Deterministic versus stochastic modelling in biochemistry and systems biology

Deterministic versus stochastic modelling in biochemistry and systems biology

Paola Lecca
Ian Laurenzi
and
Ferenc Jordan

WOODHEAD PUBLISHING

Oxford Cambridge Philadelphia New Delhi

Woodhead Publishing Limited, 80 High Street, Sawston, Cambridge, CB22 3HJ, UK
www.woodheadpublishing.com
www.woodheadpublishingonline.com

Woodhead Publishing, 1518 Walnut Street, Suite 1100, Philadelphia, PA 19102-3406, USA

Woodhead Publishing India Private Limited, G-2, Vardaan House, 7/28 Ansari Road, Daryaganj, New Delhi 110002, India
www.woodheadpublishingindia.com

First published in 2013 by Woodhead Publishing Limited
ISBN: 978-1-907568-62-6 (print); ISBN 978-1-908818-21-8 (online)
Woodhead Publishing Series in Biomedicine ISSN 2050-0289 (print); ISSN 2050-0297 (online)

British Library Cataloguing-in-Publication Data: A catalogue record for this book is available from the British Library.

Library of Congress Control Number: 2013930126

Typeset by Domex e-Data Pvt. Ltd., India

Printed in the UK and USA

Contents

Published by Woodhead Publishing Limited, 2013

Published by Woodhead Publishing Limited, 2013

Published by Woodhead Publishing Limited, 2013

List of figures

Published by Woodhead Publishing Limited, 2013

Published by Woodhead Publishing Limited, 2013

Published by Woodhead Publishing Limited, 2013

Published by Woodhead Publishing Limited, 2013

Published by Woodhead Publishing Limited, 2013

Published by Woodhead Publishing Limited, 2013

Published by Woodhead Publishing Limited, 2013

Published by Woodhead Publishing Limited, 2013

Published by Woodhead Publishing Limited, 2013

Published by Woodhead Publishing Limited, 2013

Published by Woodhead Publishing Limited, 2013

Published by Woodhead Publishing Limited, 2013

Published by Woodhead Publishing Limited, 2013

Published by Woodhead Publishing Limited, 2013

Published by Woodhead Publishing Limited, 2013

Published by Woodhead Publishing Limited, 2013

Published by Woodhead Publishing Limited, 2013

List of tables

Published by Woodhead Publishing Limited, 2013

Published by Woodhead Publishing Limited, 2013

Published by Woodhead Publishing Limited, 2013

Published by Woodhead Publishing Limited, 2013

Preface

Modeling and simulation of biological systems have received increased interest in the context of system biology. The central dogma of this re-emerging paradigm states that it is system dynamics and organizing principles of complex biological phenomena that give rise to functioning and function of cells, tissues, organisms. Similarly, it is systems dynamics and organizing principles that govern the evolution of ecological communities of individuals. Namely, cell functions, such as growth, division, differentiation and apoptosis are temporal processes. The life of an ecosystem is marked by the changes in time of the numbers of the species belonging to the system. Consequently, the processes leading the evolution of biological systems at any scale can be understood if these systems are treated as dynamic systems.

This book presents the formalisms of modelling and the algorithms of simulation of biological system dynamics both in deterministic and stochastic regime. The book describes the deterministic and the stochastic approaches to model the time evolution of the systems, and reviews the most used algorithms performing stochastic simulation. The increasing availability of experimental structural and kinetics data about the micro- and meso-scale mechanisms governing the life of cells, tissues and organisms, as well as the recent

Published by Woodhead Publishing Limited, 2013

experimental approaches to the study of complex ecological and population dynamics ask researchers and modelers to have an appropriate knowledge of the paradigms and formalisms both of the deterministic and stochastic process theory. We hope that such knowledge could provide researchers and modelers with a useful tool to make a rational choice of the more suitable modelling formalisms in the application domains of their researcher study. Deterministic and stochastic approaches are two different visions of the natural process. Nevertheless, in systems biology, both the approaches try to answer the same questions and to pursue the same objectives.

Namely, system biology focuses on an understanding of functional activity from a system-wide perspective and, consequently, it is defined by two key questions: (i) how do the components within a cell interact, so as to bring about its structure and functioning? (ii) How do cells interact, so as to develop and maintain higher levels of organization and functions? In recent years, wet-lab biologists embraced mathematical modeling and simulation as two essential means toward answering the above questions. The credo of dynamics system theory is that the behavior of a biological system is given by the temporal evolution of its state. Our understanding of the time behavior of a biological system can be measured by the extent to which a simulation mimics the real behavior of that system. Deviations of a simulation indicate either limitations or errors in our knowledge. Deterministic and stochastic modelling and simulations share the objectives in this sense but could give different answers, and the reasons for this can be found in the different theoretical formulations of these two modeling frameworks. Comparing the simulations of a deterministic model with those of a stochastic one in order to understand which is the most suitable and informative modelling approach may be very hard if the modeler does not have a background about how these two frameworks conceive and abstract a physical and biological system. For this reasons,in writing the book the authors made

Published by Woodhead Publishing Limited, 2013

use of their experience in teaching to master's and Ph. D. students and in their experience of working in a multidisciplinary research environment.

Paola Lecca, PhD
The Microsoft Research - University of Trento
Centre for Computational and Systems Biology
Rovereto, Italy

Published by Woodhead Publishing Limited, 2013

About the Authors and Contributors

Paola Lecca received a M.S. in Theoretical Physics from the University of Trento (Italy) in 1997 and a PhD in Computer Science in 2006 from the International Doctorate School in Information and Communication Technologies at the University of Trento (Italy). From 1998 to 2000 she was research assistant at the FBK-Center for Information Technologies of Trento by the research unit of Predictive Models for Biological and Environmental Data Analysis. From 2001 to 2002 she was research assistant at the Department of Physics at the University of Trento in the area of data manipulation and predictive modelling in research programs of the National Institute of Nuclear Physics (Italy). She is currently Principal Investigator at The Microsoft Research - University of Trento Centre for Computational and System Biology of Rovereto (Italy) where she coordinates the research activities in the area of model identifications and sensitivity analysis. Her research interests include stochastic modelling and simulations, biological network inference, optimal experimental design, and algorithmic systems biology. Paola started her collaboration with COSBI in December 2006. Paola Lecca is author of many conference and journal papers and belongs to the editorial board of journals in the areas of biophysics, bioinformatics, biological and medical research.

Published by Woodhead Publishing Limited, 2013

A complete list of publications can be found at the url http://www.cosbi.eu. Recently, she co-edited the book *Systemic Approaches in Bioinformatics & Computational Systems Biology: Recent Advances* published by IOGI-Global. In 2007 she won the young researchers' best-paper grant for the article *Molecular mechanism of glutamate-triggered brain glucose metabolism: a parametric model from FDG PET-scans*, Brain, Vision and Artificial Intelligence 2007, LNCS (Ed. F. Mele, G. Ramella, S. Santillo and F. Ventriglia), 4729:350-359 Springer Verlag, 2007. In 2009, in team with 11 CoSBi researchers integrated with 3 French researchers and 1 German researcher she won first prize at international modeling competition in Dagstuhl, Germany. Paola Lecca hold teaching positions at The University of Trento with the courses on **Computer Science Methods in Physics** and **Simulation of Biological Systems** for master students in Physics and Computer Science at University of Trento.

Ian J. Laurenzi Ph.D., is a chemist and chemical engineer with extensive experience in biochemical modeling, computational biology, and bioinformatics. In 1997, he received a B.S. in chemistry from the State University of New York at Albany and a B.S. in chemical engineering from Rensselaer Polytechnic Institute. He subsequently pursued research on blood coagulation and receptor-mediated phenomena at the Institute for Medicine and Engineering at the University of Pennsylvania, where he completed his Ph.D. in chemical engineering in 2002. As a PhRMA postdoctoral fellow in the Department of Molecular Biophysics and Biochemistry at Yale University, he investigated the dynamics of gene networks and the effect of sex upon mammalian gene expression. Subsequently, as a professor of chemical engineering and bioengineering at Lehigh University, he conducted research on DNA microarray design, biostatistics of single-molecule experiments, and cooperative bimolecular adhesion, among other topics. He is an expert in the areas of stochastic processes, statistics, computational biology, biological aggregation, and receptor-mediated adhesion

Published by Woodhead Publishing Limited, 2013

of human blood cells. He is currently a Senior Researcher at ExxonMobil Research and Engineering.

Ferenc Jordán is a Hungarian biologist with MSc in biology (1996) and PhD in genetics (1999), both from Etvs University, Budapest. He has also got DSc from the Hungarian Academy of Sciences (2009). His main interest is applying network analysis in biological systems, mostly in ecology (food webs, landscape graphs) but also in human and animal social networks, as well as molecular and transportation networks. He has developed several structural centrality and redundancy indices for structural network analysis but recently he is increasingly interested also in dynamical simulations (in a stochastic, individual-based framework). The key interest is better understanding the relationship between structure and functioning, improving the predictability of network models. Key applications include sustainable marine fisheries, keystone species in conservation biology and group structure and dynamics of social animals. In the framework of a recent collaborative network he is involved in exploring natural security, seeking for natural solutions for various risks. Jordán has published over 60 papers and 10 book chapters, acted as Referee for over 30 journals and is an active Editor at Ecology Letters, Ecological Complexity and Community Ecology. He was the chief organizer of the 7th European Conference on Ecological Modelling (Riva del Garda, Italy, 2011). Worked as Fellow at Collegium Budapest, Institute for Advanced Study for six years, in five of these having been holding the Branco Weiss Fellowship of the Society in Science foundation (ETH, Zrich, Switzerland). Presently, he is principal investigator at The Microsoft Research - University of Trento Centre for Computational and Systems Biology (Rovereto, Italy), where he leads the Ecology research group.

Published by Woodhead Publishing Limited, 2013

1

Deterministic chemical kinetics

Abstract: The deterministic approach to chemical kinetics is used by chemical and life scientists to characterize the time evolutions of chemical reactions in large systems. In this chapter, we introduce key concepts including the chemical rate equation, the chemical rate constant, the material balance and the relationship between the "steady state" and chemical equilibrium.

Keywords: chemical kinetics, law of mass action, material balance, determinism.

1.1 Determinism and Chemistry

Chemical reaction is not instantaneous: When one or molecules react, they do so over a period of time determined by the microphysics of the reaction, e.g. molecular collision due to Brownian motion, the transfer of electrons, etc. Those timescales determine everything from the dynamic properties of materials to the timescales of cellular and physiological response. The study of the rates of chemical reactions is called *chemical kinetics*.

Of particular interest to many researchers in academia and industry are the timescales at which (a) drugs act upon the body (pharmacodynamics) and (b) the body acts upon drugs (pharmacokinetics). However, the characterization of biochemical kinetics is essential to biotechnology, agriculture

Published by Woodhead Publishing Limited, 2013

and other industries. The question common to most investigations of kinetics is: *At some point in the future, how much will I have?* The subject will vary from study to study: *how much* drug? *how much* of a particular RNA transcript? *how much* of a particular ligand? The quantity will typically be measured in amount (moles), mass, or concentration (e.g. mol/L).

As we shall discuss later, molecules collide and react at random. Thus, the number of molecules of a particular species is a random variable. Nevertheless, the time evolution of macroscopic amounts or concentrations of molecules is usually quite reproducible, subject to the reproducibility of experimental conditions. This empirical observation has led to the development of *rate laws* for the prediction of the time rate of change of molecular concentrations. Such laws are calle deterministic, because they imply that the time evolution of the chemical process is "predestined" from the initial conditions.

Such models are common in physics, e.g. the laws of kinematics, classical electromagnetic theory, and gravity. For

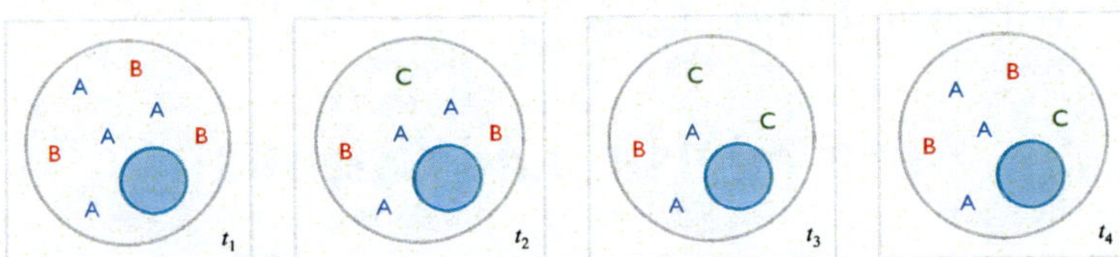

Figure 1.1 A chemical system: Biomolecular species A, B and C reside within the cytosol of a eukaryotic cell, which defines the system volume V. Two reactions may occur: A+B → C and C → A+B. Initially, concentrations of A, B and C are $4/V$, $3/V$ and zero, respectivey (left frame). As time progresses (first frame to the third), the reaction A+B → C occurs twice, yielding concentrations of A, B, and C of $2/V$, $1/V$ and $2/V$ respectively. Then, the reaction C → A+B occurs. The systems in the second and fourth frames are chemically indistinguishable from each other.

Published by Woodhead Publishing Limited, 2013

instance, the equations of motion, be they Newtonian or relativistic, specify the trajectories of each planet to infinite precision. If one knows their positions at time t, one may derive equations which specify their positions at any other time $-\infty < t' < \infty$. Likewise, a deterministic approach to chemical kinetics specifies the time evolutions of the amounts of each chemical species to infinite precision: If one knows the amounts of each species at time t, one could calculate the amounts at any time $t' > t$. Generally speaking, if a set of equations specify $A(t)$, then the time evolution of A is deterministic.

In chemical kinetics, the "material balance" and "rate law" are analogous to kinematics and dynamics of deterministic physics. Let us explore these concepts in more detail.

1.2 The Material Balance

The material balance is simply a statement of conservation of mass, although we will employ an equivalent formalism by way of conservation of species. Before writing any equations, let us define a chemical system via a boundary, e.g. a reaction vessel, cell membrane or endosome. The boundary may be permeable or impermeable, and retains a volume V within which molecular species $i = 1, 2, \ldots M$ exist at concentrations c_i. We assume that these molecules are maintained at a constant temperature T, easily achieved with sufficient mixing in a lab setting.

Having defined the system, let us consider the molecular species i within. The initial amount may have been ours to specify, or perhaps species i is generated by a cell within (or defining) the system boundary. However, the amount of i may change via reaction with other species or as a consequence of the reactions of other species. Moreover, i may enter or leave the volume. Without any loss in generality, we may express all of these changes as

$$\begin{aligned}\text{Accumulation} &= \text{Input} - \text{Output} + \text{Generation} \\ &\quad - \text{Consumption}\end{aligned} \tag{1.1}$$

Published by Woodhead Publishing Limited, 2013

where each term accounts for the rate of change of the amount (mol or number of molecules) of species *i*. "Input" and "output" refer to the rates at which species *i* enter and depart the system (e.g. via endocytosis and exocytosis), and "generation" and consumption" refer to the rates at which species *i* is generated and consumed by chemical reactions.

If we define the amount of *i* as N_i, then we may express the rate of accumulation of *i* within the system as

$$\text{Accumulation} = \frac{dN_i}{dt} = \frac{d}{dt}(Vc_i) \tag{1.2}$$

where c_i (mol/L or molar[1]) and V (L) are the concentration of species *i* and the system volume, as defined previously.

The reaction terms may be expressed generally as

$$\text{Generation} - \text{Consumption} = \sum_j \mathscr{R}_j \nu_{ij}$$

where $\mathscr{R}_j$ is the *reaction rate law* for the *j*th possible reaction among *M* possible chemical reactions in the control volume and ν_{ij} is the *stoichiometric coefficient* for species *i* in reaction *j*. The stoichiometric coefficient is a characteristic of the chemical reaction, defining the number of molecules consumed or generated by the reaction. For instance, in the reaction

$$a\text{A} + b\text{B} \rightarrow c\text{C} + d\text{D} \tag{1.3}$$

then $\nu_A = -a$, $\nu_B = -b$, $\nu_C = +c$, and $\nu_D = +d$.

In many if not most situations of relevance to life scientists, the system will be physically isolated from its surroundings, i.e. Input – Output = 0. This condition sets $V = const$ and results in the following general expression

$$\frac{dc_i}{dt} = \sum_{j=1}^{M} \mathscr{R}_j \nu_{ij} \quad \text{Material Balance in a Closed System} \tag{1.4}$$

As a matter of convention, a species consumed in a chemical reaction is called a *reactant* and has a negative stoichiometric coefficient. A species generated by a reaction is called a *product* and has a positive stoichiometric coefficient.

Published by Woodhead Publishing Limited, 2013

Example 1.2.1 (Michaelis-Menten Kinetics). The following reactions describe the enzymatic (E) conversion of a substrate S into a product P as follows

$$\begin{aligned} &E + S \rightarrow ES && j = 1 \\ &ES \rightarrow E + S && j = 2 \\ &ES \rightarrow E + P && j = 3 \end{aligned} \tag{1.5}$$

If enzyme and substrate are rapidly added to a closed volume (e.g. a well in a 384-well plate), the concentrations of each species (E, S, ES and P) will evolve in time in accordance with Eq. 1.4. Let us begin by considering the material balance for the substrate S: Since one molecule of S is consumed by the first reaction, one molecules of S is generated by the second, and S does not participate in the third, the stoichiometric coefficients for S are $\nu_{S1} = -1$, $\nu_{S2} = +1$, and $\nu_{S3} = 0$. Thus,

$$\frac{dc_S}{dt} = -\mathscr{R}_1 + \mathscr{R}_2$$

The material balance for the enzyme-substrate complex ES may be constructed likewise. In this case, one molecule of ES is generated by the first reaction, and one molecule of ES is consumed by the second and third reactions. Thus, $\nu_{ES1} = +1$, $\nu_{ES2} = -1$, and $\nu_{ES3} = -1$ and

$$\frac{dc_{ES}}{dt} = \mathscr{R}_1 - \mathscr{R}_2 - \mathscr{R}_3$$

The rate laws for the enzyme E and product P may be constructed likewise. We will explore these reactions in greater detail in Section 1.5.1.

Example 1.2.2 (Dimerization). In the preceding example, the stoichiometric coefficients were either -1 or 1 due to the nature of the reactions. For other reactions, they may take on other integers. For instance, consider the dimerization reaction,

$$\begin{aligned} &2A \rightarrow C && j = 1 \\ &C \rightarrow 2A && j = 2 \end{aligned} \tag{1.6}$$

Published by Woodhead Publishing Limited, 2013

The species A has a stoichiometric coefficient of $\nu_{A1} = -2$ for the first reaction, and and $\nu_{A2} = +2$ for the second reaction. Thus, the rate law for species A is

$$\frac{dc_A}{dt} = -2\mathscr{R}_1 + 2\mathscr{R}_2 \tag{1.7}$$

The species C, however, has stoichiometric coefficients $\nu_{C1} = +1$ and $\nu_{C2} = -1$, yielding

$$\frac{dc_C}{dt} = \mathscr{R}_1 - \mathscr{R}_2 \tag{1.8}$$

Note that the values of $\mathscr{R}_1$ and $\mathscr{R}_2$ used in Eqs. 1.7 and 1.8 are common to both and describe the rates (mol/L/s) of the two reactions independently. Let us now consider these rate laws from an empirical approach.

1.3 The Rate Law

Countless observations of chemical reactions have revealed that the rate of reaction $\mathscr{R}$ depends upon the concentrations of the reactants, and occassionally, the products. The form of this dependency is called the rate law

$$\mathscr{R} = f(c_1, c_2, \ldots c_M) \tag{1.9}$$

The functional forms of $f(c_1, c_2, \ldots c_M)$ vary considerably from one reaction to another. For instance, the aforementioned Michaelis reactions are often expressed as

$$\mathrm{E+S} \overset{\mathscr{R}}{\rightarrow} \mathrm{E+P}$$

The rate of this "net" reaction is often modeled by the following expression

$$\mathscr{R} = \frac{v_{\max} c_S}{K_M + c_S}$$

where K_M and $v_{\max}$ are constants. Rate laws of this type are often used to fit rate data for enzymatic and catalytic reactions.

Published by Woodhead Publishing Limited, 2013

An arbitrary non-catalytic reaction of the form $a\mathrm{A} + b\mathrm{B} + \cdots + z\mathrm{Z} \rightarrow$ products will often have a rate law expressed as

$$\mathscr{R} = k c_A^{\alpha} c_B^{\beta} \ldots c_Z^{\zeta} \tag{1.10}$$

where k is the *rate constant.* Since the rate of a chemical reaction must be a positive quantity, the rate constant is necessarily positive. However, the exponents may take on any real value. Their sum is called the *reaction order* and is generally an empirical quantity, as these constants, like the rate constant, are determined via curve fitting.

Generally speaking, one does not know the functional form of a rate law a priori. One posits a hypothesis regarding the functional form of the rate law, collects data, fits the data and evaluates the goodness of fit. It is possible that the data may be fit by multiple rate laws: In this case, one must choose the "best" model using techniques from information theory, e.g. the Akaike Information Criterion.

1.3.1 Elementary Reactions and Molecularity

There is an exception to this general observation for certain reactions called *elementary reactions.* A reaction is called "elementary" if the reactants directly contact each other and react. For instance, consider Eq. 1.5: If

1. E *directly* reacts with S to produce ES,
2. ES *directly* reacts to produce E and S, and
3. ES *directly* reacts to produce E and P

then all three reactions are elementary. Furthermore, if these reactions *are* elementary, then the reaction $\mathrm{E} + \mathrm{S} \rightarrow \mathrm{E} + \mathrm{P}$ is *not* elementary, as this is an aggregate of all three steps. Elementary reactions are the actual reactions that occur between molecules, and collections of such reactions constitute reaction mechanisms such as the citric acid cycle, transcription, translation, and signal transduction.

An elementary reaction of the form specified by $a\mathrm{A} + b\mathrm{B} + \cdots + z\mathrm{Z} \rightarrow$ products will have the form

Published by Woodhead Publishing Limited, 2013

$$\mathscr{R} = kc_A^{|a|}c_B^{|b|}\dots c_Z^{|z|} \qquad \text{(Elementary Reaction)} \qquad (1.11)$$

where $|a|, |b|, \dots |z|$ are the absolute values of the stoichiometric coefficients. If a reaction is elementary, its reaction order is synonymous with its *molecularity*. Hence, an elementary reaction of the form

$$A \xrightarrow{k_1} B \qquad (1.12)$$

is called *unimolecular* and has the rate law $\mathscr{R} = kc_A$. An elementary reaction of the form

$$A + B \xrightarrow{k_2} C \qquad (1.13)$$

is called *bimolecular* and has the rate law $\mathscr{R} = kc_Ac_B$. Note that we have placed the rate constants for these two elementary reactions above the "reaction arrow". We shall annotate reactions this way if they are elementary.

1.3.2 Caveats for Rate Laws

There are some important caveats to note when making use of the published literature on chemical kinetics, or conducting experimental investigations of chemical kinetics

1. Often, an empirical rate law for a reaction of the form $aA + bB \rightarrow$ products will have the form $\mathscr{R} = kc_A^{|a|}c_B^{|b|}$. This does not mean that the reaction is elementary.
2. Not infrequently, one needs to formulate rate laws for net reactions of the form $aA + bB \rightarrow$ products where $a + b > 2$. Such reactions are almost certainly non-elementary, as termolecular reactions are exceedingly uncommon. Hence, a more detailed understanding of the chemical mechanism is required for the characterization of the kinetics.
3. Occasionally, an empirical rate law will have an order greater than two or less than one. Such rate laws suggest that the reaction under study is not elementary.

Published by Woodhead Publishing Limited, 2013

4. Occasionally, one encounters an empirical rate law in the literature suggesting zeroeth order kinetics, i.e. $\mathscr{R} = const.$ Treat such rate laws with caution as they may not be predictive outside of the range of reactant concentrations employed in the experiments from which they were obtained.

Generally speaking, characterization of the kinetic rate law cannot inform a researcher as to whether a reaction is elementary, or not. Such determinations are made via other chemical investigations, e.g. spectroscopy, and are often major discoveries. Fortunately for researchers in the life sciences, mechanisms of many of the key biochemical reactions are understood; as long as one works with elementary reactions, kinetic studies become substantially simpler in terms of experimental design.

1.4 Solving the Conservation Equations

To predict the deterministic time evolution of chemical reactions, one must integrate Eq. 1.4 for each chemical species ($i = 1 \ldots N$) subject to its boundary condition, which is typically a specification of the initial concentration, i.e.

$$\frac{dc}{dt} = \boldsymbol{\mathscr{R}}\boldsymbol{\nu} \qquad \mathrm{c}(0) = \mathrm{c}_0 \tag{1.14}$$

where $\mathrm{c} = [c_1, c_2, c_3 \ldots c_N]$ $(1 \times N)$, $\boldsymbol{\mathscr{R}} = [\mathscr{R}_1, \mathscr{R}_2, \ldots \mathscr{R}_M]$ $(1 \times M)$, $\boldsymbol{\nu} = [\nu_{ij}]$ $(m \times n)$, and $\mathrm{c}_0 = [c_{10}, c_{20}, c_{30} \ldots c_{N0}]$ $(1 \times N)$. For most chemical mechanisms and pathways, the reaction rate laws $\{\mathscr{R}_j\}$ are nonlinear functions of the concentrations. Hence, no general solution to Eq. 1.14 exists. However, under certain circumstances, the system of ODEs may be solved in closed form.

Published by Woodhead Publishing Limited, 2013

1.4.1 Systems of First Order or Unimolecular Reactions

If all of the reactions in a system are unimolecular or "first order", then each reaction rate law is proportional to one and only one species concentration. In this case, Eq. 1.14 becomes a linear system of differential equations. These may be solved using standard techniques, including linear algebra and the method of Laplace transforms.

Example 1.4.1 (Spontaneous Isomerization). To begin, let us consider a spontaneous isomerization of a molecule

$$\mathrm{A} \underset{k_{-1}}{\overset{k_1}{\rightleftharpoons}} \mathrm{B} \tag{1.15}$$

The isomerization takes place in a closed volume. At $t = 0$, species A is rapidly added to the reaction vessel and mixed rapidly to yield a concentration of A is c_{A0} before any molecules are in the B state. If we apply Eq. 1.14 to these reactions, we obtain the following two ordinary differential equations

$$\frac{dc_A}{dt} = -k_1 c_A + k_2 c_B \qquad c_A(0) = c_{A0}$$

$$\frac{dc_B}{dt} = +k_1 c_A - k_2 c_B \qquad c_B(0) = 0$$

This system of equations may be rewritten as

$$\frac{d\mathbf{c}}{dt} = \mathbf{A}\mathbf{c} \tag{1.16}$$

where $\mathbf{c} = [c_A, c_B]^T$ and

$$\mathbf{A} = \begin{bmatrix} -k_1 & k_2 \\ k_1 & -k_2 \end{bmatrix} \tag{1.17}$$

To solve these equations, we apply a Laplace transform to Eq. 1.16, i.e.

Published by Woodhead Publishing Limited, 2013

$$\mathbf{C}(s) = \int_0^{\infty} e^{-st}\mathbf{c}(t)dt \tag{1.18}$$

This yields

$$s\mathbf{C}(s) - \mathbf{c}_0 = \mathbf{AC} \tag{1.19}$$

Rearranging, we obtain

$$\mathbf{C}(s) = [s\mathbf{I} - \mathbf{A}]^{-1}\mathbf{c}_0 \tag{1.20}$$

where **I** is the identity matrix.[2] In this case

$$s\mathbf{I} - \mathbf{A} = \begin{bmatrix} s+k_1 & -k_2 \\ -k_1 & s+k_2 \end{bmatrix}$$

As this is a 2×2 matrix, the inverse matrix is easily calculated

$$[s\mathbf{I} - \mathbf{A}]^{-1} = \frac{\begin{bmatrix} s+k_2 & k_2 \\ k_1 & s+k_1 \end{bmatrix}}{s\left(s+k_1+k_2\right)}$$

from which we obtain the Laplace-transformed expressions for the concentrations of A and B:

$$C_A(s) = \frac{c_{A0}}{\left(s+k_1+k_2\right)} + \frac{k_2 c_{A0}}{s\left(s+k_1+k_2\right)}$$

$$C_B(s) = \frac{k_1 c_{A0}}{s\left(s+k_1+k_2\right)}$$

These expressions may be inverted from Laplace space to time using tables of Laplace transforms (e.g. [1]). One obtains

$$c_A(t) = c_{A0}e^{-(k_1+k_2)t} + \frac{k_2}{k_1+k_2}c_{A0}\left(1 - e^{-(k_1+k_2)t}\right) \tag{1.21}$$

Published by Woodhead Publishing Limited, 2013

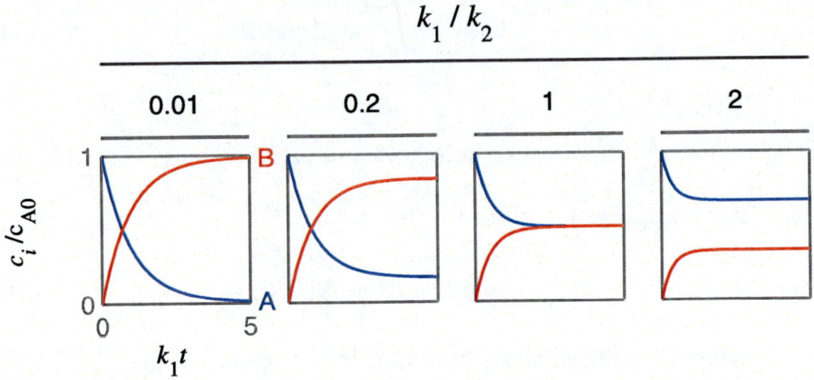

Figure 1.2 Deterministic time evolution of the isomerization reactions $A \underset{k_{-1}}{\overset{k_1}{\rightleftharpoons}} B$ initiated with A only at a concentration c_{A0}.

and

$$c_B(t) = \frac{k_1}{k_1 + k_2} c_{A0} \left(1 - e^{-(k_1+k_2)t}\right) \tag{1.22}$$

The time evolutions of species A and B are illustrated in Figure 1.2. Note that if the "forward" rate constant k_1 exceeds the "reverse" rate constant, then the "product" species B will be favored as $t \to \infty$, and vice-versa. Also note that the time evolutions of these species are monotonic. These qualities are observed for many reversible chemical reactions that occur in closed systems.

The time evolution of more complex reaction mechanisms may be calculated in exactly the same way. However, as one increases the number of species or reactions beyond two ($N > 2$ or $M > 2$), the mathematical operations become progressively more challenging.

Example 1.4.2 (The simple open ion channel block mechanism). Consider the reaction mechanism outlined in Table 1.1, which represents the transitions of states of ion channels in a cell membrane. The channels fluctuate between open and closed states $\mathscr{A}$ and $\mathscr{C}$ respectively. Moreover, the

Published by Woodhead Publishing Limited, 2013

Table 1.1 Simple open ion channel block mechanism of Colquhoun and Hawkes [16]. In this model, the ion channel blocker is at a concentration so high that its concentration does not vary, i.e. $c_B \approx const.$ Thus, this reaction is *pseudo*-first order. The parameters β', α, k_{+1} and k_{-1} are rate constants; the nomenclature is specific to Colquhoun and Hawkes.

j	Reaction Equation	$\mathscr{R}_j$	$\nu_{\mathscr{A}j}$	$\nu_{\mathscr{B}j}$	$\nu_{\mathscr{C}j}$
1	$\underset{\text{closed}}{\mathscr{C}} \xrightarrow{\beta'} \underset{\text{open}}{\mathscr{A}}$	$\beta' c_{\mathscr{C}}$	+1	0	−1
2	$\underset{\text{open}}{\mathscr{A}} \xrightarrow{\alpha} \underset{\text{closed}}{\mathscr{C}}$	$\alpha c_{\mathscr{A}}$	−1	0	+1
3	$\mathrm{B} + \mathscr{A} \xrightarrow{k_{+1}} \underset{\text{blocked}}{\mathscr{B}}$	$(k_{+1} c_{\mathrm{B}})\, c_{\mathscr{A}}$	−1	+1	0
4	$\underset{\text{blocked}}{\mathscr{B}} \xrightarrow{k_{-1}} \mathrm{B} + \mathscr{A}$	$(k_{-1})\, c_{\mathscr{B}}$	+1	−1	0

open state may be blocked if bound by the "blocker" ligand B, which, if present, would exist at much higher concentrations than the ion channels. Applying Eq. 1.14 to this reaction network we obtain

$$\frac{dc_{\mathscr{A}}}{dt} = \beta' c_{\mathscr{C}} - \alpha c_{\mathscr{A}} - (k_{+1} c_{\mathrm{B}})\, c_{\mathscr{A}} + (k_{-1})\, c_{\mathscr{B}} \quad c_{\mathscr{A}}(0) = c_{\mathscr{A}0}$$
$$\frac{dc_{\mathscr{B}}}{dt} = (k_{+1} c_{\mathrm{B}})\, c_{\mathscr{A}} - (k_{-1})\, c_{\mathscr{B}} \qquad c_{\mathscr{B}}(0) = c_{\mathscr{B}0}$$
$$\frac{dc_{\mathscr{C}}}{dt} = -\beta' c_{\mathscr{C}} + \alpha c_{\mathscr{A}} \qquad c_{\mathscr{C}}(0) = c_{\mathscr{C}0} \tag{1.23}$$

Again, this system of equations may be expressed via Eq. 1.16, but in this case $\mathbf{c} = [c_{\mathscr{A}}, c_{\mathscr{B}}, c_{\mathscr{C}}]^T$ and

$$s\mathbf{I} - \mathbf{A} = \begin{bmatrix} s + \alpha + (k_{+1} c_{\mathrm{B}}) & -k_{-1} & -\beta' \\ -(k_{+1} c_{\mathrm{B}}) & s + (k_{-1}) & 0 \\ -\alpha & 0 & \beta' \end{bmatrix} \tag{1.24}$$

The general procedure of inverting the matrix $[s\mathbf{I} - \mathbf{A}]$ is beyond the scope of this text. In this case, we may obtain it

Published by Woodhead Publishing Limited, 2013

via Cramer's Rule.[3] The Laplace space solution for the time evolution of ion channels in the $\mathscr{A}$ state is

$$C_{\mathscr{A}}(s) = \frac{\det\begin{bmatrix} c_{\mathscr{A}0} & -k_{-1} & -\beta' \\ c_{\mathscr{B}0} & s+(k_{-1}) & 0 \\ c_{\mathscr{C}0} & 0 & s+\beta' \end{bmatrix}}{\det[s\mathbf{I}-\mathbf{A}]}$$

In principle, one could calculate both the numerator and denominator of this expression, express it as a ratio of polynomials in s, and utilize tables of Laplace transforms to yield an expression for $c_{\mathscr{A}}(t)$. However, this explicitly entails calculation of the three roots of the polynomial of s defined by $\det[s\mathbf{I}-\mathbf{A}]$. The resulting expression for $C_{\mathscr{A}}(s)$ may then be inverted using tables of Laplace transforms, as previously discussed, to obtain $c_{\mathscr{A}}(t)$. Calculation of $c_{\mathscr{B}}(t)$ and $c_{\mathscr{C}}(t)$ follows from an identical procedure.

Note that in these two examples, the number of reactions corresponded to the number of roots of the characteristic polynomial $\det[s\mathbf{I}-\mathbf{A}] = 0$. Although it may be possible to find closed-form expressions for those roots, the effort required to do so will rarely enhance one's scientific understanding of the reaction dynamics.

1.4.2 Bimolecular Reactions

If bimolecular reactions are to be investigated, Eq. 1.14 becomes a nonlinear set of ordinary differential equations. In this case, there exists no general method of solving such equations. However, the rate equations may be solved for the general classes of chemical reactions listed in Table 1.2. Each of these "mechanisms" involves two and only two reactions involving the same species, which simplifies the governing equations and permits a general solution.

Example 1.4.3 (Binding of Receptors by Ligands). Let us consider an experiment designed to investigate the kinetics of the binding of receptors (A) by their ligands (B) to produce complexes (C), which is described by the reaction

Published by Woodhead Publishing Limited, 2013

Table 1.2 Reversible bimolecular reactions

Reaction	$\mathscr{R}_1$	ν_{A1}	ν_{B1}	ν_{C1}	ν_{D1}	$\mathscr{R}_2$	ν_{A2}	ν_{B2}	ν_{C2}	ν_{D2}
$A + B \underset{k_2}{\overset{k_1}{\rightleftharpoons}} C$	$k_1 c_A c_B$	−1	−1	+1	0	$k_2 c_C$	+1	+1	−1	0
$A + B \underset{k_2}{\overset{k_1}{\rightleftharpoons}} C + D$	$k_1 c_A c_B$	−1	−1	+1	+1	$k_2 c_C c_D$	+1	+1	−1	-1
$2A \underset{k_2}{\overset{k_1}{\rightleftharpoons}} C$	$k_1 c_A^2$	-2	0	+1	0	$k_2 c_C$	+2	0	−1	0
$2A \underset{k_2}{\overset{k_1}{\rightleftharpoons}} C + D$	$k_1 c_A^2$	-2	0	+1	+1	$k_2 c_C c_D$	+2	0	−1	-1
$A + B \underset{k_2}{\overset{k_1}{\rightleftharpoons}} 2A$	$k_1 c_A c_B$	+1	−1	0	0	$k_2 c_A^2$	−1	+1	0	0

$$A + B \underset{k_2}{\overset{k_1}{\rightleftharpoons}} C \tag{1.25}$$

In this experiment, a solution containing ligands is rapidly mixed with a solution containing the receptors at time $t = 0$, yielding initial concentrations of A and B and C of c_{A0}, c_{B0}, and zero respectively. The subsequent time evolution should be described by the equations

$$\begin{aligned}\frac{dc_A}{dt} &= -k_1 c_A c_B + k_2 c_C \qquad c_A(0) = c_{A0}\\ \frac{dc_B}{dt} &= -k_1 c_A c_B + k_2 c_C \qquad c_B(0) = c_{B0}\\ \frac{dc_C}{dt} &= k_1 c_A c_B - k_2 c_C \qquad c_C(0) = 0\end{aligned} \tag{1.26}$$

From Eqs. 1.26 we observe that

$$dc_A = dc_B = -dc_C$$

Intregrating this expression, we obtain

$$\begin{aligned}\int_{c_{A0}}^{c_A} dc_A = \int_{c_{B0}}^{c_B} dc_B &= -\int_0^{c_C} dc_C\\ c_A - c_{A0} = c_B - c_{A0} &= -c_C\end{aligned} \tag{1.27}$$

If one inserts these expressions for c_B and c_A into Eq. 1.26, a single ordinary differential equation results:

$$\frac{dc_C}{dt} = k_1(-c_C + c_{A0})(-c_C + c_{B0}) - k_2 c_C \quad c_C(0) = 0 \tag{1.28}$$

Similar ODEs may be written in terms of c_A or c_B. The choice of variable is subjective, as the solution of each each ODE is related to the others by Eq. 1.27.

Let us proceed with the solution of Eq. 1.28. Unlike many nonlinear differential equations, this equation may be solved as follows: For convenience, we begin by defining new variables $y = c_C/c_{A0}$ and $\tau = k_1 c_{A0} t$, and defining parameters $\xi = \frac{c_{B0}}{c_{A0}}$ and $\lambda = \frac{k_2}{k_1 c_{A0}}$. This renders Eq. 1.28 dimensionless

Published by Woodhead Publishing Limited, 2013

$$\frac{dy}{d\tau} = (1-y)(\xi - y) - \lambda y \qquad y(0) = 0, \tag{1.29}$$

and reveals that it is an ODE is of the Riccati form.[4] Thus, Eq. 1.29 may be transformed into a linear ODE for a variable u defined by

$$y = \frac{-\frac{du}{d\tau}}{u} \tag{1.30}$$

One obtains

$$\frac{d^2u}{d\tau^2} + (1+\xi+\lambda)\frac{du}{d\tau} + \xi u = 0 \tag{1.31}$$

which has the solution

$$u = \kappa_1 e^{m_1\tau} + \kappa_2 e^{m_2\tau} \tag{1.32}$$

where κ_1 and κ_2 are two arbitrary contants and

$$m_{1,2} = \frac{-(1+\xi+\lambda) \pm \sqrt{(1+\xi+\lambda)^2 - 4\xi}}{2} \tag{1.33}$$

If we substitute Eq. 1.32 into Eq. 1.30 we obtain

$$y = \frac{-(m_1 e^{m_1\tau} + Cm_2 e^{m_2\tau})}{e^{m_1\tau} + Ce^{m_2\tau}}$$

The constant $C = \frac{\kappa_2}{\kappa_1}$ may be determined from the initial condition. After some algebra one obtains

$$\frac{c_C(\tau)}{c_{A0}} = \frac{e^{m_2\tau} - e^{m_1\tau}}{\frac{1}{m_1}e^{m_1\tau} - \frac{1}{m_2}e^{m_2\tau}} \tag{1.34}$$

Published by Woodhead Publishing Limited, 2013

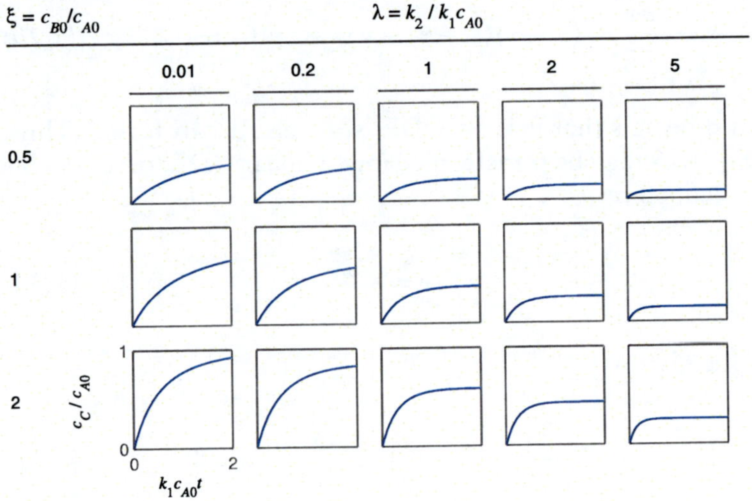

Figure 1.3 Deterministic time evolution of the reaction $A + B \underset{k_{-1}}{\overset{k_1}{\rightleftharpoons}} C$. The reaction is initiated by rapidly mixing A and B to concentrations c_{A0} and c_{B0}. These results are generated using Eq. 1.34.

where m_1 and m_2 are defined by Eq. 1.33 with $\xi = \frac{c_{B0}}{c_{A0}}$ and $\lambda = \frac{k_2}{k_1 c_{A0}}$ and $\tau = k_1 c_{A0} t$. Expressions for c_B and c_A may be expressed in terms of c_C using Eq. 1.27.

The deterministic time evolution is illustrated for various ξ and λ in Figure 1.3. Certain trends are characteristic of bimolecular equations: First, as the initial amount of ligand (B) relative to receptor (A) is increased (top row to bottom), one observes increases in both the steady state concentration of complexes and the rate of formation of those complexes. This is a manifestation of Le Chatalier's principle. Moreover, as k_2 increases with respect to k_1, fewer complexes will be present at any given time. This is a consequence of the enhanced proclivity of complexes to disssociate.

This example illustrates a general approach that may be used to predict the deterministic time evolutions of the

Published by Woodhead Publishing Limited, 2013

reversible bimolecular reactions listed in Table 1.2. The procedure may be summarized by the following steps:

1. Utilize of the reaction stoichiometry to reduce the set of ODEs for the reactant and product concentrations to a single ODE for an arbitrary species *i*.
2. Express the ODE for $c_i(t)$ as a Riccati equation, solve the corresponding linear ODE.
3. Transform the resulting solution to obtain an expression for $c_i(t)$
4. Express the concentrations of other species ($c_{k \neq i}(t)$) in terms of the newly derived expression for $c_i(t)$ via stoichiometric relationships.

That said, the procedure is only applicable to reversible reactions among chemical species that have no additional reactions. If they do, then the stoichiometric relationship between species will be disrupted, precluding the reduction of the number of ODEs to a single ODE. For instance, if the reaction $\mathrm{C} \xrightarrow{k_3} \mathrm{A} + \mathrm{D}$ is added to the preceding example, then the right hand side of the ODE for c_C will not be equal to the right hand sides of the ODEs for c_A and c_B, precluding the simplifications permitted by Eq. 1.27.

1.5 Simple Reaction Mechanisms

Nevertheless, *approximate* solutions of Eq. 1.4 are possible for simple reaction mechanisms, i.e. net reactions resulting from several elementary steps. Customarily, this is done via the *quasi-steady state approximation* (QSSA). Simply stated, one assumes that all intermediate species in a chemical pathway exist at a steady state

$$\frac{dc_{\text{Intermediates}}}{dt} = 0 \qquad \text{(QSSA)} \qquad (1.35)$$

This eliminates all of the ODEs describing the dynamics of species that do not appear in the net reaction, often leading

Published by Woodhead Publishing Limited, 2013

to an expression for the rate of the net reaction. Such expressions are occasionally of the power law type previously discussed, but are often more complex functions of the concentrations of the reactants and products. The QSSA is typically applied to specific net reactions, not complete pathways.

1.5.1 The action of enzymes – the Michaelis-Menten mechanism

A notable example is the enzymatic conversion of a substrate to a product discussed previously (Eq. 1.5). The overall reaction is

$$\mathrm{S} \xrightarrow{\mathrm{E}} \mathrm{P} \qquad \mathscr{R} = f(c_S, \ldots) \tag{1.36}$$

where the "E" above the arrow denotes the action of the enzyme. However, this masks the complexity of the mechanism

$$\mathrm{E} + \mathrm{S} \underset{k_2}{\overset{k_1}{\rightleftharpoons}} \mathrm{ES} \xrightarrow{k_3} \mathrm{E} + \mathrm{P} \tag{1.37}$$

which consists of three elementary steps. The governing equations are

$$\begin{aligned}
\frac{dc_E}{dt} &= -k_1 c_E c_S + k_2 c_{ES} + k_3 c_{ES} \qquad c_E(0) = c_{E0} \\
\frac{dc_{ES}}{dt} &= k_1 c_E c_S - k_2 c_{ES} + k_3 c_{ES} \qquad c_{ES}(0) = 0 \\
\frac{dc_S}{dt} &= -k_1 c_E c_S + k_2 c_{ES} \qquad c_S(0) = c_{S0} \\
\frac{dc_P}{dt} &= k_3 c_{ES} \qquad c_P(0) = 0
\end{aligned} \tag{1.38}$$

Since the overall reaction (Eq 1.36) is limited by the second step, $\mathscr{R} \neq k c_S$. Rather, as Eq. 1.38 shows, $\mathscr{R} = k_3 c_{ES}$.

Applying the QSSA

The enzyme-substrate complex ES is an intermediate. As such, it is often difficult if not impossible to measure owing

Published by Woodhead Publishing Limited, 2013

to its short lifetime. However, this feature often makes it an excellent candidate for the QSSA. We set

$$\frac{dc_{ES}}{dt} = 0 \qquad \text{(QSSA for ES)}$$

Applying this to the material balance for ES from Eq. 1.38, we obtain

$$c_{ES} = \frac{c_E c_S}{K_M}$$

where

$$K_M = \frac{k_2 + k_3}{k_1} \tag{1.39}$$

is known as the Michaelis constant.

The derivation of the rest of the rate law follows from additional algebraic manipulation. First, we note that the total amount of enzyme does not change:

$$c_{E0} = c_E + c_{ES} \tag{1.40}$$

Next, we combine the preceding equations to obtain an expression for the quasi steady state concentration of the intermediate, ES:

$$c_{ES} = \frac{c_{E0} c_S}{K_M + c_S}$$

The overall rate of reaction, i.e. the rate of production of the product species P, follows immediately

$$\mathscr{R} = \frac{k_3 c_{E0} c_S}{K_M + c_S} = \frac{v_{\max} c_S}{K_M + c_S} \quad \text{Michaelis-Menten Rate Law} \tag{1.41}$$

where we have recognized that the fastest rate of reaction $v_{\max} = k_3 c_{E0}$ occurs in the limit of large substrate concentrations.

Eq. 1.41 was developed by Leonor Michaelis and Maud Menten in 1913 [6] and is possibly the most influential equation in biochemical kinetics. Its predictive power holds up remarkably well for an approximation, and is routinely fit to kinetic data.

Published by Woodhead Publishing Limited, 2013

Evaluating the time evolution

Having defined the reaction rate for Eq. 1.36, the time evolution of substrate and product concentrations follow from

$$\frac{dc_S}{dt} = -\frac{v_{\max} c_S}{K_M + c_S} \tag{1.42}$$

and $c_{S0} - c_S = c_P$. Although Eq. 1.42 is nonlinear, it is variable separable. Hence,

$$\begin{aligned} \int_{c_{S0}}^{c_S} \frac{K_M + c_S}{c_S} dc_S &= -\int_0^t v_{\max} dt \\ K_M \ln \frac{c_S}{c_{S0}} + (c_S - c_{S0}) &= -v_{\max} t \end{aligned} \tag{1.43}$$

A closed form expression for $c_S(t)$ was identified in 1997 by Schnell and Mendoza [7]

$$c_S(t) = K_M W \left(\frac{c_{S0}}{K_M} \exp \left(\frac{-v_{\max} t + c_{S0}}{K_M} \right) \right) \tag{1.44}$$

where $W(x)$ is the Lambert W function[5].

The deterministic time evolution of the Michaelis-Menten mechanism is illustrated in Figure 1.4. Unlike the reactions previously discussed, it has two kinetic regimes: One in which the rate is effectively constant, and a second regime in which the substrate is fully consumed. Note that if $c_{S0} \gg K_M$ (left frame), the rate of reaction is $v_{\max}$. This has major ramifications in the study of enzymatic kinetics: Experimental designs for the quantification of kinetic parameters must be conducted over a wide range of initial substrate concentrations. If the concentration of substrate if too high, its consumption profile will be effectively linear (e.g. the cases $K_S/c_{S0} = 0.05$ and 0.1 in Figure 1.4). In such cases, one cannot estimate the Michaelis constant from the data, and careless use of data fitting software will yield nonsensical estimates (e.g. negative estimates of K_M).

Published by Woodhead Publishing Limited, 2013

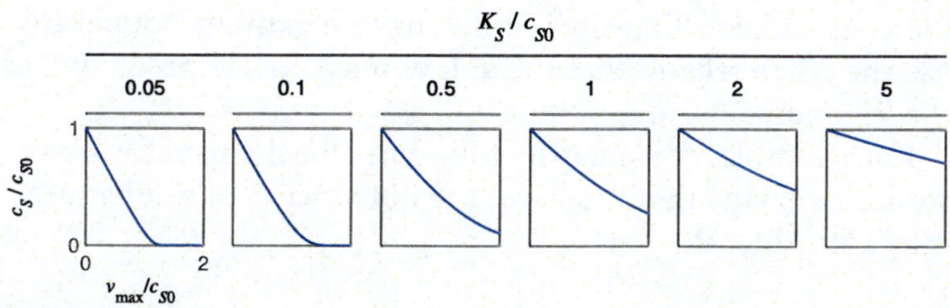

Figure 1.4 Approximate deterministic time evolution of an enzymatic reaction obeying the Michaelis Menten rate expression (Eq. 1.41). Results are generated using Eq. 1.44.

The type of reaction described by the Michaelis-Menten rate law is so common in biology that kinetic data is routinely fit to it. Since its introduction in 1913, several modified versions of rate law have been introduced to account for competitive inhibition, non-competitive inhibition, etc., in part because enzymatic inhibition constitutes a significant amount of drug development. Today, rate constants k_3 (sometimes denoted k_{cat}) and Michaelis constants, among other parameters, are collected and reported via several online databases (e.g BRENDA[6]) for a wide variety of enzymes.

1.5.2 Cooperative binding and the Hill model

Enzymes are often composed of multiple subunits, each of which has enzymatic activity. Initially, each binding site has the same kinetics of association to and dissociation from the substrate. However, as sites are bound, subtle changes in the enzyme's structure result, which in turn affects the rate constants for the other subunits. This general principle is denoted cooperativity, and may result in the enhancement or inhibition of binding with increasing level of saturation of

Published by Woodhead Publishing Limited, 2013

the enzyme's binding sites. If an enzyme exhibits cooperativity, the Michaelis-Menten rate law may not be predictive of the kinetics.

For example, consider the following mechanism for a two-subunit enzyme that mediates the conversion of a substrate S into a product P:

$$E + S \underset{k_2}{\overset{k_1}{\rightleftharpoons}} ES \xrightarrow{k_3} E + P \tag{1.45}$$

$$ES + S \underset{k_5}{\overset{k_4}{\rightleftharpoons}} ES_2 \xrightarrow{k_6} ES + P \tag{1.46}$$

The deterministic time evolution of this chemical system is described by the following differential equations

$$\frac{dc_S}{dt} = -k_1 c_E c_S + k_2 c_{ES} - k_4 c_{ES} c_S + k_5 c_{ES_2} \tag{1.47}$$

$$\frac{dc_E}{dt} = -k_1 c_E c_S + k_2 c_{ES} + k_3 c_{ES} \tag{1.48}$$

$$\frac{dc_{ES}}{dt} = k_1 c_E c_S - k_2 c_{ES} - k_3 c_{ES} - k_4 c_{ES} c_S + k_5 c_{ES_2} + k_6 c_{ES_2} \tag{1.49}$$

$$\frac{dc_{ES_2}}{dt} = k_4 c_{ES} c_S - k_5 c_{ES_2} - k_6 c_{ES_2} \tag{1.50}$$

$$\frac{dc_P}{dt} = +k_3 c_{ES} + k_6 c_{ES_2} \tag{1.51}$$

As the formation of the product P is unidirectional, the rate of the net reaction S → P is equal to the rate at which P is produced

$$\mathcal{R} = +k_3 c_{ES} + k_6 c_{ES_2} \tag{1.52}$$

Therefore, formulation of a rate law for this two-subunit enzyme requires expressions for the concentrations of the intermediates ES and ES_2 in terms of the concentrations of measureable quantities, e.g. c_S.

Published by Woodhead Publishing Limited, 2013

Once again, we employ the QSSA to do so, recognizing that the species ES and ES_2 are the intermediates of this mechanism. Eqs. 1.49 and 1.50 become

$$0 = k_1 c_E c_S - k_2 c_{ES} - k_3 c_{ES} - k_4 c_{ES} c_S + k_5 c_{ES_2} + k_6 c_{ES_2}$$

$$0 = k_4 c_{ES} c_S - k_5 c_{ES_2} - k_6 c_{ES_2}$$

After some algebra, one obtains

$$c_{ES} = \frac{k_1 c_E c_S}{k_2 + k_3} \tag{1.53}$$

and

$$c_{ES_2} = \frac{k_4 c_{ES} c_S}{k_5 + k_6} \tag{1.54}$$

Lastly, we utilize the the overall material balance for enzyme molecules

$$c_{E0} = c_E + c_{ES} + c_{ES_2} \tag{1.55}$$

to relate these quantities to the instantaneous concentration of enzyme. Combining Eqs. 1.52, 1.53, 1.54 and 1.55, one arrives at the rate law for this sequential mechanism

$$\mathscr{R} = \frac{v_{\max,1}\left(\frac{c_S}{K_{M1}}\right) + v_{\max,2}\left(\frac{c_S^2}{K_{M1}K_{M2}}\right)}{\left(1 + \left(\frac{c_S}{K_{M1}}\right) + \left(\frac{c_S^2}{K_{M1}K_{M2}}\right)\right)} \quad \text{Adair/KNF Rate Law} \tag{1.56}$$

where K_{M1} is defined by Eq. 1.39,

$$K_{M2} = \frac{k_5 + k_6}{k_4}$$

$v_{\max,1} = c_{E0} k_3$ and $v_{\max,2} = c_{E0} k_6$.

Eq. 1.56 is typical of the rate laws for cooperative enzymatic reactions. The rate expression exhibits polynomials of substrate concentration in both the numerator and denominator, with order specified by the number of binding sites. In this case, the polynomials are second order because the

Published by Woodhead Publishing Limited, 2013

enzyme was modeled as having two binding sites. Occasionally, rate laws of this type are classfied as "Adair" rate laws, as they resemble Gilbert Adair's model of the binding of oxygen to hemoglobin - an protein with four binding sites ($n = 4$) [2]. They are also denoted "KNF" rate laws, as they are derived in the manner of the model of Koshland, Némethy and Filmer [5].

Insofar as laws such as Eq. 1.56 are constructed from a putative chemical mechianism, their parameters are physically meaningful and ostensibly related to molecular energetics of binding and dissociation. However, estimation of those parameters from experimental data requires (a) exceptionally precise measurements of the rate of reaction and (b) nonlinear least squares fitting software. Thus, putatively cooperative reaction rates are often modeled via the Hill equation [4]:

$$\mathscr{R} = v_{\max}\frac{c_S^n}{K_H + c_S^n} \qquad \text{Hill Rate Law} \qquad (1.57)$$

In theory, the exponent n may be related to the number of binding sites on the multi subunit enzyme provided that

1. All subunits bind their substrate *simultaneously*, i.e. the reactions $E + nS \rightarrow ES_4$ and $ES_4 \rightarrow E + nS$ are *elementary*
2. All subunits bind their substrate instantaneously, i.e. the rates of those reactions are infinitely faster than the conversion step

In reality, neither of these assumptions are valid. The first implies that the association step is an $(n + 1)$-molecular elementary reaction - a physical impossibility. The second assumption is akin to the QSSA. In practice, the constants K_H and n are both adjustable parameters and are estimated by curve fitting software. Often, the fit is very good within the range of substrate concentrations. However, the parameter n is not equal to the number of binding sites for an enzyme.

Published by Woodhead Publishing Limited, 2013

The Michaelis-Menten (Eq. 1.41), Adair/KNF (Eq. 1.56), and Hill (Eq. 1.57 with $n = 2$) models are illustrated in Figure 1.5.

If the maximum turnover rates for both subunits of a two-subunit enzyme are equal ($\nu_{\text{max},1} = \nu_{\text{max},2}$), and the Michaelis constant for the bound enzyme (ES) is sufficiently high, then the Michaelis-Menten rate law describes the kinetics of the two-subunit enzyme. This essentially means that the binding of ES by a second substrate molecule is precluded. Although the Hill rate law presented in Figure 1.5 is incongruous with both the Adair/KNF and Michaelis-Menten rate laws, this is partly due to the choice of n and K_H. The Hill expression can often be made to fit rate-concentration data by variation of those parameters.

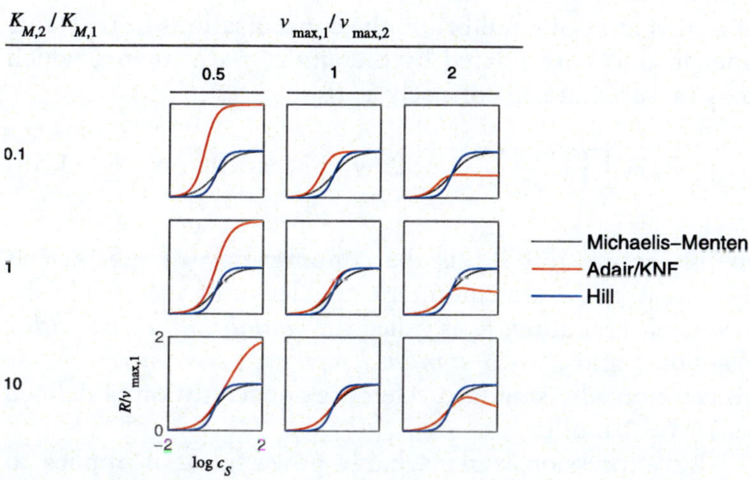

Figure 1.5 The Michaelis-Menten (Eq. 1.41), Adair/KNF (Eq. 1.56), and Hill (Eq. 1.57) rate laws. The Hill law is illustrated with $K_H = K_{M1}$ and $n = 2$. The Hill and Michaelis-Menten rate laws cannot agree unless $n = 1$; the Adair/KNF rate law agrees with the Michaelis-Menten law at low concentrations and the Hill law at high concentrations.

Published by Woodhead Publishing Limited, 2013

1.6 The Law of Mass Action

In our discussions of enzymatic catalysis, the direction of the overall reaction is irreversible; in our models, no mechanism exists by which the product P can be transformed back into S. In reality, all chemical reactions are reversible, regardless of whether or not they are facilitated by an enzyme. In accordance with experimental evidence, the theory of statistical mechanics reveals that the concentrations of reactive species will always reach a steady state at which $dc_i/dt = 0$ for all reactive species i. This limit is called *chemical equilibrium*. Strictly speaking, the reactions never stop occuring: their rates simply match each other. Fluctuations in concentration concomitant with those reactions are typically observable at very small scales, and will be discussed in Chapter 2.

In this limit of equilibrium the concentrations of reactants and products are related by the *law of mass action*, which may be summarized concisely as [8]

$$K_j = \prod_i \left(\frac{c_{i\infty}}{c^\star}\right)^{\nu_{ij}} \qquad \text{Law of Mass Action} \qquad (1.58)$$

In this expression $c_{i\infty}$ is the concentration of species i at $t = \infty$, ν_{ij} is the stoichiometric coefficient of species i in the reversible reaction j, K_j is called the *equilibrium constant* for reaction j and $c^\star$ is a *standard reference concentration*. By convention, the standard reference concentration is defined as 1 M = 1 mol/L.

This expression is remarkably powerful as it applies to *all* chemical reactions, not just elementary reactions. Moreover, it relates the concentrations of species at equilibrium to the energetics of those reactions by way of the equilibrium constant.

1.6.1 *The equilibrium constant*

The equilibrium constant is related to the energetic properties of the reactants and products as follows

Published by Woodhead Publishing Limited, 2013

$$
\begin{aligned}
-R_g T \ln K_j &= \Delta G^0_{rxn,j} \\
&= \sum_i \nu_{ij} G^0_i \\
&= \sum_i \nu_{ij} H^0_i - T \sum_i \nu_{ij} S^0_i \\
&= \Delta H^0_j - T \Delta S^0_j \qquad (1.59)
\end{aligned}
$$

where ΔG^0_j is the *Gibbs energy of reaction,* G^0_i is the Gibbs energy of formation for species i, ΔH^0_j is the *enthalpy of reaction,* H^0_i is the *enthalpy of formation* for species i, ΔS^0_j is the *entropy of reaction,* S^0_i is the *entropy of formation* for species i, T is the system temperature and R_g is the universal gas constant.[7] We refer the interested reader to Ref. [8] for more detail regarding these quantities and thermodynamics of chemical reactions. For many biochemical reactions or reactants, some or all of these parameters may be found in the published literature or via online tools. For example, the Gibbs energy of hybridization between DNA or RNA molecules may be calculated using the UNAFOLD tool.[8]

Note that the direction of the reaction is implicitly accounted for via Eqs. 1.59. If one "reverses" the direction reaction, then the stoichiometric coefficients and energies and entropies of change sign. This in turn inverts the equilibrium constant. Let us illustrate using the reactions listed in Table 1.3.

If we relate the law of mass action for Reaction 1 to the thermodynamic relationship between K_1 and the Gibbs energies of formation of the reactants and product, we obtain

$$\frac{c_{C\infty} c^\star}{c_{A\infty} c_{B\infty}} = e^{-(-G^\circ_A - G^\circ_B + G^\circ_C)/RT}$$

For Reaction 2, one obtains

$$\frac{c_{A\infty} c_{B\infty}}{c_{C\infty} c^\star} = e^{-(+G^\circ_A + G^\circ_B - G^\circ_C)/RT}$$

Careful inspection of these two expressions reveals them to be identical. This is almost to be expected, as the natural

Published by Woodhead Publishing Limited, 2013

Table 1.3 Application of the law of mass action to the association and dissociation reactions.

j	Reaction	ν_{Aj}	ν_{Bj}	ν_{Cj}	$\Delta G^\circ_{rxn,j}$	Mass Action Law	K_j
1	$A + B \rightarrow C$	−1	−1	+1	$-G^\circ_A - G^\circ_B + G^\circ_C$	$K_1 = \frac{c_{C\infty} c^\star}{c_{A\infty} c_{B\infty}}$	$e^{-(-G^\circ_A - G^\circ_B + G^\circ_C)/RT}$
2	$C \rightarrow A + B$	+1	+1	−1	$+G^\circ_A + G^\circ_B - G^\circ_C$	$K_2 = \frac{c_{A\infty} c_{B\infty}}{c_{C\infty} c^\star}$	$e^{-(+G^\circ_A + G^\circ_B - G^\circ_C)/RT}$

relationship between concentrations at equilibrium should not depend upon the subjective, human decision to write a reaction as $A + B \rightarrow C$ versus $C \rightarrow A + B$. When we employ the law of mass action to define concentrations, we typically write reactions as either $A + B \rightleftarrows C$ or $C \rightleftarrows A + B$ to explicitly denote equilibrium, leaving the redundant reaction out of our analysis.

Experimental investigations typically incorporate the standard reference concentration $c^\star$ into the equilibrium constant. For example, many investigators report the equilibrium constants for the dissociation of receptors from ligands in units of molarity (mol/L),

$$K_D = \frac{c_{A\infty} c_{B\infty}}{c_{C\infty}} \quad \text{Dissociation Constant} \tag{1.60}$$

where $K_D = c^\star K$. For this reason, it is exceptionally important to understand and interpret the units of reported and calculated equilibrium constants before employing them in biochemical models.

1.6.2 Calculating the equilibrium concentrations

To calculate the equilibrium concentrations $c_{i\infty}$, one needs to solve a mathematical system of equations including Eq. 1.58 for each reaction j as well as material conservation equations for each species i:

$$c_{i\infty} = c_{i0} + \sum_{j=1}^{M} \epsilon_j \nu_{ij} \tag{1.61}$$

In this expression, c_{i0} is the initial concentration of species i and ϵ_j is the "extent of reaction" for reaction j. As the equilibrium concentrations are related via the M equations defined by Eq. 1.58, the unknown quantities to be determined are the M extents of reaction. Therefore, the number of equations equal the number of unknowns, and the system of equations is properly specified.

Published by Woodhead Publishing Limited, 2013

Unfortunately, many if not all of the laws of mass action (Eq. 1.58) will be nonlinear equations. Hence, numerical methods are almost always required to solve for ϵ_j. Multi-variable versions of the Newton-Raphson algorithm are often suitable for this purpose. Matlab users will find the `fsolve`[9] function well suited to these calculations.

1.6.3 *Kinetics and thermodynamics*

If a reaction is elementary, and *only if* the reaction is elementary, then the rate constants are functionally related to the equilibrium constant. To illustrate, let us again use the example of the binding reaction, Eq. 1.25. As time proceeds, the system comes to steady state, i.e. equilibrium. At that point, the material balance for species A becomes

$$\lim_{t\to\infty} \frac{dc_A}{dt} = -k_1 c_{A\infty} c_{B\infty} + k_2 c_{C\infty} = 0$$

Similar expressions result for species B and C. Rearranging the variables, one obtains

$$\frac{k_1}{k_2} = \frac{c_{C\infty}}{c_{A\infty} c_{B\infty}} = \frac{K_1}{c^\star}$$

That is, the "experimental" equilibrium constant is the ratio of the "forward" and "reverse" reaction rate constants. This type of relationship holds for all elementary reactions, with the standard concentration appearing in the numerator or denominator of the right hand side of the equation subject to the stoichiometry of the elementary reactions. If the overall stoichiometry is zero, i.e. $\sum_i \nu_{ij} = 0$, then the ratio of rate constants will be identical to the equilibrium constant.

If a reaction is known to be elementary, it is possible to use a calculated or measured equilibrium constant to calculate one rate constant from the other. However, it is generally incorrect to utilize relationships such as these to relate rate constants of non-elementary reactions to their equilibrium constant.

Published by Woodhead Publishing Limited, 2013

1.7 Conclusions

The deterministic approach to chemical kinetics allows researchers to formulate mathematical expressions for the time evolution of chemical concentrations. Generally speaking, its use for simple chemical reactions yields solvable nonlinear ordinary differential equations, but for more complex reaction pathways, one must numerically integrate the resulting sets of differential material balances. Often, the complexity of these networks and the differences in magnitudes of the concentrations and rate constants will render these sets "stiff", making their numerical solution challenging and necessitating special algorithms.

Notes

1. Molarity is defined as the number of moles of a substance per liter of solution.
2. http://mathworld.wolfram.com/IdentityMatrix.html
3. http://mathworld.wolfram.com/CramersRule.html
4. http://mathworld.wolfram.com/RiccatiDifferentialEquation.html
5. The Lambert *W* function may be implemented in Matlab using the lambertw() command: www.mathworks.com/help/toolbox/symbolic/lambertw.html
6. http://www.brenda-enzymes.info/
7. In SI units, $R = 8.31441 \frac{\mathrm{J}}{\mathrm{mol\ K}}$
8. c.f. http://mfold.rna.albany.edu/
9. http://www.mathworks.com/help/toolbox/optim/ug/fsolve.html

References

1. Milton Abramowitz and Irene A. Stegun. *Handbook of Mathematical Functions with Formulas, Graphs, and Mathematical Table*. Dover Publications, 1965.
2. Gilbert S Adair. The hemoglobin system. vi. the oxygen dissociation curve of hemoglobin. *J. Biol. Chem.*, 63:529–545, 1925.

Published by Woodhead Publishing Limited, 2013

3. David Colquhoun and Alan G Hawkes. Relaxation and fluctuations of membrane currents that flow through drug-operated channels. *Proc Royal Soc Lond B, Biol Sci*, 199:231–262, 1977.
4. Archibald V. Hill. The possible effects of the aggregation of the molecules of haemoglobinon its dissociation curves. *J. Physiol.*, 40:i–vii, 1910.
5. Dan E Koshland, George Némethy, and David Filmer. Comparison of experimental binding data and theoretical models in proteins containing subunits*. *Biochemistry*, 5:365–385, 1966.
6. Leonor Michaelis and Maud Menten. Die kinetik der invertinwirkung. *Biochem Z*, 49(333-369), 1913.
7. Santiago Schnell and D Mendoza. Closed form solution for time-dependent enzyme kinetics. *J. Theor. Biol.*, 187:207–212, 1997.
8. Joseph M Smith, Hendrick C Van Ness, and Michael Abbott. *Introduction to Chemical Engineering Thermodynamics*. McGraw-Hill, 2000.

Published by Woodhead Publishing Limited, 2013

2

The stochastic approach to biochemical kinetics

Abstract: Cell-scale chemical environments often feature small populations of chemical or biochemical species, from metabolites to macromolecules. Thus, the mathematical description of the time evolutions of such "small" systems often requires an explicit accounting for the discreteness of molecular populations and the randomness of chemical reaction. The stochastic approach to chemical kinetics fills this role in the modeling of biological systems, especially those with complex dynamics hinging upon the action of a handful of bimolecular entities. In this chapter, we formulate the stochastic approach and show how it relates to the deterministic approach.

Keywords: reaction propensity, chemical master equations, rate constant.

2.1 Introduction

Biochemical reactions are fundamentally random events: At the nanoscale, Brownian motion mediates the interactions between molecules; at the atomic-scale, chemical restructure is mediated by the laws of quantum mechanics. Therefore, the populations of biochemical species are *random variables*,

Published by Woodhead Publishing Limited, 2013

and their values will always exhibit some degree of uncertainty. This is true for species that undergo an isolated chemical reaction as well as those that participate in complex pathways such as gene regulation of central metabolism.

Researchers typically investigate the kinetics of biomolecules in the "thermodynamic limit", where each biomolecular species has a large population, often in excess of billions of molecules. In this limit, the deterministic approach is often sufficient for the characterization of kinetics. However, as one investigates progressively smaller systems, e.g. the cytosol of a cell, the uncertainties associated with molecular populations present challenges to experimental reproducibility as well as measurement (Figure 2.1). Additionally, some conceptual questions arise when considering a deterministic analysis as the populations of chemical species become small:

- How does one define dc_i/dt when c_i is clearly discrete?
- Molecules collide and react at random. How does one define the "rate of reaction" in small systems?
- What is the relationship between random reaction events at the microscale and mascroscopic rate laws?

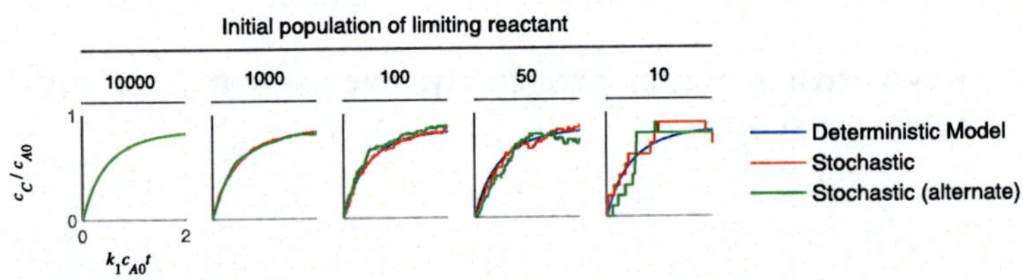

Figure 2.1 Effect of the system size upon the time evolution of chemical reactions. The deterministic time evolution and two stochastic time evolutions of the reaction A + B $\rightleftharpoons$ C are illustrated for the same reaction conditions (initial concentrations, rate constants, etc.). As the initial number of molecules decreases, the time evolution becomes progressively more random.

Published by Woodhead Publishing Limited, 2013

These questions may be answered by formulating a probabilistic description of the time evolution of chemical reactions: one that accounts for the random fluctuations of chemical populations in accordance with statistical mechanics; one that can predict the time evoutions of chemical reactions in both small and large systems. We call this the *stochastic* approach to chemical kinetics. In his landmark book, Joseph Doob defined a stochastic process as [9, 5]

a mathematical abstraction of an empirical process whose development is governed by probabilistic laws

Chemical reaction phenomena fits this description: biomolecular collisions and reactions, as discussed, are indisputably probabilistic in nature.

There are two implementations of the stochastic approach to chemical kinetics, each of which answers a different question:

1. At some point in time t, what is the probability that the system will be in a state $\mathbf{x}$?
2. Given that the system is in state $\mathbf{x}$, what is the probability that the next reaction to occur will be of type μ and will happen after an interval of time τ?

The second of these questions is related to the *stochastic simulation algorithm*, which we discuss in Chapter 3. In this chapter, we will focus on the first of these questions and the means by which it is answered: the *chemical master equation* (CME).

2.2 The chemical master equation

In Chapter 1, we defined the state of a biochemical system in terms of the populations $X_1, X_2, \ldots, X_N$, temperature T, and volume V. In experimental practice, it is often convenient to fix the temperature and volume of a mixture; in many situations it is even necessary to do so. In such cases, we may represent the states of a system by unique combinations of populations, i.e.

Published by Woodhead Publishing Limited, 2013

$$\mathbf{x} = [x_1, x_2, \ldots x_N] \qquad \text{State of a biomolecular system} \tag{2.1}$$

Note that $\mathbf{x}_k$ represents a particular set of species populations, whereas $\mathbf{X} = [X_1(t), \ldots X_N(t)]$ represents a set of variables. If $\mathbf{X}(t) = \mathbf{x}_k$, then the system is in the kth state at time t.

The time evolution of a system is a journey through these states, mediated by chemical reactions. Since these reactions occur randomly in time, the species populations $\mathbf{X} = [X_1(t), \ldots X_N(t)]$ constitute a set of random variables. As a result, it is impossible to predict the state of a biochemical system with absolute certainty *a priori*. However it is possible to know the *probability* that the system will occupy a given state.

$$P(\mathbf{x}, t) = \Pr(\mathbf{X}(t) = \mathbf{x}) \tag{2.2}$$

This probability distribution is specified via the chemical master equation, which plays a role akin to the deterministic mass balance discussed in Chapter 1.

2.2.1 Probability of chemical reaction

To specify $P(\mathbf{x}, t)$, one requires a probabilistic description of how chemical reactions occur. Therefore, let us illustrate by example: Consider the following enzymatic conversion mechanism, which is a generalization of the Michaelis Menten mechanism discussed in Chapter 1:

$$\mathrm{E} + \mathrm{S} \underset{k_2}{\overset{k_1}{\rightleftharpoons}} \mathrm{C} \tag{2.3}$$

$$\mathrm{C} \underset{k_4}{\overset{k_3}{\rightleftharpoons}} \mathrm{E} + \mathrm{P} \tag{2.4}$$

The state of the system may be expressed as $\mathbf{x} = (x_E, x_S, x_C, x_P)$, where x_E, x_S etc. are the populations of the enzyme, substrate, and other species. In Figure 2.2 we illustrate the possible states for a system initiated with two enzyme molecules and two substrate (or product) molecules. In this scenario, there are six possible states and twelve pos-

Published by Woodhead Publishing Limited, 2013

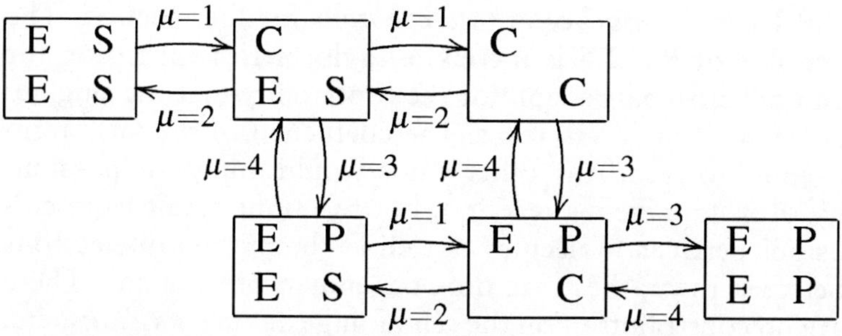

Figure 2.2 Biochemical states and transitions for the simple reaction pathway described by Eqs. 2.3 and 2.4. Here, the process is initiated with two enzyme (E) molecules and two substrate molecules (S). In this case, there are only six possible states of the system, however, the number of states grows rapidly with respect to the populations of reactive molecules.

sible transitions among them. The number of possible states strongly depends upon the reaction network structure as well as the populations of the molecules. Note also that the states are connected by way of the reaction stoichiometries: two states have three possible transitions, another two have two, and the remaining two can only transition to one other state. Many states are chemically unconnected to other states, precluding direct transitions among them.

Let us consider a scenario where there are only two molecules in the system: an enzyme E and a substate S. Without any loss in generality, we may state that

$$C_1\delta t + o(\delta t) = \Pr(\text{A specific E molecule will react with a } \textit{specific} \text{ S molecule within the imminent time interval } \delta t) \tag{2.5}$$

where

$$o(\delta t) = \kappa_2(\delta t)^2 + \kappa_3(\delta t)^3 + \kappa_4(\delta t)^4 + \ldots \tag{2.6}$$

Published by Woodhead Publishing Limited, 2013

and C_1 is the *stochastic rate constant* for the reaction. The left side of Eq. 2.5 is merely a Taylor series expansion for an undefined expression for the probability that the single E and S molecules will react. The coefficient of the $(\delta t)^0$ term is equal to zero. Any other value would imply the possibility of instantaneous reaction, but reactions result from collisions between molecules as well as the motion of electrons between them: these are time-dependent phenomena. There are no constraints upon the remaining coefficients κ_1, κ_2, etc.

Now let us consider an alternative scenario: In this case, the populations of E and S molecules, X_E and X_S, are greater than one. The probability that any specific pair of these molecules will react is still described by Eq. 2.5. However, the probability that *any* pair of molecules will react must be multiplied by the number of ways that the reaction could occur, which in this case is $X_E X_S$. Therefore,

$$C_1 X_E X_S \delta t + o(\delta t) = \Pr(\text{Any pair of E and S molecules will react within the imminent time interval } \delta t) \tag{2.7}$$

For each state, Eq. 2.7 defines a transition probability associated with this reaction: A state with $x_E = x_S = 1$ would have a transition probability defined by 2.5, and a state with $x_E = 0$ or $x_S = 0$ would have a transition probability zero, since the required molecules would not be present to react.

The logic employed for this particular reaction may be extended to any reaction. Thus, the transition probability for any elementary reaction may be defined from

$$C_\mu h_\mu(\mathbf{x}) \delta t + o(\delta t) = \Pr(\text{A reaction of type } \mu \text{ will occur within the imminent time interval } \delta t) \tag{2.8}$$

where C_μ is the stochastic rate constant for reaction μ, and $h_\mu(\mathbf{X})$ is the number of distinct combinations of reactant molecules for reaction μ. The quantity h_μ is defined for all types of elementary reactions in Table 2.1. Note that the value of $h_\mu(\mathbf{X})$ depends upon the state of the system. Hence, transition probabilities are conditional probabilities.

Published by Woodhead Publishing Limited, 2013

Table 2.1 The effect of reactant combinations (h) upon the effect upon the stochastic rate of reaction (a). The deterministic rates of reaction are provided for comparison.

Reaction	h	a (in terms of C)	a (in terms of k)	$\mathcal{R}$
$A \xrightarrow{k} B$	X_A	CX_A	kX_A	kc_A
$A+B \xrightarrow{k} C+D$	X_AX_B	CX_AX_B	$\frac{k}{VN_{Av}}X_AX_B$	kc_Ac_B
$A+B \xrightarrow{k} C$	X_AX_B	CX_AX_B	$\frac{k}{VN_{Av}}X_AX_B$	kc_Ac_B
$C \xrightarrow{k} A+B$	X_C	CX_C	kX_C	kc_C
$A+B \xrightarrow{k} 2C$	X_AX_B	CX_AX_B	$\frac{k}{VN_{Av}}X_AX_B$	kc_Ac_B
$2C \xrightarrow{k} A+B$	$\binom{X_C}{2}$	$C\binom{X_C}{2}$	$\frac{2k}{VN_{Av}}\binom{X_C}{2}$	kc_C^2
$2A \xrightarrow{k} C$	$\binom{X_A}{2}$	$C\binom{X_A}{2}$	$\frac{2k}{VN_{Av}}\binom{X_A}{2}$	kc_A^2
$C \xrightarrow{k} 2A$	X_C	CX_C	kX_C	kc_C
$A+B \xrightarrow{k} 2A$	X_AX_B	CX_AX_B	$\frac{k}{VN_{Av}}X_AX_B$	kc_Ac_B
$2A \xrightarrow{k} A+B$	$\binom{X_A}{2}$	$C\binom{X_A}{2}$	$\frac{2k}{VN_{Av}}\binom{X_A}{2}$	kc_A^2

2.2.2 *The total probability theorem*

Biochemical reaction pathways are Markov Chains: the transition probabilities from a state depend only upon the state from which the transition occurs. Therefore, given definitions for the complete set of transition probabilities, one may define $P(\mathbf{x}, t)$ via the *total probability theorem.*[1]

$$P(\mathbf{x}, t+\delta t) = \sum_{\mu=1}^{M} P(\mathbf{x}-\boldsymbol{\nu}_\mu \to \mathbf{x}, t)P(\mathbf{x}-\boldsymbol{\nu}_\mu, t) \quad (2.9)$$

$$+P(\mathbf{x} \to \mathbf{x}, t)P(\mathbf{x}, t) \quad (2.10)$$

The first term on the right hand side of Eq. 2.9 is the probability that state $\mathbf{x}$ results from a transition from every other state that is one reaction event away from $\mathbf{x}$. The second term

Published by Woodhead Publishing Limited, 2013

is the probability that the system remains in state $\mathbf{x}$ from t until $t+\delta t$. Since the set of transitions from a state, including the lack of a transition, is mutually exclusive , one may rewrite the total probability theorem as

$$P(\mathbf{x},t+\delta t)-P(\mathbf{x},t)=\sum_{\mu=1}^{M} P(\mathbf{x}-\boldsymbol{\nu}_{\mu}\rightarrow \mathbf{x},t)P(\mathbf{x}-\boldsymbol{\nu}_{\mu},t) \tag{2.11}$$

$$-\sum_{\mu=1}^{M} P(\mathbf{x}\rightarrow \mathbf{x}+\boldsymbol{\nu}_{\mu},t)P(\mathbf{x},t) \tag{2.12}$$

Next, we recognize the transition probabilities from our previous discussion (Eq. 2.8):

$$P(\mathbf{x}\rightarrow \mathbf{x}+\boldsymbol{\nu}_{\mu},t)=C_{\mu}h_{\mu}(\mathbf{x})\delta t+o(\delta t)$$

Inserting this expression into Eq. 2.12, and rearranging terms, one obtains

$$\frac{P(\mathbf{x},t+\delta t)-P(\mathbf{x},t)}{\delta t}=\sum_{\mu=1}^{M}\left[C_{\mu}h_{\mu}(\mathbf{x}-\boldsymbol{\nu}_{\mu})+\frac{o(\delta t)}{\delta t}\right] \times P(\mathbf{x}-\boldsymbol{\nu}_{\mu},t) \tag{2.13}$$

$$-\sum_{\mu=1}^{M}\left[C_{\mu}h_{\mu}(\mathbf{x})+\frac{o(\delta t)}{\delta t}\right]P(\mathbf{x},t) \tag{2.14}$$

which, in the limit $\delta t\rightarrow\infty$ becomes

$$\frac{dP(\mathbf{x},t)}{dt}=\sum_{\mu=1}^{M} C_{\mu}h_{\mu}(\mathbf{x}-\boldsymbol{\nu}_{\mu})P(\mathbf{x}-\boldsymbol{\nu}_{\mu},t)-\sum_{\mu=1}^{M} C_{\mu}h_{\mu}(\mathbf{x})P(\mathbf{x},t) \tag{2.15}$$

This is the *chemical master equation* (CME). In some regards, it is the stochastic analogue to the deterministic material balance equations discussed in Chapter 1, with the exception that it specifies the probability $P(\mathbf{x},t)$ in lieu of the concentration $c_i(t)$.

Published by Woodhead Publishing Limited, 2013

2.2.3 The stochastic rate constant

The stochastic rate of reaction

$$a_\mu(\mathbf{x}) = C_\mu h_\mu(\mathbf{x}) \qquad \text{Rate of reaction } \mu \tag{2.16}$$

is analogous to the rate of reaction $\mathcal{R}_\mu$ discussed in Chapter 1. However, there are distinctions between these rates and between their corresponding rate constants that merit some discussion. According to deterministic approach, the rate of change of molecular population resulting from reaction μ is

$$VN_{\mathrm{Av}}\mathcal{R}_\mu$$

where $N_{\mathrm{Av}} = 6.022 \times 10^{23}/\mathrm{mol}$ is Avogadro's number. According to stochastic approach to chemical kinetics this quantity is a_μ. Therefore, one would expect that

$$a_\mu = VN_{\mathrm{Av}}\mathcal{R}_\mu \tag{2.17}$$

However, this relationship reveals distinctions between the rate constants of the deterministic and stochastic approaches.

Unimolecular reactions of the form $\mathrm{A} \xrightarrow{k} \mathrm{B}$ A reaction of this type will have a stochastic rate defined by $a = CX_A$ and a deterministic rate of

$$\begin{aligned} \mathcal{R} &= VN_{\mathrm{Av}}kc_A \\ &= VN_{\mathrm{Av}}k\left(\frac{X_A}{V\mathcal{N}_A}\right) \\ &= kX_A \end{aligned}$$

Setting these expressions equal to each other, one obtains

$$C_\mu = k_\mu \qquad \mathrm{A} \rightarrow \mathrm{B} \tag{2.18}$$

That is, the stochastic rate constant for unimolecular reactions is equal to the deterministic rate constant.

Published by Woodhead Publishing Limited, 2013

Bimolecular reactions of the form $A + B \xrightarrow{k} C$ Reactions of this type will have stochastic rates defined by $a = CX_AX_B$ and deterministic rates of the form

$$\begin{aligned}\mathscr{R} &= VN_{\mathrm{Av}}kc_Ac_B \\ &= VN_{\mathrm{Av}}k\left(\frac{X_A}{VN_{\mathrm{Av}}}\right)\left(\frac{X_B}{VN_{\mathrm{Av}}}\right) \\ &= \frac{k}{VN_{\mathrm{Av}}}X_AX_B\end{aligned}$$

Therefore

$$C_\mu = \frac{k_\mu}{VN_{\mathrm{Av}}} \qquad \mathrm{A} + \mathrm{B} \rightarrow \mathrm{C} \tag{2.19}$$

Although the expression is somewhat more complex than Eq. 2.18 it reflects a fundamental difference in quantities C and k for this bimolecular reaction. Rate constants for bimolecular reactions are typically measured in units of $M^{-1}s^{-1}$, whereas C has units of s^{-1} regardless of reaction type.

Bimolecular reactions of the form $2A \xrightarrow{k} C$ This type of reaction is has a stochastic rate of

$$\begin{aligned}a &= C\binom{X_A}{2} \\ &= C\frac{X_A(X_A-1)}{2} \\ &\approx \tfrac{1}{2}CX_A^2\end{aligned}$$

The deterministic rate, however, is typically expressed as

$$\begin{aligned}\mathscr{R} &= kc_A^2 \\ &= k\left(\frac{X_A}{VN_{\mathrm{Av}}}\right)\left(\frac{X_A}{VN_{\mathrm{Av}}}\right)\end{aligned}$$

Relating these two rates via Eq. 2.17 in the limit of large X_A yields the following expression

$$C_\mu = \frac{2k_\mu}{VN_{\mathrm{Av}}} \qquad 2\mathrm{A} \rightarrow \mathrm{C} \tag{2.20}$$

Published by Woodhead Publishing Limited, 2013

The factor of two in the numerator of this expression distinguishes it from Eq. 2.19 as a consequence of the form of $\mathscr{R}$ utilized by most experimentalists. From a theoretical point of view the deterministic rate should be expressed as $\mathscr{R} = \frac{1}{2}kc_A^2$, and some practitioners have done so when studying the kinetics of dimerization. However, the common convention is to lump the factor of 1/2 into the rate constant. Before utilizing an experimentally measured rate constant in a stochastic model, it is essential to know which type of rate law was used to fit the data. If the rate constant was fit to data using a rate model of $\mathscr{R} = \frac{1}{2}kc_A^2$, then Eq. 2.19 should be used to relate the stochastic and deterministic rate constants.

2.3 Solution of the Master Equation

For a given reaction or set of reactions, the CME is often more challenging to solve than the corresponding deterministic material balance equations. The complexity increases with the number of reactions as well as the molecularity of the reactions. Common strategies for the solution of differential difference equations such as the CME include transformation to a partial differential equation and transformation to a linear system of equations. The choice of approach often depends upon the type of reaction or reaction network under inquiry.

2.3.1 Zeroeth order processes

Strictly speaking, there are no zeroeth order chemical reactions in nature. As Table 2.1 shows, the rates of chemical reactions are always at least first order in terms of the reactant concentrations. A zeroeth order reaction rate means that $h_\mu = 1$, implying that there is only one way that the reaction could happen regardless of the population of reactant molecules. Essentially, it means that *there are no reactant molecules at all* and that product molecules spontaneously and randomly appear in the system, i.e.

Published by Woodhead Publishing Limited, 2013

$$\varnothing \rightarrow \mathrm{P} \tag{2.21}$$

Clearly, chemistry does not permit the spontaneous generation of matter. Indeed, biochemists speak of unimolecular reactions and bimolecular reactions, but there is not even a proper name for zeroeth order reactions.

Nevertheless, in some circumstances, a reaction may *appear* to be zeroeth order for a limited period of time. For instance, consider the Michaelis-Menten rate of Chapter 1. If the concentration of substrate c_S is much higher than K_M, then $\mathscr{R} = v_{\max}$. However, as the substrate is consumed by this *pseudo*-zeroeth order reaction, it's concentration will decreases until c_S and K_M are relatively close in magnitude. At this point, the reaction rate ceases to behave as zeroeth order. In other words, reaction rates may appear to be zeroeth order, but will only do so for a limited amount of time. That said, there do exist physical processes that may be conceptualized as first order. Consider for instance an action performed by thousands of biochemists every day: the addition of a protein solution to a small reaction chamber (e.g. a well of a 384 well plate). We assume

1. the reaction chamber (e.g. the well) is much smaller than the reservoir of protein solution, and
2. the bulk concentration of protein in the reservoir is constant, c_0.

Although the protein solution to be added to the reaction chamber is homogeneous, a droplet of volume δV added to the chamber may or may not contain any protein molecules, as protein molecules move randomly through the media via Brownian motion. If a single protein molecule exists within the reservoir of volume V, then

$$\frac{\delta V}{V} = \Pr(\text{A protein molecule will be found within a droplet of volume } \delta V)$$

Alternatively, if there are many protein molecules in the media at a concentration c_0, then the total number of molecules in bulk is $c_0 V$. Therefore,

Published by Woodhead Publishing Limited, 2013

$$c_0 \delta V = \text{Pr(Protein molecules will be found within a droplet of volume } \delta V)$$

If the media is introduced into the reaction chamber at a constant rate $F = \delta V/\delta t$, then the average rate of addition of proteins to the reaction chamber is $\lambda = c_0 F$, and

$$\lambda \delta t + o(\delta t) = \text{Pr(A protein molecule will be introduced to the reaction chamber within the imminent time interval } \delta t) \quad (2.22)$$

From the point of view of the reaction chamber, the stochastic process resembles the non-physical reaction of Eq. 2.21.

Let us define the state of the system by the number of protein molecules within it, and the probability distribution for these states as

$$P_x(t) = \Pr(X_P = x) \quad (2.23)$$

Combining Eq. 2.15 with the "reaction rate" defined by Eq. 2.22 ($a = \lambda$), the master equation corresponding to this process is

$$\frac{dP_x(t)}{dt} = \lambda P_{x-1}(t) - \lambda P_x(t) \quad (2.24)$$

with initial condition $P_x(0) = \delta_{x0}$, where δ_{ij} is a Kronecker delta function, indicating that the reaction chamber is initially empty.

Solution of this differential difference equation is relatively straightforward. First, let us insert $x = 0$ into Eq. 2.24. Recognizing that $P_x = 0$ for $x < 0$, we obtain

$$\frac{dP_0(t)}{dt} = -\lambda P_0(t)$$

which has the solution

$$P_0(t) = e^{-\lambda t}$$

Next, let us insert $x = 1$ into Eq. 2.24. We obtain

$$\frac{dP_1(t)}{dt} = \lambda P_0(t) - \lambda P_1(t)$$

Published by Woodhead Publishing Limited, 2013

which has the initial condition $P_1(t) = 0$. This differential equation may be solved via the integrating factor method [19], yielding

$$P_1(t) = \lambda t e^{-\lambda t}$$

This procedure can be repeated for successive values of x, yielding a clear pattern that satisfies Eq. 2.24:

$$P_x(t) = \frac{(\lambda t)^x}{x!} e^{-\lambda t} \qquad \varnothing \rightarrow \mathrm{P} \qquad (2.25)$$

This is a Poisson distribution, i.e. $x \sim \mathrm{Pois}(\lambda t)$.

Statistics for the Poisson distribution are well known: The expectation value (average) for the population of protein molecules is

$$E(X_P) = \lambda t \qquad \varnothing \rightarrow \mathrm{P} \qquad (2.26)$$

And in an unusual coincidence, the variance is equal to the mean

$$V(X_P) = \lambda t \qquad \varnothing \rightarrow \mathrm{P} \qquad (2.27)$$

Therefore, as time proceeds, the "standard deviation" $\sigma = \sqrt{V(X_P)}$ will grow much slower than the mean, resulting in increased certainty in the reaction chamber's population over the course of time.

Although the population of protein molecules increases over time, the volume of the reaction chamber, v, also increases as $v = Ft$. Therefore, the average concentration is $\lambda/F = c_0$, the concentration of protein in the bulk protein solution. The statistical fluctuations are themselves quite large, however, and $P_x(t)$ admits the possibility that as much as a mole (6.022×10^{23}) or more of protein molecules could randomly enter the reaction chamber in finite time, regardless of the volume of the chamber. This type of paradox is common when reactions are treated as zeroeth order. The conclusion follows from our assumption that the reservoir solution is not "stochastically diluted" as its contents flow to the reaction chamber. By stating that the concentration of the bulk solution remains constant, one explicitly forbids this possibility; if one includes it, the "filling process" becomes first order.

Published by Woodhead Publishing Limited, 2013

2.3.2 First order reactions

Analytical solutions are known for several unimolecular reactions: The first solution of a birth-death equation describing a chemical reaction is attributed to Max Delbrück, who investigated the non-physical autocatalytic reaction $A \rightarrow 2A$ [8]. Over ten years later, following the publication of Doob's book, Anthony Bartholomay solved the CMEs for the reactions $A \rightarrow B$ and $A \rightleftharpoons B$ [3, 4]. Subsequently, Donald McQuarrie published solutions of the CME for the coupled reactions $A \rightarrow B$ and $A \rightarrow C$ [16]. In 1966, Arnold Fredrickson formulated solutions for more generalized unimolecular reaction networks [11, 10]. However, a general method of solution for arbitrary unimolecular reaction networks with arbitrary initial conditions did not exist until 2004, when Xueying Zhang and coworkers published it in the *Journal of Chemical Physics* [21].

To illustrate these methods, let us consider the "simplest" of unimolecular reactions: the elementary isomerization reactions

$$A \underset{k_2}{\overset{k_1}{\rightleftharpoons}} B \tag{2.28}$$

Each reaction represents the spontaneous and direct conversion of one type of molecule into the other without catalysis or additional reaction steps. For instance, "A" and "B" might represent a protein in two states of activity. In the most general scenario the reaction is initiated with N_A A molecules and N_B B molecules, for a total of $N = N_A + N_B$ reactive molecules.

Applying Eq. 2.15 to this set of reactions, one obtains

$$\begin{aligned} \frac{dP(x_A, x_B)}{dt} &= k_1(x_A + 1)P(x_A + 1, x_B - 1) \\ &\quad + k_2(x_B + 1)P(x_A - 1, x_B + 1) \\ &\quad - \left(k_1 x_A + k_2 x_B\right) P(x_A, x_B) \end{aligned} \tag{2.29}$$

The initial condition is concisely expressed as

$$P(x_A, x_B, 0) = \delta_{x_A, N_A} \delta_{x_B, N_B} \tag{2.30}$$

Published by Woodhead Publishing Limited, 2013

Although Eq. 2.29 appears to be a partial differential difference equation with two discrete variables and one continuous variable, it may be simplified via the reaction stoichiometry. The stoichiometry of Eq. 2.35 demands that

$$X_B + X_A = N \tag{2.31}$$

Thus, the number of discrete random variables in Eq. 2.29 may be reduced by one. If we define

$$P_x(t) = \Pr(X_A = x, X_B = N - x|t) \tag{2.32}$$

then Eq. 2.29 becomes

$$\begin{aligned}\frac{dP_x}{dt} &= k_1(x+1)P_{x+1} + k_2(N-x+1)P_{x-1}\\ &\quad -(k_1x + k_2(N-x))P_x\end{aligned} \tag{2.33}$$

which has the initial condition

$$P_x(0) = \delta_{N,x} \tag{2.34}$$

The irreversible reaction A → B

If Eq. 2.28 is irreversible, i.e.

$$\mathrm{A} \xrightarrow{k_1} \mathrm{B} \tag{2.35}$$

and the reaction process starts with N A molecules and no B molecules, then the Eq. 2.33 becomes

$$\frac{dP_x}{dt} = k_1(x+1)P_{x+1} - k_1xP_x \qquad P_x(0) = \delta_{x,N} \tag{2.36}$$

Equations such as 2.36 are often solved via the method of generating functions,[2] which is equivalent to the method of Z transforms.[3] In this approach, one transforms Eq. 2.36 from an equation in terms of the discrete variable x to a differential equation in terms of a continuous variable s. The simplifed equation in s is then solved, and the resulting solution is then inverted to yield $P_x(t)$.

To begin, let us define the generating function:

$$G(s,t) = \sum_{x=0}^{\infty} s^x P_x(t) \tag{2.37}$$

In this expression, s is a continuous variable limited to the range $[-1, 1]$. It does not depend on t. Transformation of $P_x(t)$ to $G(s,t)$ is accomplished by multiplying the CME (Eq. 2.36) by s^x and summing over the entire range of x. One obtains

$$\frac{d}{dt}G = k_1 \left(\sum_{x=0}^{N} s^x (x+1) P_{x+1} - \sum_{x=0}^{N} s^x x P_x \right) \tag{2.38}$$

The sums on the right hand side may then be expressed entirely in terms of G and s by relating them to the derivatives of G:

$$\frac{\partial G}{\partial s} = \sum_{x=0}^{\infty} x s^{x-1} P_x(t) \tag{2.39}$$

$$\frac{\partial^2 G}{\partial s^2} = \sum_{x=0}^{\infty} x(x-1) s^{x-2} P_x(t) \tag{2.40}$$

Often, this may be facilitated by defining new indices for the sums, e.g. $i = x + 1$, and recognizing that $P_{-1} = 0$ and $P_{x>N} = 0$. In this case, one obtains

$$\frac{\partial G}{\partial \tau} = (1-s)\frac{\partial G}{\partial s} \qquad G(0,s) = s^N \tag{2.41}$$

where we have introduced the dimensionless variable $\tau = k_1 t$.

This partial differential equation may be solved in the standard way via separation of variables. First, we subtitute $G(s,\tau) = T(\tau)S(s)$ into Eq. 2.41, which yields

$$\frac{1}{T(\tau)}\frac{\partial T(\tau)}{\partial \tau} = (1-s)\frac{1}{S(s)}\frac{\partial S(s)}{\partial s} \tag{2.42}$$

Published by Woodhead Publishing Limited, 2013

Since the right and left sides of Eq. 2.42 feature derivatives of different variables, and τ and s are independent, each side of the equation must be equal to a common constant:

$$\frac{1}{T(\tau)}\frac{\partial T(\tau)}{\partial \tau} = C \tag{2.43}$$

$$C = (1-s)\frac{1}{S(s)}\frac{\partial S(s)}{\partial s} \tag{2.44}$$

Insofar as these equations are of one variable, they are ordinary differential equations and may be solved using standard approaches. One immediately obtains

$$T(\tau) = A' \exp(C\tau) \tag{2.45}$$

and

$$S(s) = B'(1-s)^{-C} \tag{2.46}$$

where A' and B' are constants to be determined from the initial condition. We may express the general solution as

$$G = \sum_n A_n \left((1-s)e^{-\tau}\right)^{-C_n} \tag{2.47}$$

The constants $\{A_n\}$ and $\{C_n\}$ may be obtained by setting the s-derivatives of the initial condition for Eq. 2.41 equal to the s-derivatives of Eq. 2.47. The resulting expression is then evaluated at $\tau = 0$ and $s = 1$ to give

$$\left.\frac{(N)!}{(N-k)!} s^{N-k}\right|_{s=1} = \left.\frac{d^k G}{ds^k} \sum_n A_n (1-s)^{-C_n}\right|_{s=1} \tag{2.48}$$

For the first derivative ($k = 1$), we obtain

$$N = -\lim_{s\to 1} \sum_n (-C_n) A_n (1-s)^{-C_n - 1} \tag{2.49}$$

The only term in the sum that survives in this limit is the one for which $C_n = -1$. If we define $C_1 = -1$, then $A_1 = -N$.

Published by Woodhead Publishing Limited, 2013

A similar analysis of the higher derivatives reveals $A_2 = \binom{N}{2}$, $A_3 = -\binom{N}{3}$, etc., suggesting

$$A_n = (-1)^n \binom{N}{n} \tag{2.50}$$

which may be proven via induction. The corresponding constants $C_2 = -2$, $C_3 = -3$, etc. suggest

$$C_n = -n \tag{2.51}$$

which may also be proven via induction. Therefore, the solution to Eq. 2.41 is

$$G(s,t) = \sum_{n=0}^{N} \binom{N}{n} \left((s-1)e^{-k_1 t}\right)^n \tag{2.52}$$

Employing the binomial theorem,[4] Eq. 2.52 may be expressed the form

$$G(s,t) = \left(1 + (s-1)e^{-k_1 t}\right)^N \qquad \mathrm{A} \xrightarrow{k_1} \mathrm{B} \tag{2.53}$$

In Eq. 2.37, $P_x(t)$ serves the role of a Taylor coefficient. Thus $P_x(t)$ may be inverted from $G(s,t)$ using Taylor's theorem,

$$P_x(t) = \frac{1}{x!} \left.\frac{\partial^x G(s,t)}{\partial s^x}\right|_{s=0} \qquad \text{Inversion Formula} \tag{2.54}$$

In this case, the derivatives are easily calculated

$$\begin{aligned}
\frac{\partial G(s,t)}{\partial s} &= N\left(1 + (s-1)e^{-k_1 t}\right)^{N-1} e^{-k_1 t} \\
\frac{\partial^2 G(s,t)}{\partial s^2} &= N(N-1)\left(1 + (s-1)e^{-k_1 t}\right)^{N-2} e^{-2k_1 t} \\
\frac{\partial^3 G(s,t)}{\partial s^3} &= N(N-1)(N-2)\left(1 + (s-1)e^{-k_1 t}\right)^{N-3} e^{-3k_1 t} \\
&\vdots
\end{aligned}$$

Published by Woodhead Publishing Limited, 2013

$$\frac{\partial^x G(s,t)}{\partial s^x} = N(N-1)\cdots(N-x)\left(1+(s-1)e^{-k_1 t}\right)^{N-x} e^{-xk_1 t}$$

Therefore, the solution to the CME for the irreversible reaction $A \xrightarrow{k_1} B$ is

$$P_x(t) = \binom{N}{x}\left(1-e^{-k_1 t}\right)^{N-x} e^{-xk_1 t} \qquad A \xrightarrow{k_1} B \qquad (2.55)$$

We recognize this as a binomial distribution

$$x \sim \text{Binomial}(N, p) \qquad (2.56)$$

with $p = \exp(-k_1 t)$.

The time evolution of the distribution is illustrated in Figure 2.3 for a case with $N = 10$ molecules. Initially, there is no distribution *per se*, as the state is known with absolute certainty. However, as time proceeds the uncertainty in the population of A molecules grows. Ultimately, the irreversible nature of the reaction results in the consumption of all A molecules, once again eliminating the uncertainty in X_A.

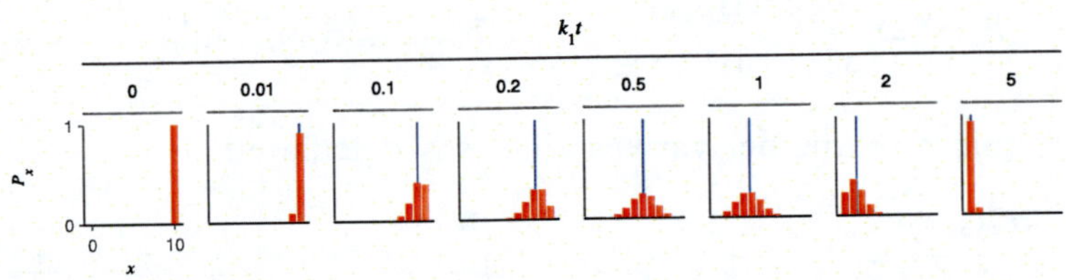

Figure 2.3 The probability density $P_x(t)$ for the reaction $A \xrightarrow{k_1} B$. This reaction is initiated with $N = 10$ A molecules (Eq. 2.55) . At $t = 5/k_1$, it is almost certain that no A molecules remain, but there is a finite probability that $X_A = 1$.

Published by Woodhead Publishing Limited, 2013

Expressions for the expectation value and variance may be obtained from the generating function using the following equations[5]

$$E(X) = \sum_{x=0}^{\infty} xP_x(t) = \left.\frac{\partial G(s,t)}{\partial s}\right|_{s=1} \tag{2.57}$$

$$V(X) = \textstyle\sum_{x=0}^{N} (x - E(x))^2 P_x(t) = \left.\frac{\partial^2 G(s,t)}{\partial s^2}\right|_{s=1} + E(x) - [E(x)]^2 \tag{2.58}$$

Aplying these, we find that the average population of A molecules is

$$E(X_A) = Ne^{-k_1 t} \qquad \mathrm{A} \xrightarrow{k_1} \mathrm{B} \tag{2.59}$$

and the variance is

$$V(X_A) = Ne^{-k_1 t}\left(1 - e^{-k_1 t}\right) \qquad \mathrm{A} \xrightarrow{k_1} \mathrm{B} \tag{2.60}$$

In Figure 2.4, we illustrate the time evolution of the population of A molecules (mean $\pm$ standard deviation) as the molecules are consumed. In the thermodynamic (large N) limit, the standard deviation of the mean $\sigma = \sqrt{V(X)}$ becomes insignificant. This is why stochastic fluctuations of molecular population are not commonly observed in bulk experiments. Moreover, it is a statistical justification of the use of the deterministic formalism for the characterization of chemical kinetics in bulk. However, systems featuring fewer than 100 reactant molecules, i.e. those that take place within nanoscopic volumes (e.g. organelles, endosomes, or viruses) will exhibit strong fluctuations, and will not have deterministic time evolutions.

The reversible reaction $\mathrm{A} \rightleftharpoons \mathrm{B}$

If the reversible isomerization process

$$\mathrm{A} \underset{k_2}{\overset{k_1}{\rightleftharpoons}} \mathrm{B} \tag{2.61}$$

Published by Woodhead Publishing Limited, 2013

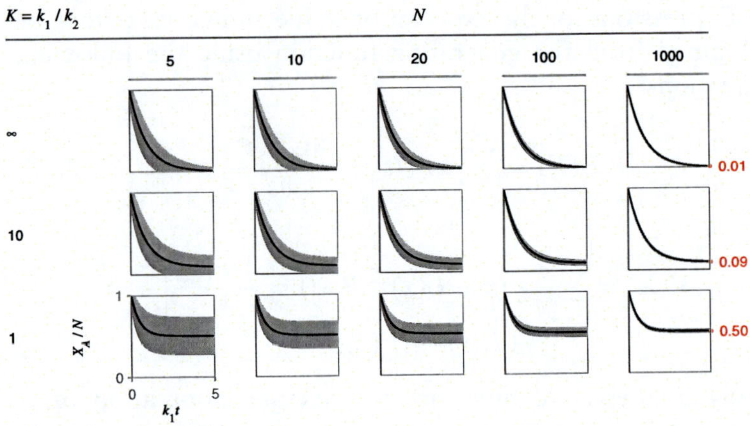

Figure 2.4 Time evolution of the reaction A $\xrightarrow{k_1}$ B. The average population of A molecules ($E(X)/N$, Eq. 2.59) is illustrated as a black line and surrounded by a region constituting one standard deviation, $\sigma = \sqrt{V(X)}/N$, Eq. 2.60). Results for irreversible reactions are presented in the top row. As the total population of molecules increases, the standard deviation decreases as $N^{-1/2}$.

is initiated with N A molecules and no B molecules, then one may utilize the generating function method to obtain $P_x(t)$. However, if the process is initiated with N_A A molecules and $N_B = N - N_A$ B molecules, the generating function will have the form [7]

$$G(s,\tau) = \left(\frac{-K(1-s)e^{-(1+K^{-1})\tau} + K + s}{(1-s)e^{-(1+K^{-1})\tau} + K + s} \right)^{N_A} \times \left(\frac{(1-s)e^{-(1+K^{-1})\tau} + K + s}{1+K} \right)^{N} \quad (2.62)$$

This expression, while differentiable, does not yield a neat Taylor series from which the probability distribution $P_x(t) = \Pr(X_A = x, X_B = N - x)$ may be inverted. Although the

Published by Woodhead Publishing Limited, 2013

solution for the CME for this reaction initiated with all A or all B molecules was published by Bartholomay in the 1950s, the general solution for the case initiated with N_A A molecules and N_B B molecules remained elusive until 2005, when a team of researchers from Petar Djurić's group at SUNY Stony Brook published a general solution of the CME for *any* uncoupled chemical reaction network, i.e any network that lacks bimolecular reactions [21]. Let us explore this powerful technique, which has applications to a wide variety biochemical reaction networks and their CMEs.

Reaction with one molecule Let us begin by considering an experiment initiated with just one molecule. As the system evolves in time, it may exist in only one of two states: $x = 0$ (no A molecules, one B molecule) and $x = 1$ (one A molecule, no B molecules). Eq. 2.33 may be expressed as two equations, one for each state:[6]

$$\frac{dP_0}{dt} = -k_2 P_0 + k_1 P_1 \tag{2.63}$$

$$\frac{dP_1}{dt} = k_2 P_0 - k_1 P_1 \tag{2.64}$$

It is convenient to express Eqs. 2.63 and 2.64 in matrix form

$$\frac{d}{d\tau}\mathbf{P} = \mathbf{A}\mathbf{P} \tag{2.65}$$

where once again we use the dimensionless time $\tau = k_1 t$, and define $\mathbf{P}$ and $\mathbf{A}$ by

$$\mathbf{P}(\tau) = \begin{bmatrix} P_0(\tau) \\ P_1(\tau) \end{bmatrix} \tag{2.66}$$

and

$$\mathbf{A} = \begin{bmatrix} -K^{-1} & 1 \\ K^{-1} & -1 \end{bmatrix} \tag{2.67}$$

Conveniently, $K = k_1/k_2$ is the equilibrium constant for the reaction.

Applying a Laplace transform to Eq. 2.65 [1]

Published by Woodhead Publishing Limited, 2013

$$P_x(s) = \int_0^{\infty} e^{-s\tau} P_x(\tau) \tag{2.68}$$

one obtains

$$s\mathbf{P}(s) - \mathbf{P}_0 = \mathbf{A}\mathbf{P}(s) \tag{2.69}$$

where

$$\mathbf{P}_0 = \lim_{t \to 0} \begin{bmatrix} P_0(t) \\ P_1(t) \end{bmatrix} \tag{2.70}$$

and

$$\mathbf{P}(s) = \begin{bmatrix} P_0(s) \\ P_1(s) \end{bmatrix} \tag{2.71}$$

Eq. 2.69 is readily expressed in a standard matrix form

$$[s\mathbf{I} - \mathbf{A}]\,\mathbf{P}(s) = \mathbf{P}_0 \tag{2.72}$$

where **I** is the identity matrix. The general solution is

$$\mathbf{P}(s) = [s\mathbf{I} - \mathbf{A}]^{-1}\,\mathbf{P}_0 \tag{2.73}$$

where in this simple case, the inverse matrix is

$$[s\mathbf{I} - \mathbf{A}]^{-1} = \frac{1}{s\left(s + (1 + K^{-1})\right)} \begin{bmatrix} s+1 & 1 \\ K^{-1} & s + K^{-1} \end{bmatrix} \tag{2.74}$$

Once expressions for $P_x(s)$ are obtained, they may be inverted back to time-space using tables of Laplace transforms. The tables provided by Abramovitz and Stegun are particularly useful for this purpose[7] [1].

Since we are considering a system with only one molecule, there are two initial conditions worth considering: The case where the molecule is initially A and the case where the molecule is initially B. Let us consider each in turn. If the process begins with a single A molecule ($N_A = 1$), then $\mathbf{P}_0 = [0, 1]^T$. Inserting this initial condition into Eq. 2.73 yields, after some algebra,

$$P_x(s) = \begin{cases} \frac{1}{s(s+(1+K^{-1}))} & x = 0 \\ \frac{1}{(s+(1+K^{-1}))} + \frac{K^{-1}}{s(s+(1+K^{-1}))} & x = 1 \end{cases}$$

Published by Woodhead Publishing Limited, 2013

The solution to the CME is obtained by performing an inverse Laplace transform upon these expressions. Utilizing the inverse transform tables, we obtain

$$P_x(t) = \begin{cases} \frac{K}{1+K}\left(1 - e^{-(k_1+k_2)t}\right) & x = 0 \\ \frac{1}{1+K}\left(1 + Ke^{-(k_1+k_2)t}\right) & x = 1 \end{cases} \qquad N_A = 1, N_B = 0 \tag{2.75}$$

This is a Bernoulli distribution, i.e. $x \sim \text{Bernoulli}(p)$, with $p = \frac{1}{1+K}\left(1 + Ke^{-(k_1+k_2)t}\right)$.

On the other hand, if the process is initiated with one B molecule ($N_B = 1$), then $\mathbf{P}_0 = [1, 0]^T$. In this case, the Laplace-space solution to Eqs. 2.63 and 2.64 is

$$P_x(s) = \begin{cases} \frac{1}{(s+(1+K))} + \frac{1}{s(s+(1+K^{-1}))} & x = 0 \\ \frac{K^{-1}}{s(s+(1+K^{-1}))} & x = 1 \end{cases}$$

Once again making use of the inverse Laplace transform tables, one obtains

$$P_x(\tau) = \begin{cases} \frac{1}{1+K}\left(K + e^{-(k_1+k_2)t}\right) & x = 0 \\ \frac{1}{1+K}\left(1 - e^{-(k_1+k_2)t}\right) & x = 1 \end{cases} \qquad N_A = 0, N_B = 1 \tag{2.76}$$

This too is Bernoulli distribution, i.e. $x \sim \text{Bernoulli}(p)$, with $p = \frac{1}{1+K}\left(1 - e^{-(k_1+k_2)t}\right)$.

Interestingly as $t \to \infty$, Eqs. 2.75 and 2.76 are exactly the same, reflecting chemical equilibrium even though there is *only* one molecule. From this we can conclude that chemical equilibrium is conceptually deeper than a mere equivalence of reaction rates, as the deterministic approach would suggest. A single reactive molecule can literally exist in a state of equilibrium in the absence of any other reactive molecules. This is a common motif in biology, as ion channels regularly transition between open (A) and closed (B) states. Patch clamp measurements of current through individual ion channels often use expressions such as 2.75 and 2.76 to characterize the distinctively random kinetics of transitions between states.

Published by Woodhead Publishing Limited, 2013

Reaction with many molecules – the method of Zhang [21] In Eq. 2.61, A molecules do not interact with B molecules. Nor, for that matter, do A and B molecules interact amongst themselves. Each molecule acts independently, for all intents and purposes in isolation from its neighbors. This quality of all unimolecular reactions permits the construction of the solution to their CMEs from the molecules up.

To illustrate, consider the A $\rightleftharpoons$ B reaction initiated with N_A A molecules and no B molecules at $\tau = 0$. Because the states of each of the N_A molecules are independent of each other, the probability that there will be ℓ A molecules at a dimensionless time $\tau = k_1 t$ is just

$$P_\ell(\tau|N_A) = \binom{N_A}{\ell} \underbrace{p \cdots p}_{\text{Pr}(\ell \text{ are A})} \; \underbrace{(1-p)\cdots(1-p)}_{\text{Pr}(N_A-\ell \text{ are not A})}$$

where $\binom{N_A}{\ell}$ is the number of ways that one could have chosen which of the molecules were in the A state and which were in the B state. Alternatively, we may write this as

$$P_\ell(t|N_A) = \binom{N_A}{\ell} p_A^\ell (1-p_A)^{N_A-\ell} \tag{2.77}$$

where

$$p_A(t) = \frac{1}{1+K}\left(1 + Ke^{-(k_1+k_2)t}\right) \tag{2.78}$$

Likewise, if the reaction is initiated with N_B B molecules and no A molecules, one may show that the probability that there will be ℓ A molecules at time t is

$$P_m(t|N_B) = \binom{N_B}{m} p_B^m (1-p_B)^{N_B-m} \tag{2.79}$$

where

$$p_B(t) = \frac{1}{1+K}\left(1 - e^{-(k_1+k_2)t}\right) \tag{2.80}$$

Now let us consider the scenario where the reaction is initiated with N_A A molecules *and* N_B B molecules. This state can occur in as many ways as one can obtain x A molecules from

Published by Woodhead Publishing Limited, 2013

the initial N_A A molecules and N_B B molecules. Since the the reactions of A and B molecules are completely independent of each other, the probabilities $P_m(t|N_B)$ and $P_\ell(t|N_A)$ are also independent. Making use of the total probability theorem, we may write

$$\begin{aligned} P_x(t) = {} & P_0(t|N_A)P_x(t|N_B) \\ & + P_1(t|N_A)P_{x-1}(t|N_B) \\ & \ldots \\ & + P_x(t|N_A)P_0(t|N_B) \end{aligned} \tag{2.81}$$

keeping in mind that $P_m(t|N_B)$ and $P_\ell(t|N_A)$ are only defined for $m \leq N_B$ and $\ell \leq N_A$.

If $N_A \geq N_B \geq x$ we may express the solution to the CME as

$$P_x(t) = \sum_{n=0}^{x} P_k(t, N_A)P_{x-n}(t, N_B) \quad (N_A \geq N_B \geq x) \tag{2.82}$$

However, if $x > N_A$ or $x > N_B$, many of these terms in Eq. 2.81 should vanish. For instance, if $\ell > N_A$ then $P_\ell(t|N_A)$ is undefined, and if $m > N_B$ then $P_m(\tau|N_B)$ is undefined. By careful selection of the limits of summation, these terms may be removed,[8] yielding the general solution to the CME

$$P_x(t) = \sum_{n=n_1}^{n_2} \binom{N_A}{n}\binom{N_B}{x-n} p_A^n(1-p_A)^{N_A-n} p_B^{x-n}(1-p_B)^{N_B-x+n} \tag{2.83}$$

where n_1 and n_2 are defined by

$$n_1 = \begin{cases} 0 & N_A \geq N_B \geq x \\ x - N_B & \text{otherwise} \end{cases}$$

and

$$n_2 = \begin{cases} x & N_A \geq x \\ N_A & \text{otherwise} \end{cases}$$

and p_A and p_B are defined by Eqs. 2.78 and 2.80. Although this solution assumes that $N_A > N_B$, it is entirely general. If

Published by Woodhead Publishing Limited, 2013

one desires to calculate $P_x(t)$ for a case with $N_A \leq N_B$, one only needs to choose the species A as the species with the larger initial population.

The time evolution of the reversible isomerization process initiated with ten A molecules and five B molecules is illustrated in Fig. 2.5. Depending on the value of the equilibrium constant $K = k_1/k_2$, the distribution of X_A will drift to the right or left as the system approaches equilibrium. The mean population of A molecules also conforms with the law of mass action, the results of which are illustrated by a blue line. We conclude that the stochastic approach to chemical kinetics conforms with the predictions of the deterministic approach to chemical kinetics as well as the macroscopic law of mass action. Additionally, it conforms with the microscopic predictions of statistical mechanics as $t \to \infty$ [12]

In the special case where the reaction is initiated with A molecules alone, i.e. $N_A = N$, then Eq. 2.83 reduces to [7]

$$P_x(\tau) = \binom{N}{x}\left(\frac{K}{1+K}\right)^N \left(e^{-(k_1+k_2)t} + K^{-1}\right)^x \left(1 - e^{-(k_1+k_2)t}\right)^{N-x} \tag{2.84}$$

After some manipulation, this expression may be rewritten as

$$P_x(t) = \binom{N}{x} p_A^x (1 - p_A)^{N-x}$$

which is a Binomial distribution with p_A given by Eq. 2.78.

Comparing this result with the grand probability distribution for the unidirectional reaction A $\rightarrow$ B (Eq. 2.55), we recognize that if a biochemical reaction of the type A $\rightleftharpoons$ B is initiated with N A molecules and no B molecules - the experimentally expedient initial condition - then the distribution of A-populations will *always* be binomial

$$x \sim \text{Binomial}(p, N), \quad p = \begin{cases} e^{-k_1 t} & K^{-1} = \frac{k_2}{k_1} = 0 \\ \frac{1}{1+K}\left(1 + Ke^{-(k_1+k_2)t}\right) & \text{otherwise} \end{cases} \tag{2.85}$$

Published by Woodhead Publishing Limited, 2013

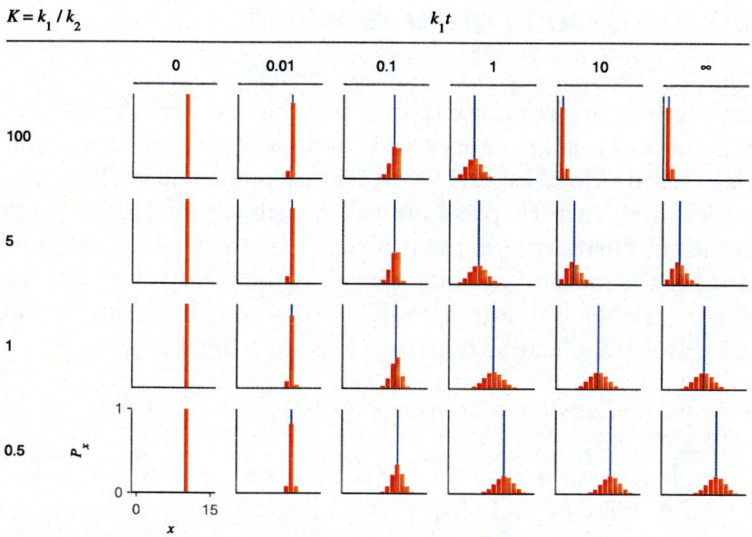

Figure 2.5 Time evolution of the grand probability distribution $P_x(\tau) = \Pr(X_A = x)$ for the reaction A $\rightleftarrows$ B. Each row represents the time evolution of a different reaction. All reaction processes are initiated with $N_A = 10$ A molecules and $N_B = 5$ B molecules. In the limit $t \to \infty$, the distribution shifts to the right as $K = k_1/k_2$ decreases. This is a microscopic reflection of the law of mass action.

In fact, almost any solution of the CME for unimolecular reaction networks will have some relationship to the binomial or multinomial distribution, as the molecules transition among many possible states *independently of each other*. This quality of unimolecular reaction networks makes them particularly well suited to experimental design and analysis, especially for experiments that investigate the dynamics of single molecules.

Published by Woodhead Publishing Limited, 2013

2.3.3 Second order reactions

The CMEs that describe "second order" reactions are substantially more complex than those that describe first order reactions. First and foremost, two or more terms on the right hand side of Eq. 2.15 will feature quadratic coefficients, which may in turn preclude the possibility of closed form solution. Furthermore, the molecules interact. Thus, we cannot use the method of Zhang to solve the CME as we did for the reversible reaction $A \rightleftharpoons B$. As a result, solutions of the CME for bimolecular reactions are often limited to

- single *irreversible* reactions, e.g. $A + B \rightarrow C$, and $2A \rightarrow B$ [20, 13, 18].
- single reversible reactions *at equilibrium*, e.g. $A + B \rightleftharpoons C$ and $A + B \rightleftharpoons C + D$ [6]

Let us take a closer look at of these classes of CMEs and their solutions.

The Irreversible Reaction A + B → C

Let us revisit the irreversible association reaction from Chapter 1,

$$A + B \xrightarrow{k_1} C \tag{2.86}$$

This class of reaction is one of the most studied in biochemistry, as A could represent a biological receptor, B could represent a ligand, and C, their complex. Other contexts include DNA hybridization and and the binding of antigens by antibodies. We consider a process initiated by the mixing of N_A A molecules and N_B B molecules to a final volume V in the absence of C molecules. Experimentally, this might be implemented by rapidly adding ligands to receptor-expressing cells in culture.

The molecular populations of A, B and C are related via stoichiometry

$$N_A - X_A = N_B - X_B = X_C \tag{2.87}$$

Published by Woodhead Publishing Limited, 2013

Therefore, any of these three populations may identify the state of the system. For convenience, we choose the population of A molecules to represent the state, and assert that this species is the limiting reactant (i.e. $N_B \geq N_A$). The probability distribution may then be defined as

$$P_x(t) = \Pr(X_A = x, X_B = Z + x, X_C = N_A - x) \qquad (2.88)$$

where $Z = N_B - N_A$. The stoichiometric relationships defined by 2.87 also allows us to express the rate of the reaction (Table 2.1) in terms of the state x

$$a_1(x) = \frac{k_1}{VN_{\text{Av}}} x(Z + x) \qquad (2.89)$$

Inserting Eqs. 2.88 and 2.89 into Eq. 2.15, one obtains the CME for the reaction $A + B \rightarrow C$

$$\frac{dP_x}{d\tau} = (x+1)(Z+x+1)P_{x+1} - x(Z+x)P_x \qquad P_x(0) = \delta_{x,N_A} \qquad (2.90)$$

Again, we express the CME in terms of a dimensionless time $\tau = k_1 t / N_{\text{Av}} V$.

Although this CME may be solved via the method of generating functions, the calculations are tedious and the resulting generating function is not easily inverted from s to x space. A better alternative is the method of Laplace transforms, also discussed previously (Section 2.3.2). First, we express Eq. 2.90 as a system of $N_A + 1$ ($x \in [0, N_A]$) linear differential equations as follows

$$\frac{d}{dt}\mathbf{P} = \mathbf{AP}$$

where

$$\mathbf{P} = [P_0(t), P_1(t), \ldots P_{N_A}(t)]^T$$

and

Published by Woodhead Publishing Limited, 2013

$$\mathbf{A} = \begin{bmatrix} 0 & u_1 & 0 & \cdots & \cdots & 0 \\ 0 & -u_1 & u_2 & \cdots & \cdots & 0 \\ \vdots & 0 & -u_2 & u_3 & \cdots & 0 \\ \vdots & \ddots & \ddots & \ddots & \ddots & \vdots \\ 0 & \cdots & \cdots & 0 & -u_{N_A-1} & u_{N_A} \\ 0 & 0 & \cdots & 0 & 0 & -u_{N_A} \end{bmatrix} \tag{2.91}$$

and $u_n = n(Z + n)$. The initial condition transforms to

$$\mathbf{P}_0 = [0, 0, \ldots 0, 1]^T$$

In our discussion of the reaction A $\rightleftharpoons$ B (Section 2.3.2), we showed that the Laplace space solution to the CME may be expressed as

$$\mathbf{P}(s) = [s\mathbf{I} - \mathbf{A}]^{-1}\mathbf{P}_0 \tag{2.92}$$

with $P_x(s)$ related to $P_x(t)$ by Eq. 2.68. However, determination of the elements of the matrix $[s\mathbf{I} - \mathbf{A}]^{-1}$ is substantially more challenging in this case because

1. $\mathbf{A}$ is an $(N_A + 1) \times (N_A + 1)$ matrix, and
2. we ultimately seek to obtain a general solution for any value of N_A.

Therefore, we employ the general expression

$$[s\mathbf{I} - \mathbf{A}]^{-1} = \frac{\mathrm{adj}[s\mathbf{I} - \mathbf{A}]}{\det[s\mathbf{I} - \mathbf{A}]}$$

where $\mathrm{adj}[s\mathbf{I} - \mathbf{A}]$ is the adjoint matrix and $\det[s\mathbf{I} - \mathbf{A}]$ is the determinant.

It is at this point that one must recognize an important aspect of the solution: Every element of $\mathbf{P}_0$ is zero with one exception: The last. Since this vector multiplies $[s\mathbf{I} - \mathbf{A}]^{-1}$ (Eq. 2.92), the only information in $[s\mathbf{I} - \mathbf{A}]^{-1}$ to survive the multiplication step resides in its last column. Therefore, we only needs to calculate the last row of the matrix of cofactors $\mathbf{C} = \mathrm{adj}[s\mathbf{I} - \mathbf{A}]^T$ and the determinant $\det[s\mathbf{I} - \mathbf{A}]$ to specify $\mathbf{P}(s)$, i.e.

$$P_x(s) = \frac{C_{N_A+1,x+1,}(s)}{\det[s\mathbf{I} - \mathbf{A}]}$$

Published by Woodhead Publishing Limited, 2013

From the definition of matrix cofactors and the structure of $[s\mathbf{I} - \mathbf{A}]$, it is straightforward to show that

$$C_{N_A+1,j} = (-1)^{N_A+j+1} D_{j-1}(s) \prod_{n=j}^{N_A} (-u_n) \tag{2.93}$$

where $D_j(s)$ is the determinant of the submatrix of $[s\mathbf{I} - \mathbf{A}]$ composed of its first j rows and columns. After some algebra, this expression may be substantially simplified, yielding the following expression for $P_x(s)$

$$P_x(s) = \frac{N_A!N_B!}{(Z+x)!x!} \frac{D_x(s)}{D_{N_A+1}(s)} \tag{2.94}$$

where $D_{N_A+1}(s) = \det[s\mathbf{I} - \mathbf{A}]$.

Development of an expression for $D_x(s)$, is relatively straightforward. The submatrix of $[s\mathbf{I} - \mathbf{A}]$ composed of its first x rows and columns, like $\mathbf{A}$, is diagonal

$$[s\mathbf{I}-\mathbf{A}] = \begin{bmatrix} s & -u_1 & 0 & \cdots & \cdots & 0 \\ 0 & s+u_1 & -u_2 & \cdots & \cdots & 0 \\ \vdots & 0 & s+u_2 & -u_3 & \cdots & 0 \\ \vdots & \ddots & \ddots & \ddots & \ddots & \vdots \\ 0 & \cdots & \cdots & 0 & s+u_{x-2} & -u_{x-1} \\ 0 & 0 & \cdots & 0 & 0 & s+u_{x-1} \end{bmatrix} \tag{2.95}$$

Thus, the determinant is the product of the diagonal elements, i.e.

$$D_x(s) = \prod_{n=0}^{x-1} [s + n(Z+n)] \tag{2.96}$$

Substituting this expression into Eq. 2.94 yields the Laplace space solution of the CME

$$P_x(s) = \frac{N_A!N_B!}{(Z+x)!x!} \frac{D_x(s)}{D_{N_A+1}(s)} \prod_{n=x}^{N_A+1} \left[\frac{1}{s+n(Z+n)}\right] \tag{2.97}$$

One may invert Eq. 2.97 via the residue theorem

Published by Woodhead Publishing Limited, 2013

$$P_x(\tau) = \sum_{n=0}^{N_A} \lim_{s \to \lambda_n} \left[(s - \lambda_n) e^{s\tau} P_x(s) \right] \tag{2.98}$$

where $\{\lambda_n\} = \{u_n\}$ are the roots of the polynomial in the denominator of Eq 2.97. The resulting expression is the solution of the CME:

$$P_x(\tau) = \frac{N_A! N_B!}{(Z+x)! x!} \sum_{n=x}^{N_A} (-1)^{n-x} \times \frac{(Z+2n)(Z+x+n-1)!}{(N_A-n)!(n-x)!(N_B+n)!} e^{-n(Z+n)\tau} \tag{2.99}$$

The stochastic time evolution of the reaction A + B → C is illustrated in Figures 2.6 and 2.7 for processes initiated with ten A molecules ($N_A = 10$) and various initial populations of B molecules, N_B. As we observed in the case of the irreversible reaction A → B discussed previously, the uncertainties in molecular populations grow and then decrease as the reactants are consumed. This growth and subsequent reduction of uncertainty occur for *any* irreversible reaction.

When the stochastic model of the kinetics of the association reaction was first studied by Renyi [20], Ishida [13], and McQuarrie [18], they concluded that the expected value of the distribution would deviate from the result predicted by the deterministic formalism. However, as both figures clearly illustrate, the expectation value $E(X_A)$ (represented by a dashed green line in 2.6 and a black line in Figure 2.7) matches the prediction of the deterministic approach almost exactly. Although there is a tiny difference in the predictions, it is so small that it cannot be measured: it always falls within the standard deviations for the species population (Fig. 2.7). Again, we find that the deterministic approach is a suitable method of predicting the time evolution of the average population of a reactant or product in a small system. However, it cannot offer any information regarding the uncertainty of the populations.

Published by Woodhead Publishing Limited, 2013

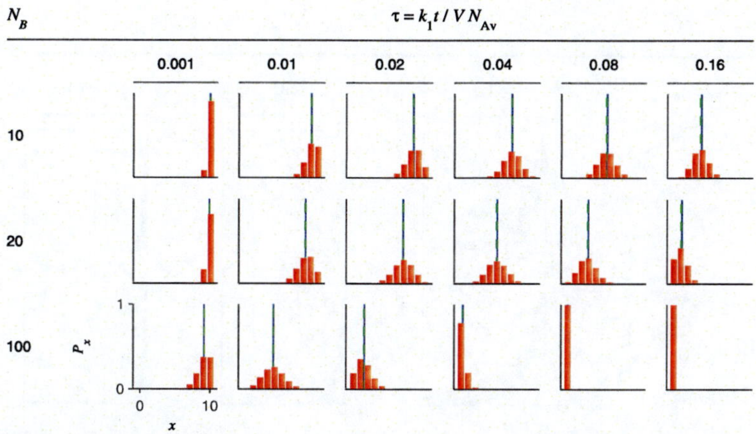

Figure 2.6 Time evolution of $P_x(t) = \Pr(X_A = x)$ for the reaction $A + B \xrightarrow{k_1} C$ (Eq 2.99). Species "A" is the limiting reactant. As the initial population of species B (N_B) increases, the distribution evolves faster, driven by Le Chatalier's principle. The predictions of the deterministic approach to chemical kinetics and the expectation value of $P_x(t)$ are illustrated as blue and green dashed lines, respectively. The deterministic approach to chemical kinetics represents the expectation value quite well, but does not offer any information regarding the uncertainty associated with molecular populations.

The distributions resulting from solution of the CMEs for zeroeth order processes and unimolecular reactions are popular, and are routinely used to characterize to a wide variety of physical processes and technologies. By contrast, Eq. 2.99 has no obvious commonality with statistical distributions from other areas of science. However, if Z is large ($N_B \gg N_A$), then Eq. 2.99 becomes a binomial distribution

$$P_x(t) = \binom{N_A}{x} \left(1 - e^{-k_1 (N_B/(VN_{Av}))t}\right)^{N-x} e^{-xk_1 (N_B/(VN_{Av}))t}$$

Published by Woodhead Publishing Limited, 2013

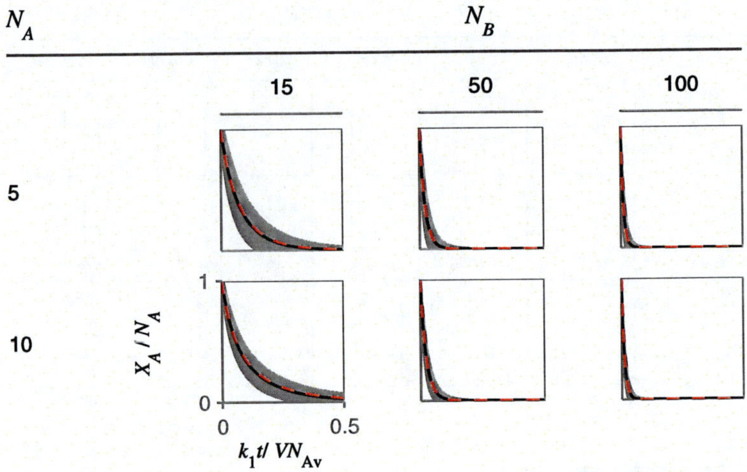

Figure 2.7 Time evolution of the population of species A (X_A) as a consequence of the reaction A + B $\xrightarrow{k_1}$ C. Species "A" is the limiting reactant. Here, X_A is normalized by its initial population, N_A. As the initial population of species B (N_B) increases, the distribution evolves faster, driven by Le Chatalier's principle.

$$N_B \gg N_A \tag{2.100}$$

This "large N_B" limiting behavior is a consequence of the reaction stoichiometry. If the population of B is so large that it is effectively constant, then the reaction A + B $\xrightarrow{k_1}$ C behaves like a unimolecular reaction of the form A $\xrightarrow{k_1'}$ C, with $k_1' = k_1 N_B / VN_{Av}$. Indeed, the only difference between Eq. 2.100 and 2.55 is the pseudo-first-order rate constant.

Outside of this extreme limit, the expectation value and variance of Eq. 2.99 are as functionally complex as the distribution from which they come. Inasmuch as x and N_A will generally be small for most investigations, it is often optimal to calculate $E(X_A)$ and $V(X_A)$ numerically from Eq. 2.99 using programs such as R, Excel® or Matlab® .

Published by Woodhead Publishing Limited, 2013

Other Irreversible Bimolecular Reactions

The CMEs for bimolecular reactions such as 2A $\rightarrow$ B and A + B $\rightarrow$ 2A may be constructed and solved similarly, and yield population distributions and statistics of equivalent complexity. We refer the reader to the literature for more detail [13, 18, 2].

The reversible reaction A + B $\rightleftharpoons$ C at equilibrium

As we discussed in Chapter 1, all chemical reactions are reversible. Therefore, Eq. 2.99 will fail to represent reality as the population of complexes increases. Even the most robust antibody-antigen associations are reversible to some extent, as a fundamental consequence of thermodynamics.

Unfortunately, time dependent master equations that describing reversible bimolecular reactions are exceptionally challenging to solve. The reversible association reaction

$$\mathrm{A} + \mathrm{B} \underset{k_2}{\overset{k_1}{\rightleftharpoons}} \mathrm{C} \tag{2.101}$$

illustrates the concepts well. Consider these reactions occurring in a fixed volume V at a fixed temperature T. The reaction is initiated with N_A A molecules and N_B B molecules, following the experimental protocol discussed in our analysis of the irreversible reaction A + B $\rightarrow$ C. Let us investigate the time evolution of P_x, once again defined by Eq. 2.88.

The two reactions of Eq. 2.101, have rates and rate constants specified in Table 2.1. If we express the dissociation reaction rate in terms of the state of the system, we obtain

$$a_2(x) = k_2(N_A - x) \tag{2.102}$$

Incorporating Eqs. 2.102 and 2.89 into Eq. 2.15, we obtain the following master equation for the reversible association reaction

$$\begin{aligned} \frac{dP_x}{d\tau} =& (x+1)(Z+x+1)P_{x+1} - x(Z+x)P_x \qquad P_x(0) = \delta_{x,N_A} \\ &+ K'(N_A - x + 1)P_{x-1} - K'(N_A - x)P_x \end{aligned} \tag{2.103}$$

Published by Woodhead Publishing Limited, 2013

where once again we have expressed the CME in terms of the dimensionless time $\tau = k_1 t / VN_{\mathrm{Av}}$. The quantity K' is defined by

$$\begin{aligned} K' &= \frac{k_2}{k_1} VN_{\mathrm{Av}} \\ &= K_D VN_{\mathrm{Av}} \end{aligned} \tag{2.104}$$

where K_D is the empirical equilibrium constant for the reaction. Note K' is dimensionless, whereas K_D is expressed in units of M.

Insofar as the association reaction couples the populations of A, B, and C, Eq. 2.103 cannot be solved via the aforementioned method of Zhang. The method of generating functions yields an equally complex partial differential equation that does not easily afford solution. And although it may be solved using the method of Laplace transforms, the resulting expression is quite complex and must be expressed in terms of exotic functions [15]. Therefore, let us reduce the scope of our investigation to the solution to the CME at equilibrium. The molecular populations fluctuate in this limit, but the distribution of states does not. Thus, P_x is invariant with time. Eq. 2.103 then becomes

$$\begin{aligned} 0 = &(x+1)(Z+x+1)P_{x+1}(\infty) - x(Z+x)P_x(\infty) \\ &+ K'(N_A - x + 1)P_{x-1}(\infty) - K'(N_A - x)P_x(\infty) \end{aligned} \tag{2.105}$$

This CME may be solved via the method of generating functions. Utilizing the generating function discussed previously (Eq. 2.37), one obtains [17]

$$s\frac{d^2G}{ds^2} + (Z + 1 + K's)\frac{dG}{ds} - K'N_A G = 0 \tag{2.106}$$

This differential equation is easily transformed into Kummer's differential equation, which has a solution of the confluent hypergeometric function [6]

$$G(s,\infty) = C_1F_1(-N_A, Z+1, -K's) \tag{2.107}$$

Published by Woodhead Publishing Limited, 2013

In this expression, C is a constant to be specified by a boundary condition. Previously, we made use of the initial condition so specify such constants. However, the time-independence of Eq. 2.105 precludes the use of the initial condition for this purpose.

However, there is an alternative way to specify C: The sum of any probability density over its state space must be one, i.e.

$$\sum_{x=0}^{N_A} P_x(\infty) = 1$$

An examination of Eq. 2.37 reveals that this normalization condition is reproduced by the equation $G(1, \infty) = 1$. Applying this equation to Eq. 2.107, we find

$$1 = C_1F_1(-N_A, Z+1, -K')$$

Therefore, the generating function for the probability distribution $P_x(\infty)$ is

$$G(s, \infty) = \frac{{}_1F_1(-N_A, Z+1, -K's)}{{}_1F_1(-N_A, Z+1, -K's)} = \frac{L_{N_A}^{(Z)}(-K's)}{L_{N_A}^{(Z)}(-K')} \qquad (2.108)$$

where $L_n^k(x)$ is a Laguerre polynomial[9]. These special functions have the following series representation

$$L_n^k(x) = \sum_{i=0}^{n} \binom{k+n}{n-i} \frac{(-x)^i}{i!}$$

Therefore,

$$P_x(\infty) = \frac{1}{\sum_{m=0}^{N_A} \binom{N_B}{N_A - m} \frac{(K')^m}{m!}} \binom{N_B}{N_A - x} \frac{(K')^x}{x!} \qquad (2.109)$$

The distribution is illustrated for various values of K' in Figure 2.8, along with its expectation value $E(X_A)$ and the corresponding prediction from the deterministic approach. As in the case of the irreversible reaction, the expectation value and deterministic approach do differ, but by an

Published by Woodhead Publishing Limited, 2013

amount that is insignificant with respect to the distribution width. Thus, we must again conclude that the deterministic approach to chemical kinetics accurately predicts the average equilibrium populations.

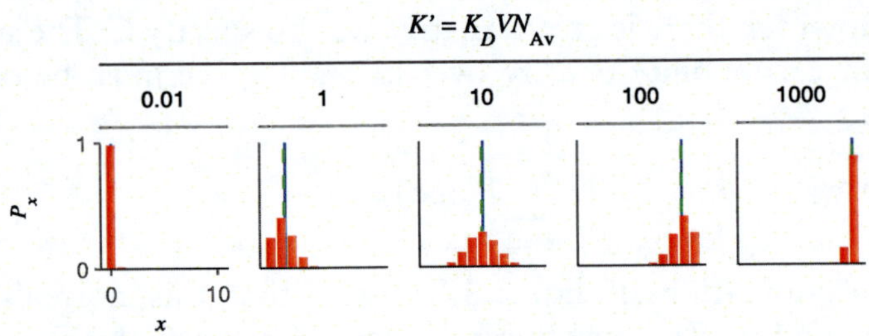

Figure 2.8 $P_x(t) = \Pr(X_A = x)$ for the reaction A+B $\rightleftharpoons$ C at equilibrium ($t \to \infty$). The prediction of the deterministic formalism is represented by a blue line, and the expectation value of the distribution is represented by a dashed green line. The average result of the stochastic approach to chemical kinetics is almost indistinguishable from the result obtained via the deterministic material balance.

Other reversible bimolecular reactions at equilibrium

The CMEs for other bimolecular reactions at equilibrium may be constructed and solved similarly, yielding generating functions of equivalent complexity. We present them all in Table 2.2. An interesting finding is that if both the "forward" and "reverse" reactions are bimolecular, then the generating function may be expressed in terms of the hypergeometric function ${}_2F_1(\ldots)$, but if only one of the reactions is bimolecular, then the generating function may be expressed in terms of the confluent hypergeometric function ${}_1F_1(\ldots)$.

Expressions for the expectation values of these distributions are provided in Table 2.3.

Published by Woodhead Publishing Limited, 2013

Table 2.2 Generating functions for $P_x = \Pr(X_A = x)$ for bimolecular reactions at equilibrium [6, 2]. The initial populations for species A, B, C, and D are N_A, N_B, N_C and N_D respectively.

Reaction	K'	Generating Function	
$A + B \underset{k_2}{\overset{k_1}{\rightleftharpoons}} C$	$\frac{k_2 V N_{Av}}{k_1}$	$\frac{{}_1F_1(-N_A-N_C, N_B-N_A+1, -K's)}{{}_1F_1(-N_A-N_C, N_B-N_A+1, -K')}$	$N_B \geq N_A$
$A + B \underset{k_2}{\overset{k_1}{\rightleftharpoons}} C + D$	$\frac{k_2}{k_1}$	$\frac{{}_2F_1(-N_A-N_C, -N_A-N_D, N_B-N_A+1, K's)}{{}_2F_1(-N_A-N_C, -N_A-N_D, N_B-N_A+1, K')}$	$N_B \geq N_A$
$2A \underset{k_2}{\overset{k_1}{\rightleftharpoons}} C$	$\frac{k_2 V N_{Av}}{k_1}$	$\frac{{}_1F_1(-\frac{1}{2}N_A-N_C, \frac{1}{2}, -\frac{1}{2}K's^2)}{{}_1F_1(-\frac{1}{2}N_A-N_C, \frac{1}{2}, -\frac{1}{2}K')}$	N_A even
		$\frac{{}_1F_1(-\frac{1}{2}(N_A-1)-N_C, \frac{3}{2}, -\frac{1}{2}K's^2)}{{}_1F_1(-\frac{1}{2}(N_A-1)-N_C, \frac{3}{2}, -\frac{1}{2}K')}$	N_A odd
$2A \underset{k_2}{\overset{k_1}{\rightleftharpoons}} C + D$	$\frac{k_2}{k_1}$	$\frac{{}_2F_1(-\frac{1}{2}N_A-N_C, -\frac{1}{2}N_A-N_D, \frac{1}{2}, \frac{1}{2}K's^2)}{{}_2F_1(-\frac{1}{2}N_A-N_C, -\frac{1}{2}N_A-N_D, \frac{1}{2}, \frac{1}{2}K')}$	N_A even
		$\frac{{}_2F_1(-\frac{1}{2}(N_A-1)-N_C, -\frac{1}{2}(N_A-1)-N_D, \frac{3}{2}, \frac{1}{2}K's^2)}{{}_2F_1(-\frac{1}{2}(N_A-1)-N_C, -\frac{1}{2}(N_A-1)-N_D, \frac{3}{2}, \frac{1}{2}K')}$	N_A odd
$A + B \underset{k_2}{\overset{k_1}{\rightleftharpoons}} 2B$	$\frac{k_2}{k_1}$	$\frac{\left(1+\frac{1}{2}Ks\right)^{N_A+N_B} - \left(\frac{1}{2}Ks\right)^{N_A+N_B}}{\left(1+\frac{1}{2}K\right)^{N_A+N_B} - \left(\frac{1}{2}K\right)^{N_A+N_B}}$	$x \in [0, N_A + N_B)$

Table 2.3 Expectation values for the populations of A molecules at equilibrium [6, 2]. The stochastic approach to chemical kinetic yields complex expressions which are distinct from those calculated via the law of mass action. However, the predictions of the two approaches are very close. The initial populations for species A, B, C, and D are N_A, N_B, N_C and N_D respectively.

Reaction	K'	Expectation Value $E(X_A)$
$A + B \underset{k_2}{\overset{k_1}{\rightleftharpoons}} C$	$\frac{k_2 V N_{Av}}{k_1}$	$K' \frac{(N_A+N_C)}{(N_B-N_A+1)} \frac{{}_1F_1(-N_A-N_C+1, N_B-N_A+2, -K')}{{}_1F_1(-N_A-N_C, N_B-N_A+1, -K')}$
$A + B \underset{k_2}{\overset{k_1}{\rightleftharpoons}} C + D$	$\frac{k_2}{k_1}$	$K' \frac{(N_A+N_C)(N_A+N_D)}{(N_B-N_A+1)} \frac{{}_2F_1(-N_A-N_C+1, -N_A-N_D+1, N_B-N_A+2, K')}{{}_2F_1(-N_A-N_C, -N_A-N_D, N_B-N_A+1, K')}$
$2A \underset{k_2}{\overset{k_1}{\rightleftharpoons}} C$	$\frac{k_2 V N_{Av}}{k_1}$	$K'(N_A - 2N_C) \frac{{}_1F_1(-\frac{1}{2}N_A-N_C+1, \frac{3}{2}, -\frac{1}{2}K')}{{}_1F_1(-\frac{1}{2}N_A-N_C, \frac{1}{2}, -\frac{1}{2}K')}$
		$\frac{1}{2}K'(N_A - 1 + 2N_C) \frac{{}_1F_1(-\frac{1}{2}(N_A-1)-N_C+1, \frac{5}{2}, -\frac{1}{2}K')}{{}_1F_1(-\frac{1}{2}(N_A-1)-N_C, \frac{3}{2}, -\frac{1}{2}K')}$
$2A \underset{k_2}{\overset{k_1}{\rightleftharpoons}} C + D$	$\frac{k_2}{k_1}$	$\frac{1}{2}K'(N_A - 2N_C)(N_A - 2N_D) \frac{{}_2F_1(-\frac{1}{2}N_A-N_C+1, -\frac{1}{2}N_A-N_D+1, \frac{3}{2}, \frac{1}{2}K')}{{}_2F_1(-\frac{1}{2}N_A-N_C, -\frac{1}{2}N_A-N_D, \frac{1}{2}, \frac{1}{2}K')}$
		$\frac{1}{2}K'(N_A - 1 + 2N_C)(N_A - 1 + 2N_D) \frac{{}_2F_1(-\frac{1}{2}(N_A-1)-N_C+1, -\frac{1}{2}(N_A-1)-N_D+1, \frac{5}{2}, \frac{1}{2}K')}{{}_2F_1(-\frac{1}{2}(N_A-1)-N_C, -\frac{1}{2}(N_A-1)-N_D, \frac{3}{2}, \frac{1}{2}K's^2)}$
$A + B \underset{k_2}{\overset{k_1}{\rightleftharpoons}} 2B$	$\frac{k_2}{k_1}$	$N\frac{K}{2} \frac{\left(1+\frac{1}{2}Ks\right)^{N_A+N_B-1} - \left(\frac{1}{2}Ks\right)^{N_A+N_B-1}}{\left(1+\frac{1}{2}K\right)^{N_A+N_B} - \left(\frac{1}{2}K\right)^{N_A+N_B}}$

2.3.4 Higher order reactions

Occasionally, one will encounter biochemical reactions of the following form in the literature:

$$nA \longrightarrow A_n \tag{2.110}$$

If we assume that that this reaction is elementary with rate constants k, then its rate would be

$$a = \frac{k}{V^{n-1}} \binom{X_A}{n} \tag{2.111}$$

because the number of ways that n A molecules could come together is $\binom{X_A}{n}$. In the thermodynamic limit, the reaction rate would scale as $a = Vk_1 c_A^n/n!$, where c_A is the concentration of species A.

If $n = 3$, we would call this elementary reaction "termolecular". And indeed, there is some evidence for the existence of termolecular (third order) reactions in Nature (e.g. combustion). However, at present there is no evidence to support the existence of termolecular reactions in biochemistry. Furthermore, there is no evidence of higher order ($n > 2$) elementary reactions of any kind. Therefore, modeling of reactions such as Eq. 2.110 as elementary is physically invalid.

Generally speaking, reactions of the form specified by 2.110 are the result of a series of bimolecular reactions, e.g.

$$\begin{aligned} A + A &\longrightarrow A_2 \\ A_2 + A &\longrightarrow A_3 \\ A_3 + A &\longrightarrow A_4 \end{aligned} \tag{2.112}$$

If one is required to construct a stochastic model of the kinetics of a process with a net reaction such as Eq. 2.110, then one should model the kinetics using reactions such as Eqs. 2.112.

Published by Woodhead Publishing Limited, 2013

2.4 The relationship between the deterministic and stochastic formalisms

In many of the Figures of this chapter (e.g. 2.8, 2.7, 2.6, and 2.5) we have illustrated how the deterministic and stochastic approaches to chemical kinetics yield similar if not identical results *on average*. This is not a coincidence: the deterministic rate laws have their physical basis in the CME.

For example, consider the CME for the reaction A → B (Eq. 2.36).

$$\frac{dP_x}{dt} = k_1(x+1)P_{x+1} - k_1 x P_x$$

This equation, like all CMEs, was formulated from first principles. If one multiplies it by x and sums the resulting equation over all x, one obtains

$$\sum_{x=0}^{\infty} x\frac{dP_x}{dt} = \sum_{x=0}^{\infty} xk_1(x+1)P_{x+1} - \sum_{x=0}^{\infty} xk_1 x P_x$$

$$\sum_{x=0}^{\infty} x\frac{dP_x}{dt} = \sum_{y=1}^{\infty} (y-1)k_1 y P_y - \sum_{x=0}^{\infty} xk_1 x P_x$$

$$= \sum_{y=0}^{\infty} (y-1)k_1 y P_y - \sum_{x=0}^{\infty} xk_1 x P_x$$

$$= -\sum_{y=0}^{\infty} k_1 y P_y$$

Since $\sum_{x=0}^{\infty} xP_x = E(X_A)$, we obtain

$$\frac{d}{dt}E(X_A) = -k_1 E(X_A) \qquad (2.113)$$

which is equivalent to the deterministic material balance for this reaction

$$\frac{dc_A}{dt} = -k_1 c_A$$

Published by Woodhead Publishing Limited, 2013

and demonstrates that the deterministic and stochastic approaches are consistent on average. Such arguments can be made for any system of first order reactions, demonstrating that the deterministic rate law has a mathematical basis in the stochastic approach to chemical kinetics.

Recall, however, that the stochastic and deterministic treatments of bimolecular reactions yield average results that are close but not exact. The reason for this difference may also be gleaned from consideration of the CME. For example, if one takes the CME for the reaction A + B → C (Eq. 2.90) and applies the aforementioned operations upon it, one obtains

$$\begin{aligned}\sum_{x=0}^{N_A} x\frac{dP_x}{d\tau} &= \sum_{x=0}^{N_A} \frac{k_1}{VN_{\mathrm{Av}}} x(x+1)(Z+x+1)P_{x+1} \\ &\quad - \sum_{x=0}^{N_A} \frac{k_1}{VN_A} xx(Z+x)P_x \\ &= \sum_{y=0}^{N_A} \frac{k_1}{VN_{\mathrm{Av}}} (y-1)y(Z+y)P_y \\ &\quad - \sum_{x=0}^{N_A} \frac{k_1}{VN_A} xx(Z+x)P_x \\ &= -\sum_{x=0}^{N_A} \frac{k_1}{VN_{\mathrm{Av}}} x(Z+x)P_y\end{aligned}$$

Recalling that $X_B = Z+x$ (Eq. 2.88), this expression may be expressed as

$$\begin{aligned}\frac{d}{dt}E(X_A) &= -\frac{k_1}{VN_{\mathrm{Av}}}E(X_A X_B) \\ &= -\frac{k_1}{VN_{\mathrm{Av}}}E(X_A)E(X_B) + \frac{k_1}{VN_{\mathrm{Av}}}\mathrm{Cov}(X_A, X_B)\end{aligned} \tag{2.114}$$

By contrast, the deterministic material balance is

Published by Woodhead Publishing Limited, 2013

$$\frac{dc_A}{dt} = -k_1 c_A c_B$$

Therefore difference between the two expressions is the covariance of X_A and X_B in Eq. 2.114. Fortunately, this term is negligible for almost all chemical and biochemical pathways.

Ultimately, these results demonstrate that the rate laws of "classical" chemical kinetics have their origin in the stochastic approach, which is why the two approaches yield consistent results on average. The question might then be posed: *Why not use the mathematically simpler deterministic approach for all investigations?* As we shall explain in Chapter 3, the stochastic approach to chemical kinetics may be implemented in a way that is simpler to use than both the deterministic mass balance and the CME, offering a powerful method for the prediction of the time evolution of any biochemical pathway.

Notes

1. http://mathworld.wolfram.com/TotalProbabilityTheorem.html
2. http://mathworld.wolfram.com/GeneratingFunction.html
3. http://mathworld.wolfram.com/Z-Transform.html
4. http://mathworld.wolfram.com/BinomialTheorem.html
5. Eqs. 2.57 and 2.58 may be derived from Eqs. 2.39 and 2.40.
6. Note that we have implicitly set P_2 and P_{-1} to zero, since $X_A \in [0, 1]$.
7. An online version is available at http://apps.nrbook.com/abramowitz_and_stegun/index.html
8. See [14] for a detailed explanation
9. http://mathworld.wolfram.com/AssociatedLaguerrePolynomial.html

References

1. Milton Abramowitz and Irene A. Stegun. *Handbook of Mathematical Functions with Formulas, Graphs, and*

Published by Woodhead Publishing Limited, 2013

Mathematical Table. Dover Publications, 1965.
2. Erdem Arslan and Ian J Laurenzi. Kinetics of autocatalysis in small systems. *J Chem Phys*, 128:015101, Jan 2008.
3. Anthony F Bartholomay. On the linear birth and death processes of biology as markoff chains. *Bull Math Biophys*, 20:97–118, 1958.
4. Anthony F Bartholomay. Stochastic models for chemical reactions: Ii. the unimolecular rate constant. *Bull Math Biophys*, 21:363–373, 1959.
5. Anthony T. Bharucha-Reid. *Elements of the Theory of Markov Processes and Their Applications*. McGraw-Hill, 1960.
6. Ivan G Darvey, B W Ninham, and P J Staff. Stochastic models of second-order chemical reaction kinetics. the equilibrium state. *J Chem Phys*, 45(6):2145–2155, 1966.
7. Katrien De Cock, Xueying Zhang, Monica F Bugallo, and Peter M Djuric. Comment on "stiffness in stochastic chemically reacting systems: The implicit tau-leaping method" [j. chem phys 119,12784 (2003)]. *J Chem Phys*, 121:3347–3348, 2004.
8. Max Delbruck. Statistical fluctuations in autocatalytic kinetics. *J Chem Phys*, 8:120–124, 1940.
9. Joseph L Doob. *Stochastic processes*. Wiley, 1953.
10. Arnold G. Fredrickson. Stochastic models for sterilization. *Biotech. Bioeng*, 8:167–182, 1966.
11. Arnold G. Fredrickson. Stochastic triangular reactions. *Chem Eng Sci*, 21:687–691, 1966.
12. Terrell L. Hil. *Statistical Mechanics: Principles and Selected Applications*. Dover, 1987.
13. Kenji Ishida. Stochastic model for bimolecular reaction. *Bull Chem Soc Jap*, 41:2472–2748, 1964.
14. Michael R King and David Gee, editors. *Multiscale Modeling of Particle Interactions: Applications in Biology and Nanotechnology*. Wiley, 2010.
15. Ian J Laurenzi. An analytical solution of the stochastic master equation for reversible bimolecular reaction kinetics. *J Chem Phys*, 113(8):3315–3322, 2000.

16. Donald A McQuarrie. Kinetics of small systems. *J Chem Phys*, 38(2):433–436, 1963.
17. Donald A McQuarrie. Stochastic approach to chemical kinetics. *J Appl Prob*, 4:413–467, 1967.
18. Donald A McQuarrie, C J Jachimowski, and M E Russell. Kinetics of small systems. ii. *J Chem Phys*, 40(10):2914–2921, 1964.
19. Earl D. Rainville and Phillip E. Bedient. *Elementary Differential Equations*. Macmillan, 7 edition, 1989.
20. Alfred Renyi. A discussion of chemical reactions using the theory of stochastic processes. *MTA Alk Mat Int Kozl*, 2:83–101, 1953.
21. Xueying Zhang, Katrien De Cock, Monica F Bugallo, and Peter M Djuric. A general method for the computation of probabilities in systems of first order chemical reactions. *J Chem Phys*, 122:104101, 2005.

Published by Woodhead Publishing Limited, 2013

3

The exact stochastic simulation algorithms

Abstract. When used to model chemical and biological systems, the stochastic master equation and deterministic material balances constitute large sets of complex differential equations. These mathematical descriptions are often unwieldy for use in research and offer approximate solution at best. However, an alternative exists in the stochastic simulation algorithms (SSAs), which allow one to numerically simulate the time evolution of a chemical system without the introduction of approximations or formulation of complex sets of differential equations. In this chapter we introduce the seminal SSAs and illustrate their use by way of case studies. We also highlight key caveats regarding the simulation of living systems with these algorithms.

Keywords: Gillespie algorithms.

3.1 Introduction

In Chapters 1 and 2, we investigated the time evolutions of individual reactions. However, computational biologists are typically more interested in the dynamics of complex reaction networks such as the one illustrated in Figure 3.1. This mechanism crudely represents the regulation of a gene by its own gene product: G represents the gene of interest, M is its mRNA, Π is a polymerase, R is a ribosome, and T is the

Published by Woodhead Publishing Limited, 2013

protein encoded by G, which in this case serves an an activator of its own transcription. Although the actual mechanism is substantially more complicated, this mechanism illustrates the scope of the problem faced by the computational biologist: How does one predict the time evolution of such a process?

The deterministic approach to chemical kinetics provides one option. The rate laws yield equations such as

$$\frac{dc_G}{dt} = -k_1 c_G c_T + k_2 c_C$$

$$\frac{dc_C}{dt} = k_1 c_G c_T - k_2 c_C - k_3 c_C + k_4 c_{C'}$$

$$\vdots$$

However, there may be complications: Since the network contains several bimolecular reactions, the system of ODEs

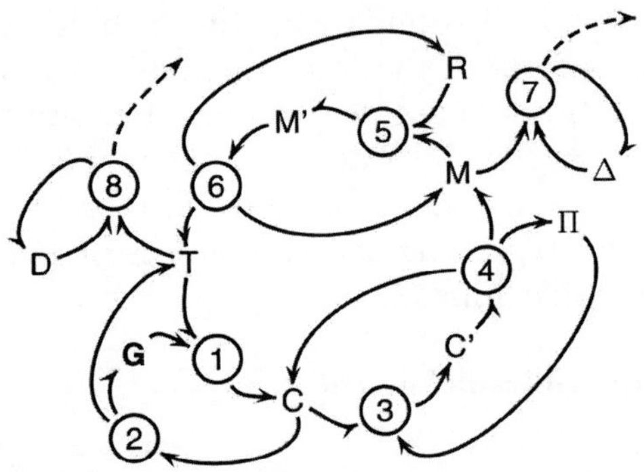

Figure 3.1 Coarse mechanism of a gene regulated by its own gene product: G represents the gene of interest, M is its mRNA, Π is a polymerase, R is a ribosome, and T is the protein encoded by G, which in this case serves an an activator of its own transcription. The species Δ and D represent degradation enzymes for the mRNA transcript and transcription factor T, respectively.

Published by Woodhead Publishing Limited, 2013

must be solved numerically, e.g. via Euler's method[1]. Such solutions are always approximate because they require approximation of the differential time interval dt by a finite interval Δt. Moreover, since the population of G (X_G) is at most two, a deterministic approach may not be able to represent the dynamics.

Alternately, one might consider using a chemical master equation to describe the stochastic time evolution of the chemical system. This approach addresses the matter of the gene population, but the resulting master equation is exceptionally complex – far more so than the aforementioned system of deterministic ODEs! And like the deterministic ODEs, it too must be solved numerically. However, the immensity of the size of the state space precludes the possibility due to considerations of computer memory alone.

Issues such as these have led many researchers to a third option: the *stochastic simulation algorithms* (SSAs). The SSAs were originally developed by Daniel T Gillespie in the mid 1970s at a time when Fortran 66 was the lingua franca of computation and computers were programed via punch cards [1, 2]. Although very powerful, the algorithms were developed before most of the researchers in the applied chemical and life sciences were prepared to use them. In 1998, several researchers independently utilized the SSA to investigate the dynamics of a variey of biological processes, including the fate of lambda-phage infected *E coli* [3]. These discoveries reintroduced SSAs to a new generation of researchers, and have changed the way that computational biology research is conducted. Today, these algorithms have become the methods of choice for researchers studying the dynamics of biological pathways in silico, and a number of user-friendly and publicly-available software packages exist for simulation of chemical processes [4, 5, 6, 7, 8, 9, 10, 11].

3.2 The reaction probability density function

Consider the following questions:

Published by Woodhead Publishing Limited, 2013

1. At time t, what is the state of a biochemical system?

2. Given that the system is in the state $\mathbf{x}$, how much time will pass before the next reaction event, and which event will it be?

The stochastic approach to chemical kinetics answers both questions probabilistically: the first via the CME, and the second via the *reaction probability density function*

$$\mathscr{P}(\mu, \tau|\mathbf{x})d\tau = \text{Pr(A reaction of type } \mu \text{ will occur in the imminent time interval } (t+\tau, t+\tau+dt)) \tag{3.1}$$

where μ is one of M reactions that may occur. Note that $\mathscr{P}(\mu, \tau|\mathbf{x})$ is a joint probability density for two random variables, where

$$\sum_{\mu=1}^{M} \int_0^\infty \mathscr{P}(\mu, \tau|\mathbf{x})d\tau = 1 \tag{3.2}$$

The reaction probability density function is the basis of the stochastic simulation algorithms: By sampling this distribution via Monte Carlo (MC), one may obtain the time until the next event and the corresponding reaction to come, adjust the system accordingly, and repeat the process until an entire realization of the process has been completed.

First, one requires an expression for $\mathscr{P}(\mu, \tau|\mathbf{x})$. To begin, let us express the distribution as

$$\mathscr{P}(\mu, \tau|\mathbf{x})d\tau = \mathscr{P}_0(\tau) \times a_\mu(\mathbf{x})d\tau \tag{3.3}$$

Published by Woodhead Publishing Limited, 2013

where

$$\mathscr{P}_0(\tau|\mathbf{x}) = \text{Pr(Given that the system is in state } \mathbf{x}, \text{ no reactions will occur between times } t \text{ and } t+\tau)$$

and

$$a_\mu(\mathbf{x})d\tau = \text{Pr(Given that the system is in state } \mathbf{x}, \text{ a reaction of type } \mu \text{ will occur within the imminent time interval } d\tau)$$

The stochastic rates of reaction a_μ may be expressed in terms of the system volume V, species populations x_i $(i = 1 \ldots N)$ and rate constants k_μ as discussed in Chapter 2.

It remains to define $\mathscr{P}_0(\tau|\mathbf{x})$. Gillespie's approach was to break up the forthcoming time interval τ into N subintervals of length ϵ

$$\tau = \epsilon N \tag{3.4}$$

Assuming that the system is in the state $\mathbf{x}$, the probability that a reaction $\nu \in [1, M]$ does not occur in the first subinterval is

$$1 - a_\mu(\mathbf{x})\epsilon + o(\epsilon)$$

where $o(\epsilon)$ is defined as discussed in Chapter 2[2]. However, reaction ν is only one of the M possible reactions. Recalling the independence of the reactions from Chapter 2, the probability that no reactions occur between $t = 0$ and $t = \epsilon$ is

$$\prod_{\nu=1}^{M}(1 - a_\nu(\mathbf{x})\epsilon + o(\epsilon)) = 1 - \sum_{\nu=1}^{M} a_\nu(\mathbf{x})\epsilon + o(\epsilon)$$

That is,

$$\mathscr{P}_0(\epsilon) = 1 - \sum_{\nu=1}^{M} a_\nu(\mathbf{x})\epsilon + o(\epsilon) \tag{3.5}$$

At this point in time (ϵ), no events have occurred, Therefore, the system is still in its initial state, and the probability that it will not transition from this state over the next subinterval ϵ is still defined by Eq. 3.5. Thus,

Published by Woodhead Publishing Limited, 2013

$$\mathscr{P}_0(2\epsilon) = \left(1 - \sum_{\nu=1}^{M} a_\nu(\mathbf{x})\epsilon + o(\epsilon)\right) P_0(\epsilon)$$
$$= \left(1 - \sum_{\nu=1}^{M} a_\nu(\mathbf{x})\epsilon\right)^2 + o(\epsilon)$$

Repeating the argument for the remaining subintervals, one obtains

$$\mathscr{P}_0(\tau) = \left(1 - \sum_{\nu=1}^{M} \frac{a_\nu(\mathbf{x})\tau}{N}\right)^N + o\left(\frac{\tau}{N}\right)$$

At this point, note that our discretization of the interval τ is entirely arbitrary. In the limit $N \to \infty$, we obtain

$$\lim_{N\to\infty} \left[\left(1 - \sum_{\nu=1}^{M} \frac{a_\nu(\mathbf{x})\tau}{N}\right)^N + o\left(\frac{\tau}{N}\right)\right] = \exp\left(-\sum_{\nu=1}^{M} a_\nu(\mathbf{x})\tau\right)$$

Thus

$$\mathscr{P}_0(\tau) = \exp\left(-\sum_{\nu=1}^{M} a_\nu(\mathbf{x})\tau\right) \tag{3.6}$$

Inserting this expression into Eq. 3.3, we obtain

$$\mathscr{P}(\mu, \tau|\mathbf{X})d\tau = a_\mu(\mathbf{x}) \exp\left(-\sum_{\nu=1}^{M} a_\nu(\mathbf{x})\tau\right) d\tau \tag{3.7}$$

The convenience of this expression is that it needn't be re-derived every time one performs a stochastic simulation. It may be treated as a simple formula.

3.3 The stochastic simulation algorithms

The stochastic simulation algorithms (SSAs) use Eq. 3.7 to predict the stochastic time evolutions of chemical processes. The general procedure is conceptually quite simple:

Published by Woodhead Publishing Limited, 2013

1. **Initialize** the populations of all molecular species $\{X_i\}$ and reaction rates $\{a_\nu\}$, and the "initial" time t.
2. **Select** the time preceding the next reaction event to come (τ) and the specific reaction event to come (μ) via Monte Carlo (MC).
3. **Update**
   (a) the populations of the reactants and products of reaction μ
   (b) the rates of all other reactions $\{\nu\}$ that depend upon these populations
   (c) the time t, by τ
4. **Iterate** by returning to Step 2..

Over many iterations, the time and populations of the reactive species evolve in tandem, progressing until the user terminates the simulation or the process reaches a terminal point where all reaction rates are zero. Note that the simulation procedure does not require the solution of any equations: differential or otherwise. However, implementations of the algorithms vary in how they implement the Monte Carlo step, and may require the use of object oriented programming for optimal performance, i.e. speed of calculation. Differences in the implementations of steps 2 and 3 result in tradeoffs between computer memory and simulation speed.

3.3.1 The Direct Method

Of the two original SSAs developed by Gillespie, the "Direct Method" (DM) is the most efficient in terms of the speed of calculation: For a system of M reaction channels, it requires $O(M)$ operations per iteration for both the reaction selection and update steps, and memory allocation for M reaction rates and rate constants.

The DM selects τ and μ independently, requiring two MC selection steps per iteration. However, this requires two probability distributions for sampling. They are defined as $\mathscr{P}_2(\mu|\tau,\mathbf{x})$ and $\mathscr{P}_1(\tau)$, defined as

Published by Woodhead Publishing Limited, 2013

$$\mathscr{P}(\mu,\tau|\mathbf{x})d\tau = \mathscr{P}_2(\mu|\tau,\mathbf{x}) \times \mathscr{P}_1(\tau)d\tau \tag{3.8}$$

The distributions may be obtained via conditioning of the reaction probability density function as follows

$$\begin{aligned}\mathscr{P}_1(\tau)d\tau &= \sum_{\mu=1}^{M} \mathscr{P}(\mu,\tau|\mathbf{x})d\tau \\ &= \sum_{\mu=1}^{M} a_\mu(\mathbf{x}) \exp\left(-\sum_{\nu=1}^{M} a_\nu(\mathbf{x})\tau\right) d\tau \\ &= a_0 \exp(-a_0\tau)\, d\tau \end{aligned} \tag{3.9}$$

where we have defined

$$a_0 = \sum_{\mu=1}^{M} a_\mu(\mathbf{x}) \tag{3.10}$$

As the sum of all reaction rates, the quantity a_0 is the total rate of reaction in the system. We immediately recognize that the quiescence interval τ is exponentially distributed, i.e. $\tau \sim$ Exponential (a_0).

Inserting Eq. 3.9 into Eq. 3.8 immediately yields

$$\mathscr{P}_2(\mu|\tau) = \frac{a_\mu}{a_0} \tag{3.11}$$

One may implement a Monte Carlo selection of a random variable by setting its cumulative distribution equal to a uniform random number. For the quiescence time, one obtains

$$r_1 = \int_0^\tau \mathscr{P}_1(\tau)d\tau \tag{3.12}$$

$$= 1 - \exp(-a_0\tau) \tag{3.13}$$

where r_1 is a uniform random number between zero and one, i.e. $r_1 \sim U(0,1)$. Rearranging, one obtains the selection rule for the quiescence interval preceding the forthcoming event

$$\tau = \frac{1}{a_0}\ln\left(\frac{1}{1-r_1}\right) \quad \text{Quiescence time selection, DM} \tag{3.14}$$

The procedure for the selection of the imminent event follows from similar argument. However, since μ is a discrete random variable, one sums over $\mathscr{P}_2(\mu|\tau)$ in lieu of integration, choosing μ by the point when the sum exceeds a different uniform random number $r_2 \sim U(0,1)$. One obtains

$$\sum_{\nu=1}^{\mu-1} \mathscr{P}_2(\nu|\tau) < r_2 \leq \sum_{\nu=1}^{\mu} \mathscr{P}_2(\nu|\tau)$$

or more concisely,

$$\sum_{\nu=1}^{\mu-1} a_\nu < r_2 a_0 \leq \sum_{\nu=1}^{\mu} a_\nu \quad \text{Reaction event selection, DM} \quad (3.15)$$

Equations 3.14 and 3.15 are entirely general - they do not need to be re-derived for each reaction pathway. Together, they constitute Step 2 of the SSA. Note that the calculation of a_0 involves M operations per iteration, and the reaction event selection may require as many as M additions. Hence, we say that the Direct Method is an $O(M)$ algorithm.

Step 3 of the DM is relatively straightforward: Once the event to come and quiescence time are selected, one must update the populations of the affected species. The reaction rates are then updated based on the newly updated populations. If one chooses to update the rates of *all* reactions after the update to the reaction populations, this becomes the speed-limiting step of the algorithm. However, if one restricts the reaction rate updates to those reactions which have a reactant whose population was just modified, the performance of the DM increases dramatically. Generally speaking, simulations of "small" reaction networks may often be performed in a few seconds with current computer technology.

3.3.2 The "First Reaction" method

The "First Reaction" Method (FRM) was published along with the Direct Method in Gillespie's seminal works on SSAs,

Published by Woodhead Publishing Limited, 2013

and is mathematically equivalent to the DM. However, it uses a radically different MC technique for the selection of the event to come and the quiescence time. Moreover, it requires the generation of M random numbers per iteration, causing it to be slower than the DM when used for the simulation of large biochemical reaction networks.

The logic of the FRM's selection criterion is subtle. Imagine that the chemical system transitions to the state $\mathbf{X} = \mathbf{x}$ at time t. It is fairly straightforward to show that the time until each reaction νth *might occur* will be distributed exponentially, i.e.[3]

$$\mathscr{P}_\nu(\tau) = a_\nu e^{-a_\nu \tau} d\tau \quad \nu = 1 \ldots M \tag{3.16}$$

However, once one of these reactions *actually occurs*, the populations of certain species will also change. This, in turn, changes the rates of other reactions, and their distributions. Thus, Equation 3.16 defines the distributions of the "tentative" times τ_ν preceding the forthcoming reaction event μ. The tentative times may be selected via MC as follows

$$\tau_\nu = \frac{1}{a_\nu} \ln\left(\frac{1}{1 - r_\nu}\right) \quad \text{"tentative" times for the FRM} \tag{3.17}$$

where $r_\nu \sim U(0,1)$ is a unique uniform random number; a new random number must be generated for each τ_ν to maintain the independence of the reactions.

Clearly, the time until the next event is the smallest of these reaction times, i.e.

$$\tau = \min \tau_\nu \quad \text{Quiescence time selection, FRM} \tag{3.18}$$

Moreover, the event to come is the reaction whose tentative time is τ_μ,

$$\tau_\mu = \tau \quad \text{Reaction event selection, FRM} \tag{3.19}$$

Therefore, both the event to come and quiescence interval are selected based on the identification of the "first reaction" that will occur. The process of identifying the minimum

Published by Woodhead Publishing Limited, 2013

time requires $O(\log(M))$ operations, but since the quiescence times must be recalculated after each iteration, M recalculations of Eq. 3.17 are required. Conceptually, this means that the FRM is an $O(M)$ algorithm, like the DM. However, the M recalculations of the tentative times $\{\tau_\nu\}$ require the generation of M uniform random numbers, which are computationally much more expensive than arithmetic operations. For this reason, the process of selecting the event to come and quiescence time tends to be much slower than the corresponding steps of the DM.

3.3.3 The "Next Reaction" method

The FRM and DM were the only SSAs for general reaction pathways until 2000, when Michael Gibson and Jehoshua Bruck of Caltech published a new SSA entitled the "Next Reaction Method" (NRM) [12]. Substantially improving upon the FRM, the NRM requires $O(\log M)$ operations per iteration for sparse reaction networks, and $O(M)$ operations for highly-coupled networks. The improvement is achieved by storing reaction information as a directed graph , and utilizing an "absolute time" selection criterion in lieu of quiescence time criterion. The data structure facilitating improvements in speed may, however, come at a significant cost of computer memory for large and highly coupled reaction networks.

To help us illustrate the algorithm, let us consider the reaction pathway of Figure 3.1. The process begins with the system in a state **x** defined by the populations of species G, C, etc. at time t. After some time interval τ_μ corresponding to the occurrence of some reaction $\mu \in [1, 8]$, the system will transition to a new state. The time at which reaction μ would occur in the absence of other reactions is tentatively $T_\mu = t + \tau_\mu$, which has the distribution

$$\mathscr{P}_\nu(T) = \begin{cases} a_\nu e^{-a_\nu(T-t)} d\tau & T > t \\ 0 & \text{otherwise} \end{cases} \qquad \nu = 1 \ldots M$$

Published by Woodhead Publishing Limited, 2013

We may select this "tentative time" *at which* the next reaction might occur via the following MC formula

$$T_\nu = t + \frac{1}{a_\nu} \ln\left(\frac{1}{1 - r_\nu}\right) \quad \text{"tentative" times for the NRM} \tag{3.20}$$

where again, $r_\nu \sim U(0, 1)$ is a unique uniform random number, and a new random number must be generated for each τ_ν.

As with the FRM, we recognize that the next reaction will occur at

$$t = \min_\nu T_\nu \text{Absolute time at which the next reaction occurs, NRM} \tag{3.21}$$

with the event to come being defined by the smallest of these times, i.e.

$$T_\mu = t \quad \text{Next reaction to occur} \tag{3.22}$$

Now let us consider the same system immediately following reaction event $\mu = 4$. Time has increased by τ_4 and the populations of species C', C, M and Π have changed. These changes, in turn, affect the rates of Reactions 2, 3, 4, 5, and 7, since

$$\begin{aligned} a_2 &= k_2 X_C \\ a_3 &= k_3 X_C \\ a_4 &= k_4 X_{C'} \\ a_5 &= k_5 X_M X_R \\ a_7 &= k_7 X_M \end{aligned}$$

Note, however, that the rates of Reactions 1, 6, 8, and 9 are completely unaffected by the preceding reaction event. Therefore, their rates a_1, a_6, a_8, and a_9 and their reaction times T_1, T_6, T_8, and T_9 *need not be updated.* Furthermore, T_2, T_3, T_5, and T_7 may be updated without the generation of a random variable via the following formula

$$T_{\alpha,\text{new}} = (T_{\alpha,\text{old}} - t)\frac{a_{\alpha,\text{old}}}{a_{\alpha,\text{new}}} + t \quad \text{Update for reaction} \, \alpha \neq \mu, \text{NRM} \tag{3.23}$$

Published by Woodhead Publishing Limited, 2013

If the rate of some reaction α (a_α) was previously zero, then the old reaction time would be infinite. In this case, one should use Eq. 3.20 to calculate the absolute reaction time. And since reaction 4 just occurred, the next occurrence will result from a new random process. Thus,

$$T_\mu = t + \frac{1}{a_{\mu,\text{new}}} \ln\left(\frac{1}{1-r_\mu}\right) \quad \text{Update for reaction } \mu\text{, NRM} \tag{3.24}$$

where $r_\mu \sim U(0,1)$ is a newly generated random number. In most cases, this will be the only random number generated in a given iteration.

Having selected the event to come and the time t at which it does, one must update the populations of the reactants and products of reaction μ as in the other SSAs. Gibson and Bruck suggest the use of a Dependency Graph to do so, which is encoded much like the schematic of Figure 3.1. Implementation of such structures in a dynamic fashion is best left to those who have experience with object-oriented programming languages such as C++ or Java, and which are beyond the scope of this book. Nevertheless, it is critically important to ensure that there is a connection between the reactions and their reactants and products that participate in other reactions, so as to easily identify the species whose populations are affected by each reaction. It is also helpful to connect each species to the reactions in which it serves as a reactant, so as to facilitate the update of reactions after a reaction event occurs *in silico*.

In summary, the NRM may be described as follows:

1. **Initialize** the populations of all molecular species $\{X_i\}$ reaction rates $\{a_\nu\}$, "tentative" times $\{T_\nu\}$ and the "initial" time t.
2. **Select**
   (a) the time in the simulation via MC using Equation 3.21.
   (b) the reaction to come μ, by $T_\mu = t$.
3. **Update**

Published by Woodhead Publishing Limited, 2013

   (a) the populations of the reactants and products of reaction μ
   (b) the rates of all other reactions that depend upon these populations
   (c) the tentative times associated with all other reactions via Equation 3.23
   (d) The tentative time associated with reaction μ via Equation 3.24

4. **Iterate** by returning to Step 2.(a).

Like the FRM, the NRM requires on average $O(\log(M))$ operations for the identification of the smallest T_ν. However, the use of absolute times and complex data structures to relate reactants to products essentially reduces the number of operations for updating the system after an event to $O(1)$. Thus, the NRM is theoretically a substantial improvement over both the DM and FRM in terms of speed. However, it comes at a cost of computer memory: the directed graph requires storage for all of the relationships between reactants and products, and requires the calculation of M random numbers before the simulation begins. For very large and coupled reaction networks, these features become memory and speed prohibitive. Therefore, enhancements of the Direct Method have been proposed for the simulation of such processes in recent years. In 2004, Cao and coworkers in the Petzold group developed an "Optimized Direct Method" illustrating that the Direct Method could be implemented in such a way as to make it as fast as the Next Reaction Method [13] without introducing approximations in the methodology.

3.4 Case studies

Thus far, we have discussed the SSAs at a high level. The simplicity of the process may be illustrated via some practical examples.

Published by Woodhead Publishing Limited, 2013

3.4.1 The Association Reaction $A + B \rightleftharpoons C$

To begin, let us perform a simulation of the complex formation reaction

$$A + B \underset{k_2}{\overset{k_1}{\rightleftharpoons}} C$$

that we studied in Chapters 1 and 2. We consider a reaction in a closed volume V initiated with N_A A molecules and N_B B molecules, which could represent biochemical receptors and their ligands. The stochastic rates of the reactions are

$$a_1 = \frac{k_1}{N_{Av} V} X_A X_B$$

and

$$a_2 = k_2 X_C$$

where $N_{Av} = 6.022 \times 10^{23}$/mol and k_1 and k_2 are measured in units of $M^{-1}s^{-1}$ and s^{-1} respectively. The solutions to the deterministic material balance and stochastic master equation are provided in Chapters 1 and 2. The First Reaction Method, by contrast, may be implemented as follows:

Published by Woodhead Publishing Limited, 2013

```
% Matlab implementation of the FRM for the reaction A + B <=> C
function [xA xB xC t] = ABC_MC(t_end, NA, NB, V, k1_VNAv, k2)

XA = NA;     % Initial population of A
XB = NB;     % Initial population of B
XC = 0;      % Initial population of C
t = 0;       % Initial time, t = 0

% MC loop
while (t < t_end)

    % Define/Update Rates
    a1 = k1_VNAv*xA*xB; % k1_VNAv is k1/V/N_Av
    a2 = k2*xC;
    a0 = a1+a2;  % Total rate of all reactions

    tau(1) = log(1/rand)/a1;      % Time until a forward reaction
    tau(2) = log(1/rand)/a2;      % Time until a reverse reaction

    [Tau,mu] = min(tau);      % Select tau and mu, First Reaction Method

    % Update time
    t = t + Tau;

    if (mu == 1)
        xA = xA - 1;
        xB = xB - 1;
        xC = xC + 1;
    else
        xA = xA + 1;
        xB = xB + 1;
        xC = xC - 1;
    end

end

return
```

If one wishes to return an entire .time series from the simulator, one may do so by adding the values of t and xA to other vectors after each iteration.

The results of the SSA (FRM), the CME, and the deterministic formalism are illustrated in Figure 3.2 for systems of various sizes. Ten simulation time series are illustrated in the upper frames: The red lines are results of specific representative simulations. The results of the deterministic material balances are illustrated as blue lines. In the lower frames, the corresponding results of the CME are presented: the black line is the normalized expectation value $E(X_A)$, the dark gray region represents the range of $E(X_A) \pm \sigma$, with $\sigma = \sqrt{V(X_A)}$, and the light gray region represents the range of $E(X_A) \pm 2\sigma$.

Clearly, all three approaches give equivalent results, especially as the size of the system increases (e.g. N_A becomes larger). However, the results illustrate an application of the SSA other than the prediction of $X_A(t)$. As the figure shows, the results of several simulations reproduce the range of uncertainty of X_A as well. The spread of as few

Published by Woodhead Publishing Limited, 2013

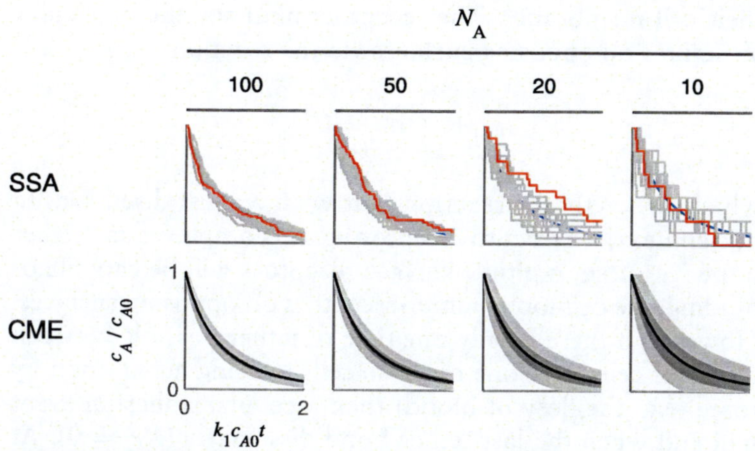

Figure 3.2 The association reaction $A + B \xrightarrow{k_1} C$ with $\xi = c_{B0}/c_{A0} = 2$. The CME and SSAs yield the same results, including distributions of results about the average time evolution. Results of ten simulations are illustrated for each N_A, with one highlighted in red.

as *ten* simulated .time series closely approximates the range predicted by the CME. Results of even more simulations give much more detailed information. Given enough simulations, one may estimate the probability distribution $P_x(t) = \Pr(X_A(t) = x)$, as well as its statistics. As each simulation requires moments, not minutes, it may be possible to computationally construct $P(\mathbf{x}, t)$ for a relatively small reaction mechanism (say $M < 10$) in a few minutes. This is something that the deterministic approach to chemical kinetics is completely incapable of doing.

3.4.2 Receptor-Mediated Adhesion

An example of where this type of information may be exceptionally useful is in the area of biological adhesion. Eukaryotic cells typically adhere to other cells, extracellular matrix, etc. via surface receptors, which are embedded in

Published by Woodhead Publishing Limited, 2013

their cell membranes. The receptors bind specific molecular structures on opposing surfaces via the reaction

$$\mathrm{R} + \mathrm{L} \underset{k_2}{\overset{k_1}{\rightleftharpoons}} \mathrm{C}$$

which is exactly the reaction that we have just discussed. In this context, we refer to receptor-ligand complexes as "tether bonds". Often, multiple surface receptors will be capable of binding one or more counter-receptors on opposing surfaces. However, if there is only one type of tether possible between a pair of cells, and one of those cells is being acted upon by force (e.g. the flow of blood) then a cellular adhesion event will end when the last tether bond dissociates ($X_C = 0$). At that point, one of the cells will depart from the other, precluding the formation of new tethers. This is somewhat different than our previous example in that this process ultimately terminates. If we represent the state of the system by the number of tether bonds (N), then the stochastic process may be represented as

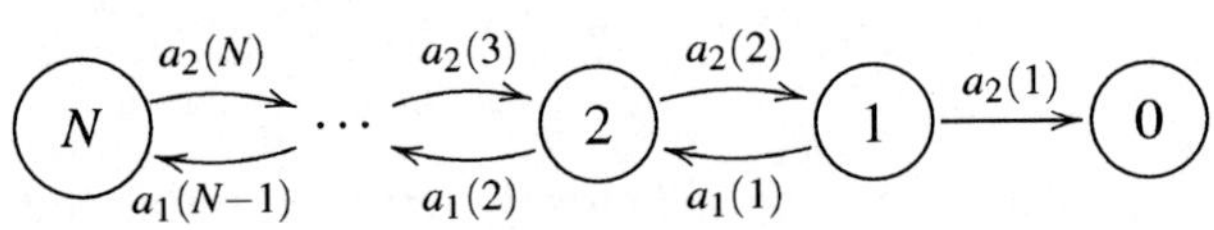

where the process is initiated with $N = 1$ (one bond formed between cells) and terminates with $N = 0$. The adhesion time is simply the time until the last bond has broken.

In principle one may formulate a distribution for the lengths of adhesion times using the CME. However, it is easily constructed via any of of the SSAs by running I simulations until they terminate, yielding I adhesion times. The empirical distribution function – an estimator for the actual adhesion time distribution, may then be calculated as

$$F_I(t) = \frac{\text{number of simulated adhesion times less than } t}{I}$$

As I increases, the closer I will be to the actual distribution.

The following Matlab program implements a Direct Method simulation of receptor mediated adhesion

Published by Woodhead Publishing Limited, 2013

```
function [X t] = Adhesion_Direct(k1, k2, NA, NB)

TMAX = 50; % Upper limit for pause times (censoring) in s

% Define reaction stoichiometry
nu(:,1) = [-1 -1 1]; % R + L -> C
nu(:,2) = [1 1 -1];  % C -> R + L

% Define initial populations of receptors, ligands and complexes
X(1) = NA - 1;  % Pop. of receptors (e.g. P-Selectin)
X(2) = NB - 1;  % Pop. of counter-receptors (e.g. PSGL-1)
X(3) = 1;       % Pop. of tether bonds at t = 0 (one)
X = X';         % transpose

% Calculate the initial reaction rates
a(1) = k1*X(1)*X(2);
a(2) = k2*X(3);

% Initialize the time
t = 0;

% MC Loop
while (t<TMAX)
    a0 = sum(a); % total reaction rate (a0)

    r1 = rand; % URN
    tau = (1/a0)*log(1/r1); % quiescence interval (tau)

    t = t + tau; % Increment time by the quiescence interval

    % Choose the event to come & modify the system state
    r2a0 = rand*a0;
    acc = 0; % accumulator of a(i)
    for i=1:size(nu,2)
        acc = acc + a(i);   % sum rates...
        if (acc > r2a0)
            mu = i;
            break           % until r2a0 is exceeded
        end
    end
    X = X + nu(:,mu);     % Modify the system state

    if (X(3) == 0)  % No bonds? (Cell detached?)
        return; % cell detached, end the simulation
    end

    % Update the reaction rates (in this sim, both are modified)
    a(1) = k1*X(1)*X(2);
    a(2) = k2*X(3);
end

return
```


Published by Woodhead Publishing Limited, 2013

By contrast, the following Matlab program implements a First Reaction Method simulation of receptor mediated adhesion

```
function [X t] = Adhesion_First(k1, k2, NA, NB)

TMAX = 50; % Upper limit for pause times (censoring) in s

% Define reaction stoichiometry
nu(:,1) = [-1 -1 1]; % R + L -> C
nu(:,2) = [1 1 -1];  % C -> R + L

% Define initial populations of receptors, ligands and complexes
X(1) = NA - 1;  % Pop. of receptors (e.g. P-Selectin)
X(2) = NB - 1;  % Pop. of counter-receptors (e.g. PSGL-1)
X(3) = 1;       % Pop. of tether bonds at t = 0 (one)
X = X';         % transpose

% Calculate the initial reaction rates
a(1) = k1*X(1)*X(2);
a(2) = k2*X(3);

% Initialize the time
t = 0;

% MC Loop
while (t<TMAX)
    % Calculate the "tentative" quiescence intervals (tau)
    r_nu = rand(1,size(nu,2)); % one URN for each reaction
    tau_nu = log(1./r_nu)./a;

    % Choose quiecence interval and reaction event to come
    [tau, mu] = min(tau_nu);

    t = t + tau; % Increment time by the quiescence interval

    X = X + nu(:,mu);     % Modify the system state

    if (X(3) == 0)   % No bonds? (Cell detached?)
        return; % cell detached, end the simulation
    end

    % Update the reaction rates (in this sim, both are modified)
    a(1) = k1*X(1)*X(2);
    a(2) = k2*X(3);
end

return
```

Finally, the following Matlab program implements a Next Reaction Method simulation of receptor mediated adhesion Note that these short programs are all similar to each other

Published by Woodhead Publishing Limited, 2013

as well as the aforementioned program for the reaction $A + B \rightleftharpoons C$, but they all terminate when $X_C = 0$. Their implementation requires only a handful of lines of code, and most of these lines could apply equally to far larger reaction networks.

```
function [X t] = Adhesion_Next(k1, k2, NA, NB)

TMAX = 50; % Upper limit for pause times (censoring) in s

% Define reaction stoichiometry
nu(:,1) = [-1 -1 1]; % R + L -> C
nu(:,2) = [1 1 -1];  % C -> R + L

% Define initial populations of receptors, ligands and complexes
X(1) = NA - 1;  % Pop. of receptors, R
X(2) = NB - 1;  % Pop. of counter-receptors, L
X(3) = 1;       % Pop. of tether bonds at t = 0 (one)
X = X';         % transpose

% Calculate the initial reaction rates
a(1) = k1*X(1)*X(2);
a(2) = k2*X(3);

% Calculate the time intervals (tau)
r_nu = rand(1,size(nu,2)); % one URN for each reaction
t_nu = log(1./r_nu)./a;

% Initialize the time
t = 0;

% MC Loop
while (t<TMAX)
    % Choose quiecence interval and reaction event to come
    [tau, mu] = min(t_nu);

    t = tau;    % Set the time in the simulation

    Xnew = X + nu(:,mu);     % Define the new system state

    % If the cell detached, end the simulation
    if (Xnew(3) == 0)  % No bonds? (Cell detached?)
        return;
    end

    % Calculate the updated reaction rates (in this sim, both are modified)
    anew(1) = k1*Xnew(1)*Xnew(2);
    anew(2) = k2*Xnew(3);

    % Update the reaction times; No directed graph needed for two reactions
    if (mu == 2)
        t_nu(1) = ( a(1)/anew(1) ) * (t_nu(1) - t) + t;
        t_nu(2) = t + log(1/rand)/anew(2);
    else
        t_nu(2) = ( a(2)/anew(2) ) * (t_nu(2) - t) + t;
        t_nu(1) = t + log(1/rand)/anew(1);
    end

    % Update rates and populations
    a = anew;
    X = Xnew;
end

return
```

The results of thousands of simulations using these programs are illustrated in Figure 3.3. As one would expect from the rigorous derivations of the MC selection rules, all three SSAs yield indistinguishable results. If one desires to construct probability distributions for very tightly binding

Published by Woodhead Publishing Limited, 2013

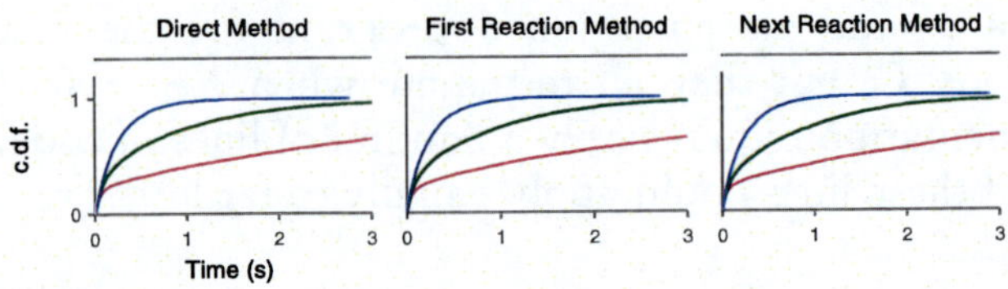

Figure 3.3 Distributions of the durations of receptor-mediated adhesion events. The cumulative pause time distributions are generated from 5000 MC-generated pause times, using $K_D/c_{A0} = 10$ (magenta), 20 (green) and 100 (blue) with $k_d = 5\ s^{-1}$, $N_A = 20$ and $N_B = 50$. All three simulation algorithms yield the same results.

receptor-ligand pairs one should not censor the simulations: Many of such adhesion events will appear to be "permanent" to the experimental researcher, as the average bond lifetimes will exceed the cell lifetimes.

The adhesion time distributions cannot be constructed using a deterministic approach to chemical kinetics, and often result from a de facto unsolvable CME. For this reason, SSAs are indispensable in this area of research. Additionally, the agreement between the results of the SSA and the variability of experimental measurements of adhesion times provides a useful empirical confirmation of the validity of the stochastic approach to chemical kinetics. We refer the reader to references [14] and [15] for more on the subject.

3.4.3 Ion Channel dynamics

The current through a specific ion channel states may be observed via the patch clamp technique, which was developed by Erwin Neher and Bert Sakmann in the late 1970s. This was one of the first "single molecule" experiments in the biological sciences and revolutionized electrophysiology, leading to the award of the Nobel Prize in Medicine to Neher and Sakmann in 1991.

Published by Woodhead Publishing Limited, 2013

The SSA can be used to simulate the patch clamp experiments with one or more ion channels. Let us consider the "simple open ion channel block mechanism" of Colquhoun and Hawkes [16], which address the actions of "channel-unlocking" and "channel-blocking" drugs. The mechanism is

$$\underset{\text{closed}}{\mathscr{C}} \underset{\alpha}{\overset{\beta'}{\rightleftharpoons}} \underset{\text{open}}{\mathscr{A}} \tag{3.25}$$

$$\mathrm{D} + \mathscr{A} \underset{k_{-1}}{\overset{k_{+1}}{\rightleftharpoons}} \underset{\text{blocked}}{\mathscr{B}} \tag{3.26}$$

where $\mathscr{A}$, $\mathscr{B}$ and $\mathscr{C}$ are states of the ion channel and D is a ligand that binds the ion channel in state $\mathscr{A}$ and precludes the flow of ions through the channel. The rate constants are denoted β', α, k_{+1} and k_{-1} by convention: They could easily be referred to by k_1, k_2, k_3 and k_4. The blocker D is typically present at such a high population relative to the ion channel that a single blocking event does not change its population appreciably. Therefore, the Eq. 3.26 may effectively be treated as a first order reaction with rate constant $k_{+1}c_D$.

The following Matlab program simulates this mechanism, yielding the .time series for the $\mathscr{A}$ state and by extension, the measured current.

This program generates the results in the upper frame of Figure 3.4. In a typical patch clamp experiment, the corresponding time series of the current would look like the lower frame.

The simulation results may be utilized to develop distributions for the "open times" and "closed/blocked times" of the ion channel, much like the aforementioned distributions of adhesion times. These distributions, in turn, may be used to estimate rate constants for a mechanism from experimental measurements.

3.4.4 The Lotka Reactions

Let us now consider the time evolution of a reaction network with more complicated dynamics than the complex formation reaction discussed earlier. The Lotka reactions do not

Published by Woodhead Publishing Limited, 2013

```
function [XA T t_close t_open] = Ion_Sim(k1, cD, k_1, alpha, beta, t_end)

X = [1 0 0];        % Populations for [A, B, C] (open, blocked, closed)
t = 0;              % Initialize the simulation time
n = 1;              % Data counter for I and T vectors
T(n) = t;           % Vector of times for output
XA = X(1);          % Vector of A populations for output

t_close = [];       % vector of 'closed intervals' for output
t_open = [];        % vector of 'open intervals' for output

nu(1,:) = [+1 0 -1]; % C -> A
nu(2,:) = [-1 0 +1]; % A -> C
nu(3,:) = [-1 +1 0]; % D + A -> B (large excess of D)
nu(4,:) = [+1 -1 0]; % B -> A + A (large excess of D)

% MC Simulation via Gillespie's "First Reaction Method" (Gillespie, 1976)
while (t < t_end)
    tau(1) = (1/(beta*X(3))) * log(1/rand);      % Opening (C -> A)
    tau(2) = (1/(alpha*X(1))) * log(1/rand);     % Closing (A -> C)
    tau(3) = (1/(k1*cD*X(1))) * log(1/rand);     % Blocking (D + A -> B)
    tau(4) = (1/(k_1*X(2))) * log(1/rand);       % Dissociation (B -> D + A)

    [Tau, mu] = min(tau);

    % Increment the time in the simulation
    t = t + Tau;

    % Update the system state
    X = X + nu(mu,:);

    % Record XA after the transition
    n = n + 1;
    T(n) = t;
    XA(n) = X(1);      % zero if closed, one if open

    % Record the length of the preceding interval
    if (X(1) == 1)   % If the channel is now open ...
        t_close(length(t_close)+1) = Tau;    % ... it was previously closed
    else             % Otherwise ...
        t_open(length(t_open)+1) = Tau;      % ... it was previously open
    end

end

return;
```

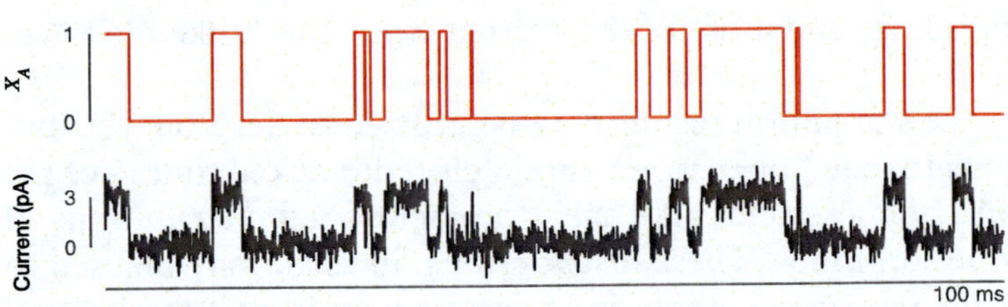

Figure 3.4 Results of a stochastic simulation of the simple ion channel blocking mechanism. These results are generated with $k_{+1} = 5.0 \times 10^7$ M^{-1}s^{-1}, a blocker concentration of $c_D = 1.0 \times 10^{-7}$ M, $k_{-1} = 2000\,\text{s}^{-1}$, $\alpha = 500\,\text{s}^{-1}$ and $\beta' = 150\,\text{s}^{-1}$.

Published by Woodhead Publishing Limited, 2013

describe a specific biochemical process *per se*, however, they do exhibit dynamics that are often observed among coupled autocatalytic reactions. The reactions are

$$X + Y_1 \xrightarrow{k_1} 2Y_1 \tag{3.27}$$

$$Y_1 + Y_2 \xrightarrow{k_2} 2Y_2 \tag{3.28}$$

$$Y_2 \xrightarrow{k_3} Z \tag{3.29}$$

The deterministic formalism discussed in Chapter 1 yields the following ordinary differential equations for this mechanism:

$$\frac{dc_{Y_1}}{dt} = k_1 c_X c_{Y_1} - k_2 c_{Y_1} c_{Y_2} \tag{3.30}$$

$$\frac{dc_{Y_1}}{dt} = k_2 c_{Y_1} c_{Y_2} - k_3 c_{Y_2} \tag{3.31}$$

$$\frac{dc_Z}{dt} = k_3 c_{Y_2} \tag{3.32}$$

Inasmuch as Eqs. 3.27 and 3.28 are bimolecular reactions, the deterministic equations may not be solved analytically. They may, however, be integrated straightforwardly using Euler's method or Runge Kutta, which approximate the derivatives in terms of finite differences, e.g $dt \rightarrow \Delta t$. By contrast, the following Matlab program implements an SSA for these reactions without any approximations:

We illustrate the stochastic and deterministic time evolutions of this reaction mechanism in Figure 3.5. The oscillations are a characteristic of this reaction network, and vary with the specific initial conditions and rate constants. The concentrations of species Y_1 and Y_2 may not be expressed in terms of the trigonometric functions, but are concisely out of phase with each other out of phase due to the structure of the mechanism.

Published by Woodhead Publishing Limited, 2013

```
function [t Y1 Y2 Z] = Lotka_sim(k1X, k2, k3, Y10, Y20, Z0, t_end, V, N)

X(1) = Y10*V;     % initial population of Y1 molecules
X(2) = Y20*V;     % initial population of Y2 molecules
X(3) = 0;         % initial population of Z molecules

% Stoichiometry matrix
stoichiometry(1,:) = [+1 0 0]; % X + Y1 -> 2Y1
stoichiometry(2,:) = [-1 +1 0]; % Y1 + Y2 -> 2Y2
stoichiometry(2,:) = [0 -1 +1]; % Y2 -> Z

t = 0;      % initial time

% MC Loop until time exceeds "t_end"
while (t < t_end)

    % Calculate reaction rates, k1X - k3 are stochastic rate constants
    a(1) = k1X*X(1);
    a(2) = k2*X(1)*X(2);
    a(3) = k3*X(2);
    a0 = sum(a);

    r1 = rand;
    tau = (1/a0)*log(1/r1); % Quiescence Interval

    t = t + tau;            % Update time

    % Choose reaction & modify system state
    r2 = rand;
    acc = 0;
    if (a(1) > r2*a0)
        X = X + stoichiometry(1,:);
    elseif ((a(1) + a(2)) > r2*a0)
        X = X + stoichiometry(2,:);
    else
        X = X + stoichiometry(3,:);
    end

end

Y1 = X(1);
Y2 = X(2);
Z = X(3);

return
```

The deterministic and stochastic results presented in Figure 3.5, initially agree, but the agreement diverges over time. If the divergence is a consequence of the numerical method used to solve Equations 3.30 -3.32 (e.g. insufficient discretization of time), then one might expect the agreement beween the deterministic formalism and the SSA to improve with increasing discretization.

However, as the results in Figure 3.6 show, the SSA is the real cause of the divergence. The deterministic results illustrated here are identical to those presented in Figure 3.5, but

Published by Woodhead Publishing Limited, 2013

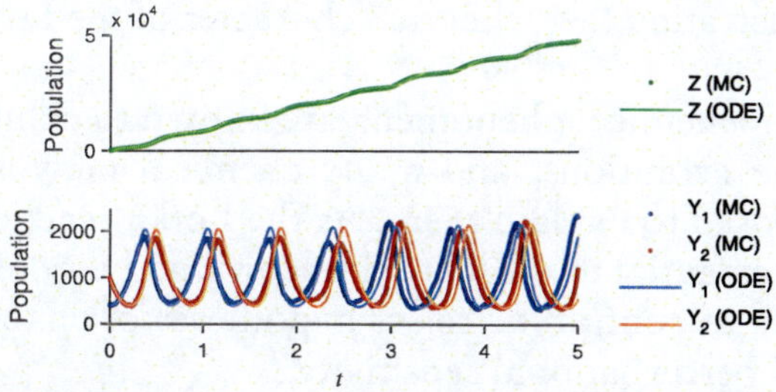

Figure 3.5 MC Simulation and deterministic results for the Lotka reactions with $k_1 c_X = 10$, $k_2/V = 0.01$, $k_3 = 10$, $X_{Y1}(0) = 400$, $X_{Y2}(0) = 1000$, $X_Z(0) = 0$. The deterministic ODEs for the Lotka reactions were integrated numerically using Euler's method with 10,000 abscissae.

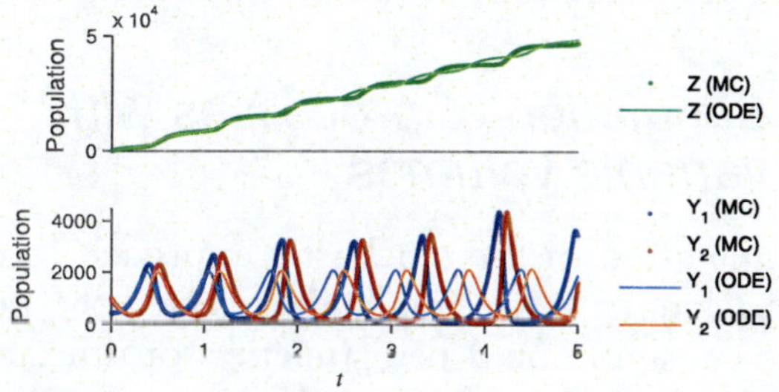

Figure 3.6 Another simulation of the Lotka reactions using conditions specified in Figure 3.5. The deterministic results are unchanged, but the stochastic time evolution is substantially different. The difference is a consequence of the sensitivity of the reaction network to perturbations.

the stochastic time evolution is considerably different. The difference between the stochastic and deterministic results is a consequence of the sensitivity of the reaction network to perturbations.The natural probabilistic perturbations in the SSA may result in dramatically different .time series from run

Published by Woodhead Publishing Limited, 2013

to run, illustrating the "chaotic" character of the Lotka reactions.

Many biochemical phenomena are known to exhibit oscillatory time evolutions, and many chemical networks have been proposed to explain them. As the Lotka reactions illustrate, it is essential to evaluate the *reproducibility* of simulation results to confirm or reject hypotheses of chaotic oscillation and perturbational sensitivity.

3.5 Caveats regarding the modeling of living systems

Reproducibility is but one of many aspects of biochemical models that must be evaluated in the course of an in silico investigation of a biological phenomenon. Physical aspects merit equal consideration – at all time and lengthscales.

3.5.1 Simulation of processes with variable volumes

In our formulation of the stochastic approach to chemical kinetics in Chapter 2, we implicitly rejected the possibility that a rate constant could be a function of time. However, from the very beginning of the stochastic approach to chemical kinetics, many researchers have investigated the time evolutions of processes with time-dependent reaction rate constants. There is no paradox: Recall that stochastic rate constants for bimolecular reactions are defined as

$$C = \frac{k}{V}$$

where V is the volume of the system. If V changes with time, then the stochastic rate constant will also be time-dependent. When performing simulations of chemical processes within a cell, V will be the volume of the compartment (e.g. the cytosol). If the cell is growing or dividing at the same timescale as the reactions, then one should account for

Published by Woodhead Publishing Limited, 2013

changes in cell volume over time. In the SSA, this will modify the MC selection step for the quiescence interval (DM and FRM) or reaction time (NRM).

The modification will be different for each of the SSAs. The direct method is the most challenging to modify owing to the form of $\mathscr{P}_1$, which is derived based on a constant value of a_0. If the volume of the cell changes, then the contributions of the bimolecular reactions to a_0 must be separated from the remaining reactions in Equation 3.13. One obtains a fairly unwieldy formula that may evade analytical integration, subject to the complexity of the time dependence of the cell volume.

By contrast, the FRM and NRM calculate each tentative reaction time or quiescence time independently. This simplifies the procedure of simulating time dependent processes somewhat. For the FRM, the MC selection rule for tentative reaction times is

$$r_\mu = \int_0^\tau f(\tau)d\tau \qquad (3.33)$$

where

$$f(\tau) = a_\mu(\tau)\exp\left(-\int_{t_0}^{\tau} a_\mu(t)dt\right) \qquad (3.34)$$

and t_0 is the actual time immediately preceding the event to come.

3.5.2 Simulation of processes with variable temperatures

Most investigations of biochemical phenomena are conducted at a single temperature that represents the "natural environment" of a reaction, e.g. body temperature. However, in some cases it may be advantageous to account for the energy released by chemical reactions as they occur, which manifests as a rise or drop in system temperature.

Rate constants almost always increase with the temperature of their environment. Often, this increase may be expressed using the Arrhenius equation

Published by Woodhead Publishing Limited, 2013

$$k_\mu = A_\mu \exp\left(-\frac{E_\mu}{R_g T}\right) \tag{3.35}$$

where $R_g = 8.31441$ J/mol/K and T is the system temperature. The parameter A_μ is denoted the "pre exponential factor", and the quantity E_μ is denoted the "activation energy". If a reaction is elementary, then both quantities may be related to statistical mechanical properties of the reaction. However, they are best considered as "free" parameters for fitting rate-temperature data. The Arrhenius equation is often sufficient for the modeling of the temperature dependence for both unimolecular and bimolecular reaction rates.

If one has Arrhenius constants for a set of biochemical reaction rates and wishes to perform a simulation, one must add a step to the SSA. After updating the populations of the just-reacted species but before recalculating the reaction rates, one must update the temperature of the system. If the system is energetically closed, i.e. heat is not provided from an external source (e.g. media), then the temperature may be calculated from the standard relationship

$$mC_p(T_{\text{old}} - T_{\text{new}}) = \Delta H_{rxn,\mu} \tag{3.36}$$

Thus, one requires the enthalpy of each reaction ($\Delta H_{rxn,\mu}$) as well as the heat capacity (C_p) and mass (m) of the system.

Once the system temperature has been updated, *every chemical rate constant must also be updated*, in turn requiring the recalculation of all tentative quiescence times in the FRM, tentative reaction times in the NRM, and total reaction rate a_0 in the DM.

In practice, very few biochemical researchers have investigated reactions in such detail that Arrhenius constants and enthalpies of reaction will be available. Absent these parameters, any simulation of the temperature dependence of a process must involve assumptions, which will introduce artifacts.

Published by Woodhead Publishing Limited, 2013

3.5.3 Mixing and diffusion

On this note, simulations are only as accurate as the rate constants they employ. Practitioners of stochastic simulation tend to prefer the use of rate constants that were carefully measured under well defined laboratory conditions. However, these conditions often entail the running of reactions in aqueous media with vigorous mixing.

The cell, by contrast, is not "well mixed" and does not possess rheological properties of water. The cytosol, for instance, is packed with proteins, nucleic acid species, and a variety of other macromolecules and metabolites. This environment will not permit the free motion of biomolecules that occurs in aqueous media, and precludes widespread mixing. In reducing their ability to diffuse and mix, it reduces their ability to contact their co-reactants and thus, every bimolecular rate constant.

This problem is shared by both the deterministic and stochastic formalisms, but it is worthwhile considering when forecasting the time evolution of any biochemical reaction that takes place within the cell. At present, there does not exist a robust science to relate *in vitro* rate constants to *in vivo* rate constants. Such a discovery will have a profound impact in computational and systems biology.

Notes

1. http://mathworld.wolfram.com/EulerForwardMethod.html
2. This is called "small o" notation, http://mathworld.wolfram.com/LandauSymbols.html
3. Here, ν refers to a reaction $\nu \in [1, M]$ whereas the subscript "1" in Equation 3.9 does not refer to any reaction whatsoever.

Published by Woodhead Publishing Limited, 2013

References

1. Daniel T Gillespie. A general method for numerically simulating the stochastic time evolution of coupled chemical reactions. *J Comput Phys*, 22:403–434, 1976.
2. Daniel T Gillespie. Exact stochastic simulation of coupled chemical reactions. *J Phys Chem*, 81(25):2340–2361, 1977.
3. Adam Arkin, John Ross, and Harley H McAdams. Stochastic kinetic analysis of developmental pathway bifurcation in phage lambda-infected escherichia coli cells. *Genetics*, 149:1633–1648, 1998.
4. Andrzej M Kierzek. Stocks: Stochastic kinetic simulations of biochemical systems with gillespie algorithm. *Bioinformatics*, 18:470–481, 2002.
5. Peter D. Karp, Suzanne Paley, and Pedro Romero. The pathway tools software. *Bioinformatics*, 18(S225-S232), 2002.
6. Lingchong You, Apirak Hoonlor, and John Yin. Modeling biological systems using dynetica - a simulator of dynamic networks. *Bioinformatics*, 19:435–436, 2003.
7. David Adalsteinsson, David McMillen, and Timothy C Elston. Biochemical network stochastic simulator (bionets): software for stochastic modeling of biochemical networks. *BMC Bioinformatics*, 5:24, 2004.
8. Stephen Ramsey, David Orrell, and Hamid Bolouri. Dizzy: stochastic simulation of large-scale genetic regulatory networks. *J Bioinform Comput Biol.*, 3:415–436, 2005.
9. Stefan Hoops, Sven Sahle, Ralph Gauges, Christine Lee, Jurgen Pahle, Natalia Simus, Mudita Singhal, Liang Xu, Pedro Mendes, and Ursula Kummer. Copasi—a complex pathway simulator. *Bioinformatics*, 22:3067–3074, 2006.
10. Ravishankar Rao Vallabhajosyula and Herbert M. Sauro. Stochastic simulation gui for biochemical networks. *Bioinformatics*, 23:1859–1861, 2007.

Published by Woodhead Publishing Limited, 2013

11. Andre S. Ribeiro and Jason Lloyd-Price. Sgn sim, a stochastic genetic networks simulator. *Bioinformatics*, 23:777–779, 2007.
12. Michael A Gibson and Jehoshua Bruck. Efficient exact stochastic simulation of chemical systems with many species and many channels. *J Phys Chem A*, 104:1876–1889, 2000.
13. Yang Cao, Hong Li, and Linda Petzold. Efficient formulation of the stochastic simulation algorithm for chemically reacting systems. *J Chem Phys*, 121:4059–4067, 2004.
14. Anne Pierres, Anne-Marie Benoliel, and Pierre Bongrand. Measuring bonds between surface-associated molecules. *J. of Immunological Methods*, 196:105–120, 1996.
15. Teresa A. Doggett, Gaurav Girdhar, Avril Lawshe, David W. Schmidtke, Ian J. Laurenzi, Scott L. Diamond, and Thomas G. Diacovo. Selectin-like kinetics and biomechanics promote rapid platelet adhesion in flow: The gpibα-vwf tether bond. *Biophys J*, 83:194–205, 2002.
16. David Colquhoun and Alan G Hawkes. Relaxation and fluctuations of membrane currents that flow through drug-operated channels. *Proc Royal Soc Lond B, Biol Sci*, 199:231–262, 1977.

Published by Woodhead Publishing Limited, 2013

4

Modelling in systems biology

Abstract Mathematical and computational frameworks for modelling and simulating biological process at the system level have to face with the main feature of such systems: the complexity. The actors of biochemical interactions have a complex structure, and the network of the interaction among them is complex. Complex behaviour occurs when many interactions at the local scale collectively lead to unpredictable larger-scale outcomes. In this chapter, we will adopt the perspective of a modeler trying to identify the expressions of biological systems complexity and searching for a suitable language to formally describe them.

Keywords: systems biology, complexity, stochastic modelling and simulation, process algebras.

4.1 What is biological modeling

Modeling is an attempt to describe an understanding of the elements of a system of interest, their states, and their interactions with other elements. The model should be sufficiently detailed and precise so that it can in principle be used to simulate the behavior of the system on a computer. In the context of molecular cell biology, a model may describe the mechanisms involved in transcription, translation, cell regulation, cellular signaling, DNA damage and repair processes, the cell cycle or apoptosis. At a higher level, modeling may be used to describe the functioning of a tissue, organ, or even an entire

Published by Woodhead Publishing Limited, 2013

organism. At still higher level, models can be used to describe the behavior and time evolution of populations of individual organisms. At the beginning of a modeling project, the first issue to confront is to decide on which feature to include in the model and the level of detail the model is intended to capture. So, for example, a model of an entire organism is unlikely to describe the detailed functioning of every individual cell, but a model of a cell is likely to include a variety of very detailed description of key cellular processes. Even then, however, a model of a cell is unlikely to contain details of every single gene and protein.

In order to show how it is possible to think about a biological process at different scales and different levels of detail, let us consider the trascription and translation process of DNA. It can be summarized by two chemical reactions in sequence.

$$\text{DNA} \dashrightarrow \text{mRNA} \dashrightarrow \text{DNA}$$

This simple reaction chain is a summary of the overall effect of the process. DNA transcription and translation consist of many reactions. The above equation to describe them globally represents the process at higher level than the more detailed description that biologists often prefer to work with. Whether a single overall equation or a full breakdown into component reactions is necessary depends on whether intermediate reagents are elements of interest to modeler. In general, we can state that the "art" to build a good model consists in the ability of capturing the essential features of the biology without burdening the model with non-essential details. However just because of the omission of the details, every model is to some extent a simplification of the biology. Nevertheless, models are valuable because they take ideas that might have been expressed verbally or diagrammatically, and make them more explicit, so that they can begin to be understood in a *quantitative* rather than purely *qualitative* way.

The features of a model depend very much on the aims of the modeling. Modeling and simulation appeared on the scientific horizon before the emergence of molecular and cellular biology. Their genesis is in the physical sciences and

Published by Woodhead Publishing Limited, 2013

engineering. In the physical sciences, besides theoretical and experimental studies, modeling and simulation are considered as the third indispensable approach because not all hypotheses are amenable for confirmation or rejection by experimental observations. In biology, researchers are facing the same or maybe even worse situation. On one hand experimental studies are unable to produce a sufficient amount of data to support theoretical interpretations; on the other hand, due to data insufficiency, theoretical research cannot provide substantial guidance and insights for experimentation. Therefore computational modeling takes a more important role in biology by integrating experimental data, facilitating theoretical hypotheses, and addressing what if questions.

Another important aim of modeling is to make clear the current state of knowledge regarding a particular system, by attempting to be precise about the elements involved and the interactions between them. Doing this can be an effective way to highlight gaps in understanding. Our understanding of the experimental observations of any system can be measured by the extent to which a simulation, we create, mimics the real behavior of that system. Behaviors of computer-executable models are at first compared with experimental values. If at this stage inconsistency is found, it means that the assumptions, that represent our knowledge on the system, are at best incomplete, or that the interpretation of the experimental data is wrong. Models survived to this initial validation can then be used to make predictions to be tested by experiments, as well to explore configurations of the system that are not easy to investigate by *in vitro* or *in vivo* experiments. Creation of predictive models can give opportunities for unprecedented control over the system. In contrast to physics, biology still lacks knowledge of the fundamental laws on which it is based. Modeling can provide valuable insights into the workings and general principles of organization of biological systems.

Modeling, simulation, and analysis of the simulation outcomes are therefore perfectly positioned for integration into the experimental cycle of cell biology (Fig. 4.1). Although

Published by Woodhead Publishing Limited, 2013

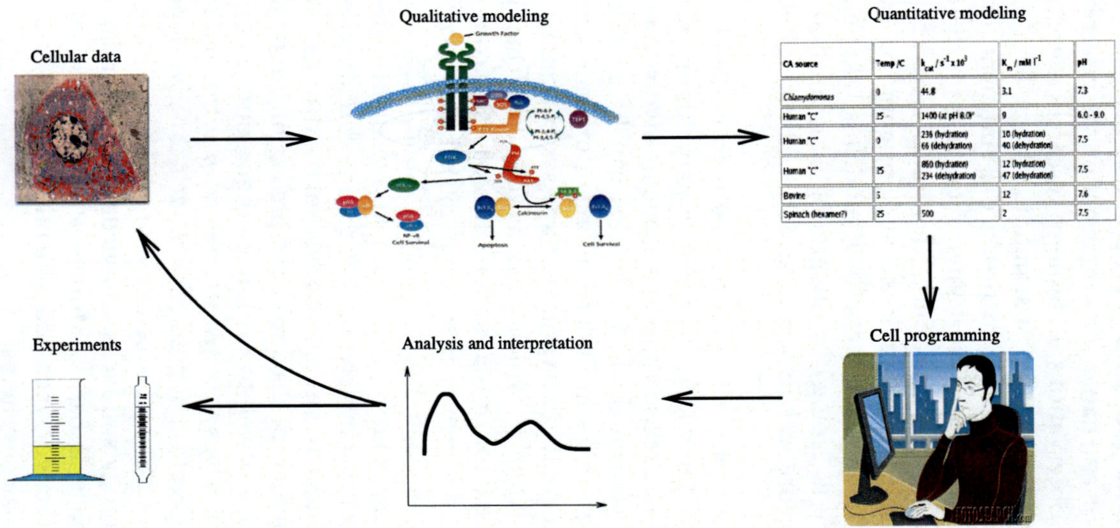

Figure 4.1 The cell biology research cycle.

we will always need real experiments to advance our understanding of biological processes, conducting *in silico*, or computer-simulated experiments can help guide the wet-lab process by narrowing the experimental search space.

4.2 System Biology

More than fifty years ago, Watson and Crick [56] identified the structure of DNA, thus paving the way for the molecular biology and genetics. Grounding the biological phenomena on molecular basis made it possible to describe the different aspects of biology, such as heredity, diseases and development, as the result of the coherent interactions between sets of elements, that are either functionally different or most often multifunctional. Grounding biological phenomena on a molecular basis made it possible to include biology in a consistent framework of knowledge based on fundamental law of physics. Since then, the field of molecular biology has emerged and enormous progress has been made. Molecular biology enables us to understand biological systems as molecular machines. Large numbers of genes and the function of transcriptional products have been identified. DNA sequences have been fully identified for various organisms such as *mycoplasma*, *Escherichia coli* (*E. coli*), *Caenorhabditis elegans* (*C. elegans*), *Drosophila melanogaster*, and *Homo sapiens*. Measurements of protein level and their interactions is also making progress [19, 49]. In parallel with such efforts, new methods have been invented to disrupt the transcription of genes, such as loss of-function knockout of specific genes and RNA interference that is particularly effective for C. elegans and is now being applied for other species. Nevertheless, such knowledge is not sufficient to provide us a complete understanding of biological systems *as systems* [24]. Cells, tissues and organs, and organisms as well as ecological webs are systems of components whose specific interactions have been defined by evolution; so a system-level understanding should be the prime goal of biology.

Published by Woodhead Publishing Limited, 2013

System-level understanding requires a set of principles and methodologies that links the behaviors of molecules to system characteristics and function. These principles and methodologies should be developed in the following four areas of investigation.

1. System structures. These include the network of gene interactions and biochemical pathways, as well as the mechanisms by which such interactions modulate the physical properties of intracellular and multicellular structures.
2. System dynamics.[1] How a system behaves over time under various conditions can be understood through metabolic analysis, sensitivity analysis, and dynamic analysis methods such as portrait and bifurcation analysis. Specifically, the system behavior analysis aim at addressing the following questions: how does a system respond to changes in the environment? How does it maintain robustness against potential damage, such as DNA damage and mutations? How do specific interaction pathways exhibit functions observed? It is not a trivial task to understand the behaviors of complex biological networks. Computer simulation and a set of theoretical analysis are essential to provide in-depth understanding of the mechanisms behind the pathways.
3. Control methods. the individuation of mechanisms that systematically control the state of the cell is necessary for two reasons: 1. their understanding can be exploited to modulate them to minimize malfunctions, and 2. they involve potential therapeutic targets for treatments of diseases.
4. The design method. Strategy to modify and constructing biological system with desired properties can be developed on definite design principles and simulations, instead of blind trial-and-error.

Any progress in each of the above areas requires breakthroughs in our understanding not only of molecular biology, but also of measurement technologies and computational sciences. Although advances in accurate, quantitative

Published by Woodhead Publishing Limited, 2013

experimental approaches will doubtless continue, insights into the functioning of biological systems will not result from purely intuitive assaults. The reason of this stays in the intrinsic complexity of biological systems, that a combination of experimental and computational simulation approaches is expected to solve.

At the present, identification of gene-regulatory logic and biochemical network is a major purpose. Nowadays, biological modeling aims at uncovering mechanisms at the fine-grained level that are internally consistent with molecule-level biological programs and at reproducing observed phenomena. Since it is hard to continuously and systematically monitor the *parallel* activities in molecular networks, molecule-level modeling [14, 36] has become and indispensable tool to bridge experimental and theoretical studies and to link system behaviors with molecular reactions.

Due to the distinctive differences between biological and physical systems, modeling a network of interacting molecules comes with additional challenges and calls for new strategies and tools. The early objective of modeling was to explore the feature of complex biological systems treated as black boxes. In such a scenario, the goal was to understand and predict the behavior of a system without knowing the microscopic details. The strategy was to reproduce observed phenomena at high level with a simplified description of the internal structures. Two methodological features emerged at this stage. First, since biological systems were approximated as structure-less entities, many methods and tools were directly borrowed from engineering fields such as the Finite Element Method[2] and Boundary Element Method [4]. The second methodology was a high-level abstraction based on the inverse approach to modeling. As consequence, numerical techniques for the solution of ordinary differential equation (ODE) and partial differential equations (PDE) were applied. Both black box assumption and inverse modeling, though suitable for modeling mechanical systems, suffer from major problems when applied to biological systems. The black box conjecture assumes that the internal structure of the system is static and thus it can not hold when the

Published by Woodhead Publishing Limited, 2013

system evolves in time as, for instance, in growth process. Complex internal structure and evolution are key features that differentiate biological systems from mechanical systems. The inverse modeling suffers from generality loss and many inverse problems are mathematically ill-posed. Even if adequate and precise data are available, unique solution is no always guaranteed and special techniques are employed specifically to the problem in hand [20].

The dynamic context in which genes operate is much more complex than the static composition of genes and genomes. Though sequence alignment can help us find homologues, the exact functions of genes still need to be confirmed experimentally. For example during embryonic development, different ectopic [9] and failed gene expression events can lead to different phenotypes. The problem is encountered by creating various knock-ins and knock-outs. The semantics of the genetic program cannot be modeled by using the black box conjecture. However, more generally, how the interaction among molecules produces the complexity of a biological system has no clear answer. Knowledge of biological complexity can lead to design of better or more efficient systems, and also for understanding of pharmacological effects for drug discovery. Because of these reasons also, a set of simulations, each of which coming from a perturbation of the parameters of an original model, are helpful in understanding the dynamic context in which gene, products and molecules operate.

A perturbed system is one in which the system's behavior is forced out of its 'normal' state by disturbances coming, for example, from external influences. This definition applies to theoretical physics or a biological system, and in both cases perturbation offers a means to study and understand a system. Furthermore, applying perturbation theory to biology may eventually allow prediction and treatment of pathological perturbations (diseases) such as exist in the clinical setting.

Perturbation analysis studies the behavior of systems forced out of their normal state. It is often the case that the behavior of a system under such perturbations is much more

Published by Woodhead Publishing Limited, 2013

amenable to theoretical analysis than the general (i.e., normal) behavior of the system. The main reason is that, mathematically, the behavior of a system close to its 'normal' state can often be described by linear equations, whose theory is very well developed. In addition, beyond linear perturbation theory, there is a well-developed theory for describing the behavior of systems as one moves away from a reference state. This theory aims, for instance, to predict under what perturbations a system will return to its reference state and which perturbations will destabilize the system. This way of thinking also applies to biology. The perturbation of a biological system by means of genetic mutation or small molecules (chemical genetics) greatly aids the understanding of the fundamental principles underlying such a system or process. Through genetic dissection biologists learned that basic cellular processes such as cell growth and cell division (as well as developmental processes depending on the interaction of groups of cells and tissues) have been highly conserved throughout evolution. Therefore, perturbations by small molecules or by targeted or random mutations in individual genes in simple model organisms such as yeast, *Drosophila*, *C. elegans*, *Arabidopsis*, and the mouse have provided, and will in the future provide important insight into the function of complex systems. Perturbation theory can also be applied biologically in a more controlled, reiterative manner. One can imagine taking some biological system of interest, defining its normal behavior, and then investigating in a general and methodological way which perturbations destabilize the system (in the sense that it will no longer return to its normal state). Examples could be regulatory systems of various kinds, such as those that keep the concentrations of different metabolites within the cell at fixed levels and restore these levels after a perturbation. One would then aim to identify what kind of perturbations would destabilize these regulatory systems. One would go back and forth between performing perturbation experiments to see how the system behaves in response to various perturbations, and building theoretical and computational models. One would start with 'small' perturbations that can be described with

Published by Woodhead Publishing Limited, 2013

linear models, and would use those to predict, and subsequently test, the behavior in response to larger perturbations.

4.2.1 What future for System Biology

Biologists are getting enthusiastic about mathematical modeling, as modelers are getting exited by biology. The *complexity* of molecular and cellular biological systems makes it necessary to consider dynamic systems theory for modeling and simulation of intra- and inter-cellular processes. To describe a system as "complex" has become a common way to either motivate new approaches or to describe the difficulties in making progress. Currently, before we can fully explain and understand the functioning and the functions of cells organs or organisms from the molecular level upwards, the major difficulties to overcome are technological and methodological. Nevertheless, whatever time is required, the complexity of these systems ensures that there is no way around mathematical modeling in this endeavor. A mathematical pathway model does not represent an objective reality outside the modeler's mind. The model is no more, and no less, a complement of biologist's reasoning. Mathematics is the handicraft of the natural sciences.

The risk in this exciting endeavor is that the following thoughts from the beginnings of System Biology will remain true in the years to come:

"In spite of the considerable interest and efforts, the application of systems theory has not quite lived up to expectations. One of the main reasons for the existing lag is that systems theory has not been directly concerned with some of the problems of vital importance in biology." [34]

The challenge is for both the theoreticians and experimentalists to change their ways:

"The real advance in the application of systems theory to biology will come about only when the biologists start asking

Published by Woodhead Publishing Limited, 2013

questions which are based on the system-theoretic concepts rather than using these concepts to represent in still another way the phenomena which are already explained in terms of biophysical or biochemical principles. Then we will not have 'the application of engineering principles to biological problems' but rather a field of System Biology with its own identity and in its own right." [34].

System biology has succeeded when it is widely accepted that there is nothing more practical than a good theory [57].

It is now necessary to clarify what complexity means in the context of system biology. A complete definition of complexity should be given with respect to

- the model: the large number of variables that can determine behavior
- the natural system: the connectivity and non-linearity of relationships
- the technology: the limited precision and accuracy measurements
- the methodology: the uncertainty arising from the conceptual framework chosen (e.g. the choice of automata instead of differential equations).

In the next section we will go through the concept of complexity in sistems biology.

Published by Woodhead Publishing Limited, 2013

4.3 Complexity of a biological system

It is often said that biological systems, such as cells, are *complex systems*, and that the grand challenge of the 21st Century is to understand and model the complexity of biological systems. Though complexity has been extensively discussed at different levels [30, 58, 60, 50], there is no operational definition for biological systems [2]. The common notion of complex systems is of very large numbers of simple and identical elements interacting to produce 'complex' behaviors. However, the reality in biology is somewhat different. In biological system large numbers of functionally different, and often multifunctional, sets of elements interact selectively and non-linearly to produce *coherent* rather than *complex* behaviors. A biological system is not equal to the sum of its parts [38], in which functions emerge from the properties of the networks rather than from any specific element. On the contrary in biological systems, functions rely on a combination of the network and the specific element involved [25]. A typical example is the p53 interactions pathway. This protein, known as 'the guardian' of the genome, acts as tumor suppressor. It is activated, inhibited and degraded by reactions as phosphorilation, de-phosphorilation, and proteolytic degradation, while its targets are selected by the different modification patterns that exist; these are properties that reflect the complexity of the element itself. Just considering this example, Kitano [25] highlighted that biological system are better characterized as *symbiotic* systems.

Beside the inherent complexity, some hallmarks of complexity, such as linearity and non-linearity, number of parameters, order of equations and evolution of network, come out only when a system is formalized in specific ways for a linear formalization of a two-component signal transduction model. Moreover, we can distinguish two types of complexity both encountered in modeling biological systems: functional and structural, or dynamic and static. The operative

Published by Woodhead Publishing Limited, 2013

definition and the identification of the complexity in biological system is not the only hard task, but also its quantitative measure by experimental biologists. The popular measure of complexity for dynamical system is the computational complexity. For instance, the complexity of a sequence can be inferred from what a finite state machine can produce. Although this measure characterizes the amount of information necessary to predict the future state of the machine, it fails to address its meaning in the world of molecular and modular cell biology [2].

Since the topological structure of a molecular network undergoes significant evolution within cells in biological development, to measure both static and dynamic complexity according to such evolution may be a practical way, namely it is easier to identify and abstract information from it [5, 26]. Furthermore, features in topological structure, such as the existence of organized biological compartments, are also helpful in identifying modularity of molecular interaction. We will return on this point in section 4.5.

Finally, there are other two important indexes of complexity in biological systems. The first is non-linearity, including parameter sensitivity and initial values sensitivity. The second, on which we will focus in this thesis is the existence of stochasticity. The noise increases the complexity of the systems even further by introducing issues of robustness, noise resonance and bi-modal behavior.

4.4 Stochastic modeling approach

An important aspect of modeling of biological networks is the handling of random events that occur inside a cell. Arguments for the application of *stochastic* models to address these random qualities of chemical and biochemical reactions are motivated by three factors:

1. they may account for random and discrete time evolutions of key variables, e.g. molecules or cell populations

Published by Woodhead Publishing Limited, 2013

2. they may conform with the laws of thermodynamics in ways that facilitate experimental designs, and
3. they are appropriate for the description of small systems, including those featuring dynamic instabilities

Many studies have reported occurrence of stochastic fluctuations and noise in living systems. Observations of gene expression in individual cells clearly illustrate the stochastic nature of transcription [1, 32]. Other studies in eukaryotic gene expression show that the messenger RNA (mRNA) production is quantal [18] and is produced in random pulses [47, 55]. It has been proposed that proteins are produced in short 'bursts' at random time intervals rather than in a continuous manner [8]. Furthermore, another clear evidence of the stochasticity of the biological phenomena at the molecular level is the existence of qualitatively and quantitatively different outcomes in the temporal behavior of a system starting from the same initial conditions. A classic example is the lysis/lysogenic switch of bacteriophage λ infected E. Coli. Due to noise, the network may randomly evolve into one of these two bistable regions [16, 15]. Role of noise has also been seen in bacterial chemotaxis [29] and cellular selection [53]. At the level of cellular population, the most important implication of noise in critical cellular processes is that in spite of identical initial conditions, with time, different cells may evolve along distinct pathways. population measurements typically show that the level of expression from the same gene vary significantly across cells with the same genetic material. The origin of such variability among isogenic population is largely attributed to stochastic phenomena [13].

In Chapters 2 and 3, we observed how the intrinsic noise of a chemical process arises from the randomness of individual chemical reaction events. In a network of molecular interactions there exists an *extrinsic* component of noise too. The extrinsic component of randomness is due to the external environmental conditions. For example, a transcription factor for a given gene is often the protein product of another gene and thus its production is also random. Such situations,

Published by Woodhead Publishing Limited, 2013

where a protein product of a stochastic triggering of a gene leads to the switching of another gene, are characterized by a cascade of stochastic events. The timings of such triggers can result in different outcomes [31].

As we discussed in Chapters 2 and 3, the formulation of the theory of stochastic kinetics does not reduce the importance of deterministic kinetics, because there exist a class of phenomena for which the stochastic model is only slightly "better" than the deterministic approach, while the mathematics of the stochastic model is much more complicated. ODE description has been practically used in many quantitative models. As we discussed in Chapter 1 the general form of an ODE model of a biochemical process may be written as

$$\frac{d[X_i]}{dt} = f_i(x) \tag{4.1}$$

where $i = 1, 2, \ldots, N$ are biochemical species and $[X_i(t)]$ is the population of the i-th species.

There have been several platforms for ODE based modeling. Among them, the most known are Gepasi [33] and E-CELL [54], which share a number of features in common, e.g. for chemical reactions simulation. Tools of mathematical analysis like metabolic control analysis and linear stability analysis of steady state, and parameter fitness have also been implemented. However, though metabolic reactions can be simulated by these tools, signaling activities may not be well supported [11]. Furthermore, signaling networks are non static and undergo evolution [5, 59]. Thus, modeling of the context-dependent cellular processes merits a different approach. A typical example is Presenilin, a protein responsible for cleaving Notch/Delta complex. It can selectively cleave a large group of membrane proteins in different contexts [28, 51]. Thus, to describe its behavior with ODEs is infeasible, because the biochemical equations would be very complex with the addition of new gene or protein into the model many equations must be re-written. This is an arduous work that greatly slows the modeling process itself.

Published by Woodhead Publishing Limited, 2013

Another example of gene with complex function is the Notch gene itself, that takes part in intercellular communication process. The semantics or function of its interaction with other proteins depends on its partners and the timing of interaction [42, 21]. In addition, in any practical model, to get complete quantitative data on gene and protein activity, such as the rate of transcription, translation and degradation of proteins, is extremely difficult. Thus, only small or medium sized models have been reported.

This brief introduction to reaction rate equations allows us to understand more deeply the meaning of the expression "intrinsically stochastic", that in this section we have used to define the character of a biological phenomenon at the molecular scale. Although the great importance and usefulness of the differential reaction rate equations approach to chemical kinetics cannot be denied, we should not lose sight of the fact that the *physical basis* for this approach is meaningless. This approach assumes that the time evolution of a chemical reacting systems both *continuous* and *deterministic*. However, since the molecular population levels can change only by discrete integer amounts, the time evolution of a chemical reacting system is no a continuous process. The time evolution is not a deterministic process either. Even ignoring quantum mechanical effects and regarding the molecular motions to be governed by the equations of classical mechanics, it is impossible even in principle to predict the dynamics of the system unless we have a complete knowledge of its state. Knowledge about the state of the system includes the details about the position, the orientation, and the momentum of every single molecule under consideration, together with a complete knowledge of the chemistry of interacting molecules. If we leave out such details of the state of the system in favor of a higher level view, the dynamics of the system is not deterministic but intrinsically stochastic. In other words, although the temporal behavior of a chemically reacting system of classical molecules is deterministic in the full position-momentum phase space, it is stochastic in the N-dimensional subspace of the molecular population levels, as Eqs. 4.1 imply.

Published by Woodhead Publishing Limited, 2013

To conclude this section, we point out some of the roles played by the stochasticity in biological phenomena. Interestingly, some living systems have noise-suppressing mechanisms. An example is *genetic redundancy* [37]. The theory of feedback loop control states that the noise is also a stabilizer and a driver of molecular motors. Moreover noise is also responsible for the phenomenon of *stochastic resonance*, that is the phenomenon in which noise enhances the detection of weak signals and helps improve the biological information processing [17].

Noise is involved in the so-called *stochastic focusing*, in which cells exploit it to reduce the random variation in regulated processes, by tuning a mechanism to a threshold [39]. Finally, stochasticity plays a crucial role in the differentiation by establishing initial asymmetries leading to different evolutive categories of different parts of a system. An example of a role of noise in differentiation can be found in many processes regarding the immune systems, such as the clonal amplification of cells expressing an antigen, but also in many processes driving the rhythm of biological oscillators such as those involved in circadian rhythm mechanism.

4.4.1 Stochastic simulation algorithms

As we discussed in Chapter 3, the Stochastic Simulation Algorithms have been applied to many *in silico* investigations of biochemical dynamics in recent yeras. Kastner et al. in [22] applied the algorithm for simulation of Hox cis-regulatory mechanisms. The simulation was successful in reproducing key features of the wild-type pattern of gene expression and *in silico* experiments yielded results similar to that of *in vivo* experiments. Besides that, Kierzek et al. in [23], applied the algorithm to model lacZ gene expression and discovered the influences of the frequencies of transcription and translation initiation on random fluctuations in gene expression. McAdams and Arkin in [31], also studied the transcription initiation and translation mechanisms in the cellular regulatory network using Gillespie's algorithms and found several

Published by Woodhead Publishing Limited, 2013

stochastic phenomena like the fluctuation in protein production and switching delay for genetically coupled links.

4.5 Formalizing complexity

The invention of conceptual and technological tools are the building blocks of any scientific revolution and paradigm shift. Such conceptual and technological tools are now emerging at the intersection of computer science, mathematics, biology, chemistry and engineering.

The main three concepts that revolutionized the approaches of the researcher to the system biology can be summarized as in the following:

- A living cell is an information processing device. Cells naturally process internal and environmental information in complex fashions and interact with neighboring cells to achieve coordinated behavior.
- Cellular information processing and passing are carried out by networks of interacting molecules.
- A better understanding of the cell requires an information processing model.

Computers have similar characteristics to the cell. Like software, cells affect, prescribe, cause, program and blueprint other behavior. All computers have an essentially similar core design and basic functions, but address a wide range of tasks. Similarly, all cells have a similar core design, yet can survive in radically different environments or fulfill widely differing functions.

Computer science also has much to offer biology in its understanding of complex systems. Indeed, the cartoons and block diagrams used by biologists to represent metabolic, signaling or regulatory pathways are qualitative models that lay out the connectivity of elements important to the phenomenon under investigation. Such models throw away details (e.g., about kinetics) whose omission in many cases renders the model irrelevant. To gain insight into the biochemical interactions that are responsible for the behavior

Published by Woodhead Publishing Limited, 2013

of the cell, a quantitative formal abstraction of the cell processes must be developed. A formal abstraction is essentially a model specified in a formal language mathematically well founded. The biologists themselves recognize the necessity of a formal language for describing the biochemical processes. With regard to this, we report two significant assertions

> *"We have no real algebra for describing regulatory circuits across different systems..."* (T. F. Smith, 1998 [52]) (TIG 14:291-293, 1998)

> *"The data are accumulating and the computers are humming, what we are lacking are the words, the grammar and the syntax of a new language ..."* (D. Bray, 1997 [6])

A formal language will be more precise than present systems of notation used by biologists. One of the challenges is making sure that the language is not at too low a level of abstraction, which might mean getting lost in a mess of details. However, if you start too high, too many details will be ignored. Neglecting these two extremes, there is a need to be able to model different levels of abstraction.

A wide variety of formal models, languages and methods have been developed in the last two decades for supporting the specification, design, verification, implementation and testing of computer networks and distributed software systems. These include CCS, π-calculus, timed and stochastic process algebra, Petri Nets, Statecharts, Logics, formal Object-Oriented approaches, and others.

Formal specification languages have been designed to support the description of system structure and behavior in terms of concepts such as event occurrence, observation and experiment, temporal ordering, causality, cooperation and synchronization among entities, non determinism, concurrency and parallelism, state changes and invariants. While considerable experience has been gained in the application of formal methods to the areas for which they were initially conceived, the high abstraction level of these concepts suggests that they could play an important role in several other disciplines such as chemistry, biology, and physics.

Published by Woodhead Publishing Limited, 2013

4.5.1 Disadvanges in using ODEs for system biology: two alternatives

The implications of thinking of the biology in terms of systems are many. The most evident consists in rediscovering those concepts typical of the system theory that can be applied to the phenomenological world of biology. The studies presented in this thesis has led to the identification of four main conceptual areas subdivided as follows:

- modularity and compositionalty
- combinatorial behavior of the chemical interactions
- multifunctionality, specificity and complementary of the interactions

Here in the following we discuss each of these concepts.

The concept of *modularity* for example, that has served engineers and systems theorists well for some time, has been rediscovered for biology. Classical biology already had this concept on a rather macroscopic scale, without explicitly calling it by this name. Now researchers see a *modular framework* for biology, "treating subsystems of complex molecular networks as functional units that perform identifiable tasks perhaps even able to be characterized in familiar engineering terms" [27].

Modularity is a design principle in engineering that allows construction of complex systems from simple components. In systems theory the concept of modularity is inherent in the definition of systems, where systems can be connected together to form higher-level systems. Cells have organelles, organs are built from cells, complex organisms have distinct functional units - i.e. organs - which are connected to form a higher level of organization. The concept can even be extended to cover populations, which consist of single organisms and organize into sub-populations.

Newer biological applications of the modeling principles of modularity and compositionality (encapsulating subsystems) are already part of developments in systems biology. Recently, an application in sub-cellular dimensions has been suggested. A work of Han et al. [12], published in *Nature*

Published by Woodhead Publishing Limited, 2013

two years ago discovered the existence of dynamically organized modularity in the yeast protein-protein interaction network. In apparently scale-free protein-protein interaction networks, or "interactome networks", most proteins interact with few partners, whereas a small but significant proportions of proteins (called "hubs") interact with many partners. The biological scale-free networks exhibit a particularly strong resistance to random node removal, but are extremely sensitive to the targeted removal of hubs. A link between the potential scale-free topology of interactome networks and genetic robustness seems to exist, because knockouts of yeast genes encoding hubs are threefold more likely to likely to confer lethality than those of non-hubs. Han et al. investigated how hubs might contribute to robustness and other cellular properties for protein-protein interactions dynamically regulated both in time and space. Both *in silico* studies of network connectivity and genetic interactions described *in vivo* support a model of organized modularity in which some classes of hubs organize the proteome, connecting biological processes - or modules - to each other, whereas some other classes of hubs function inside modules [12].

The differential equations hardly look like a modularity model even they in simple linear phenomena they can provide such a model. However if modularity is associated to compositionality, that is the norm in biology, the differential equations are not recommended.

The combinatorial and context-sensitive behavior of molecular agents raises a fundamental problem in representing systems of such agents. A description in terms of ordinary differential equations becomes inappropriate for reasoning about the dynamics of complex molecular networks, because it represents agents only in terms of the interactions that exist in a particular context. It does not represent agents in terms of potential interactions that do not exist now, but could exist (and become relevant) if other components were to appear in the system through processes such as gene expression, genetic change, or exogenous intervention. To define a system of differential equations, all the chemistry must have already happened. The differential equations are

Published by Woodhead Publishing Limited, 2013

concerned only with kinetics (rates of concentration changes) and require a complete enumeration of all reaction possibilities to begin with.

Finally, the multifunctionality of a biological entity involved in a network of chemical interaction is a common situation. Genetic polymorphism and protein conformational plasticity, for instance are two ways to promote multifunctionality. The formalism of differential equation treats each molecular species as a monolithic entity and thus it is not able to express the multifunctionality due to the presence of specific structures of interactions, protein domains, ligands, receptors, specific site of interaction that allow to the molecule or cell to have multiple simultaneous interactions with different partners.

The theories developed for the mathematical expression of all these concept are useful to further understanding of biological systems. On a more technical note, their application would enable researchers to tune abstraction levels of models to their purpose of research. This, in many cases, might reduce the requirements for computing power and speed up simulations.

Many alternative formalisms, derived from Computer Science, have been used recently to overcome the difficulties typical of the application of the differential equations to the specification of biological system, especially at molecular level.

In this thesis, we will focus on two formalisms, whose recent use in system biology modeling and simulation revealed particularly expressive and efficient: the π-calculus and the Beta binders. In the context of biological simulation the first is used in its variant of stochastic biochemical π-calculus, whereas the second, inspired to π-calculus, is deliberately formulated for describing of biological interactions. The novelty of the application of such formalisms to the modeling of biochemical pathways is twofold. First both of them integrate dynamics, molecular and biochemical details. As consequence, we can say that the use of these formalism allow specifications of *generalized dynamics systems*.

Published by Woodhead Publishing Limited, 2013

Untill now, from the standpoint of the physicist, biology is concerned with a class of extremely special systems. What makes organisms special is that their specification within the paradigm of Newtonian mechanics requires a plethora of special constraints and conditions that must be superimposed on the universal canons of system description and reduction. The determination of these special conditions is an empirical task; essentially someone else's task. Moreover, till now the biologists embraced this perspective by using Newtonian mechanics, and thus differential equations, to model and simulate the molecular interactions. It is important to recall that essential feature of the paradigm of mechanics is the employment of a mathematical language with an inherent duality, we may express as the distinction between *states* and *dynamical laws*. The states are represented by points in some manifold of phases, and the dynamical laws represent the internal or impressed forces. The resulting mathematical image is thus what, in the Newtonian context, we call *dynamical system*. From a formal point of view, the dynamical systems arising in mechanics are mathematically rather special ones, because of the way phases are defined (they have a symplectic structure). Through the work of Poincaré, Birkhoff, Lotka, and many others, however, this dynamical system paradigm has come to be regarded as the universal means for the representation of systems which could not be described in terms of mechanics (i.e. systems of interacting chemicals, organisms, ecosystems, etc.). Even the most radical changes occurring within physics itself, such as quantum physics, maintain this framework. In the formalism of ordinary differential equations (formalism of Newton's physics) the states of the agents do not appear as explicit variables, but only implicitly through the integration of these equations. On the contrary in biology we need an explicit specification of the states, because a state of a biological entity may be not representable only with a point in the phase plane, but it can represent a structurally different chemical with interaction capabilities different from those of the reagents from which it derives. The specification of

Published by Woodhead Publishing Limited, 2013

a model of a biochemical network should contain such information together with the roles defining the system dynamics. Dividing the specification into state variables plus dynamical laws corresponds to divide the world into state variables plus parameters, and this is much less than a specification of dynamical and molecular details. In Eqs. (4.1), the role of the parameters is to determine the form of the functions f, which in turn define the dynamical laws. The state variables are the arguments of these functions, while the parameters are coordinates in function spaces. There are two classes of parameters, the *intrinsic* parameters and the *extrinsic* parameters. Those which are intrinsic (called also the constitutive parameters) are connected with the specific nature of character of the system. The values they assume might, for instance, tell whether we are dealing with oxygen, carbon dioxide or any other chemical species; therefore they cannot change without our perceiving that a changes of species has occurred. The environmental parameters can change without affecting the species of the system. These distinctions cannot be accommodated with the language of differential equations; that language is too abstract. We can only recapture these distinctions in two ways. The first is introducing new equations describing the time evolution of the parameters. For instance, for the parameters $a_1, a_2, \ldots, a_r$ of the function

$$f_i = f_i(X_1, X_2, \ldots, X_N; a_1, a_2, \ldots, a_r)$$

we could introduce the following equations

$$\frac{da_i}{dt} = 0$$

for those parameters that are independent of time and

$$\frac{da_i}{dt} = g(t)$$

for those parameters whose rate of changing is a function g of the time. This way to handle the distinction complicates the form of the model and could increase the computational cost of the solution.

Published by Woodhead Publishing Limited, 2013

The second way to manage this distinction is to change the specification language itself. In this thesis we choose this second way by proposing the use of π-calculus and Beta binders. This languages move the focus from the parameters to the systems components and their changes of states. These formalisms represents biological entities as computational processes interacting with each other by synchronized pairwise communication on complementary communication channels, and modify each other by transmitting channels from one process to another. This feature is known as *mobility* , allows the network structure to change with interaction. The specification of the components structure in terms of potential reactions that they can undergo is a compact way to model a biochemical pathway without going through all of the details of the time behavior of kinetic parameters. At the same time both *pi*-calculus and Beta binders are able to handle three key feature of molecular system, such as concurrency, compositionality, mobility and hierarchical structure. In order to show the expressiveness of these language we give here a brief introduction of the syntax together with some essential modeling use-cases.

4.5.2 The π-calculus

The connection between the computer and biological world is very well described by the metaphor *cells as computations* [45]. Biological components at various levels of abstractions are represented as computational processes and their interactions result to be communications between that processes. Relying on calculi for mobility, the effect of a communication can change the future interaction of processes as it happens on the biological side for interacting components. The problem is thus to find a formal language able to represent:

- the actors of the system (molecules, genes, proteins, etc.)
- the qualitative evolution of the system in terms of its reactions
- all the quantitative aspects of the pathways (quantity of reagents, reaction rates, absolute time of simulation, etc)

Published by Woodhead Publishing Limited, 2013

The importance of formal languages to represent both structure and dynamic behavior of biological processes

Recently, Regev, Silverman and Shapiro [46] proposed the π-calculus [35] as a qualitative model of biochemical pathways, seen as network of interacting molecules, then use a stochastic variant of the π-calculus [11], yielding a language that permits to model also quantitative aspects of biochemical pathways. We now briefly introduce how the π-calculus is exploited to model networks of biochemical processes. The π-calculus for modeling biological pathways can be classified in the category of language-based approaches.

At the microscopic scale, biological processes are carried out by networks of interacting molecules. The interaction between molecules causes their biochemical modification. These modifications affect the potential of the modified molecules to interact with other molecules. The biochemical stochastic π-calculus represents the molecules as computational processes and the network of interacting molecules as a mobile concurrent system. This kind of systems is composed by a community of co-existing computational process that communicate with each other and that change their interconnection structure at execution time. Each computational process is defined by its potential communication activity. The communication between processes, namely the abstraction of chemical interaction, occurs via channels, denoted by their names (ranged over by $x, y, \ldots$). The basic communication primitives are *input* and *output* actions, denoted respectively by $x(y)$ and $\overline{x}\langle z\rangle$. Two concurrent processes can communicate only if they share a common channel name. Executing the input $x(y)$ means being ready to receive a name for y along the channel x, and executing $\overline{x}\langle z\rangle$ stays for being able to send the name z along channel x. In what follows, when the parameter of the communication is not relevant, we shortly denote an output action and an input action on channel x by $\overline{x}$ and x, respectively.

Processes (ranged over by capital letters $P, Q, \ldots$) are given by the following BNF-like syntax:

$$P ::= 0 \mid (\pi, r).P \mid (\nu y)P \mid P|P \mid P + P \mid A(y_1, \ldots, y_n). \quad (4.2)$$

Published by Woodhead Publishing Limited, 2013

The simplest process is the empty process 0, that can do nothing (deadlock process). A process P may be prefixed by (π, r) where π is either an input or an output action and r is the single parameter of an exponential distribution that characterizes the stochastic behavior of the activity corresponding to the prefix π. If π is the input action $x(y)$, then it is a binder for the name y with scope P. The restriction operator (νy) in $(\nu y)P$ is another binder for y with scope P: it declares that y is a private resource of P, as opposed to a global (or public) name. The infix operator | denotes the parallel composition of the two processes, and + the choice between the possible actions of the two operands. Finally, $A(y_1, \ldots, y_n)$ is a constant definition. Hereafter, $\tilde{y}$ denotes $y_1, \ldots, y_n$, and we use $(\nu\tilde{y})$ as a shorthand for $(\nu y_1)..(\nu y_n)$. Each agent identifier A has a unique defining equation of the form $A(y_1, \ldots, y_n) = P$ where the y_is are distinct names, $fn(P) \subseteq \{y_1, \ldots, y_n\}$, and possible occurrences of agent identifiers in P are action-guarded (i.e. are nested within an action prefix). An action π that is not prefixed by any other action is called *unguarded* action. Intuitively, an unguarded action is the action currently available to be realized, while a guarded action represents the next communication ready to occur once the action preceding it has been performed. Thus in a process of form $P = \pi_1\pi_2 \ldots pi_n.P'$, π_1 is the unguarded action, π_2 is the next action ready to be performed after the occurrence of action π_1, π_3 will be able to occur after the realization first of π_1 and then of π_2, and so on.

The formal semantics of the calculus, which makes use of a congruence relation to state the structural equivalence of syntactically different processes, is given by an *operational semantics* consisting of *reduction rules*. Those rules allow the description of the behavior of the system, namely of the transitions from one state to another state.

A significant extension of the π was realized by Priami [11], by developing a stochastic variant of the original operational semantics of the calculus. Here we briefly introduce the main concept. In the stochastic π to each channel the single parameter $r \in (0, \infty]$ of an exponential distribution describing the stochastic delay of the communication on

Published by Woodhead Publishing Limited, 2013

that channel is associated. The time evolution if a system of processes (atoms, molecules, complexes, etc) is driven by a race condition defined in a probabilistic way to reflect the stochastic nature of chemical reactions. This condition states that all the communications that are enabled in a state compete and the fastest succeeds. The speed of the communication and thus the delay of the communications are derived from the rate of corresponding channels.

The general principles outlined here allow to formally represent detailed information on complex pathways, molecules and biochemical events. In this section we illustrate these capabilities with different essential modeling use-cases representing the most common aspects of molecular systems such as complex binding and unbinding, enzymatic catalysis and competitive inhibition.

In all the following examples of this and the next session, we will omit to explicitly indicate the deadlock process, so that a channel c prefixing nothing denotes the process $c.0$.

4.5.2.1 Ionic bonding in π-calculus

Consider as first example a solution of sodium (Na) and chlorine (Cl). The solution is seen as the following system

$$System := Na|Cl \tag{4.3}$$

where the parallel symbol | stands for the fact that the two atoms Na and Cl exist at the same time and are ready to react. The chemical reaction they will undergo is

$$\mathrm{Na} + \mathrm{Cl} \rightleftharpoons \mathrm{Na}^{+} + \mathrm{Cl}^{-} \tag{4.4}$$

The chlorine has a high affinity for electrons, and the sodium has a low ionization potential. Thus the chlorine gains an electron from the sodium atom as shown in Fig. 4.2. In this figure the arrow indicates the transfer of the electron from sodium to chlorine to form the Na^{+} metal ion and the Cl^{-} chloride ion. Each ion now has an octet of electrons in its valence shell.

Published by Woodhead Publishing Limited, 2013

In the π calculus the passage of the electron e from sodium to chlorine is modeled as communication between the processes Na and Cl that abstract the respective atoms. This ionic reaction produces the processes Na^+ and Cl^-, that represent the reaction residual. The communication (that is, reaction) between processes Na and Cl occurs on a shared communication channel c. Sodium can release an electron e by the send on channel c, that can be caught by chlorine (by the input on channel c yet). The simple reaction in (4.3) is thus seen as

$$\overline{c}\langle e\rangle.Na^+ \mid c(e).Cl^- \longrightarrow Na^+ \mid Cl^- \tag{4.5}$$

where

$$Na := \overline{c}\langle e\rangle.Na^+$$
$$Cl := c(e).Cl^-$$

After this reaction, the two processes behave like Na^+ and Cl^-, that, by the reversibility of reaction (4.4), are defined as

$$Na^+ := c(e).Na$$
$$Cl^- := \overline{c}\langle e\rangle.Cl$$

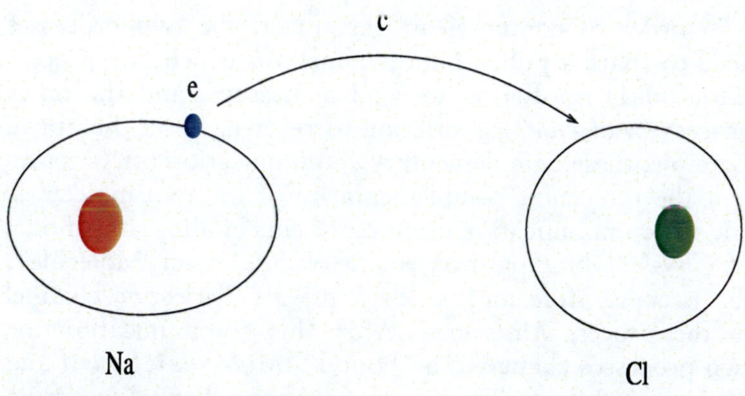

Figure 4.2 Ionic bonding between sodium and chlorine atoms. Na sends a message e on channel c to Cl that received it on the same channel c. After this communication Na becomes Na^+ and Cl becomes Cl^-.

Published by Woodhead Publishing Limited, 2013

4.5.2.2 Molecular complexation and de-complexation in π-calculus

Consider now as second example: the formation and the breakage of a complex between two molecules Molecule1 and Molecule2. In this example any molecule has some private "information", the *backbone*, that determines its identity. The interaction between two molecule to form a complex can be seen had the sending and receiving this *backbone*. In such a model a backbone is a private name of a process, declared by the scope restriction operator ν. So interaction between molecules results also in a scope change, in particular in a scope enlargement.

$$Molecule1 + Molecule2 \rightleftharpoons Molecule12$$

Each molecules is represented by a process *Molecule1* and *Molecule2*.

$$Molucule1 := (\nu backbone).\overline{bind}\langle backbone\rangle.Molecule1\_Bound$$
$$Molecule2 := bind(backbone).Molecule2\_Bound$$

In order to communicate (i.e. react) the two processes need to share a public *bind* channel, on which one process (*Molecule1*) is offering to send a message, and the other process (*Molecule2*) is offering to receive. As in the above example, these complementary communication offers represent the molecular complementarity of the two molecules, and the communication event represents binding. In a model of *physical* binding between Molecule1 and Molecule2, the process *Molecule1* sends a private *backbone* channel to the process *Molecule2*. After this communication the two processes change to a "bound" (*Molecule1_Bound* and *Molecule2_Bound*). The private *backbone* channel sent from *Molecule1* to *Molecule2* represents the formed complex. The molecules continue to exclusively communicate on the *backbone* channel, and are thus physically linked. Therefore, only a communication between the two "bound" processes on

Published by Woodhead Publishing Limited, 2013

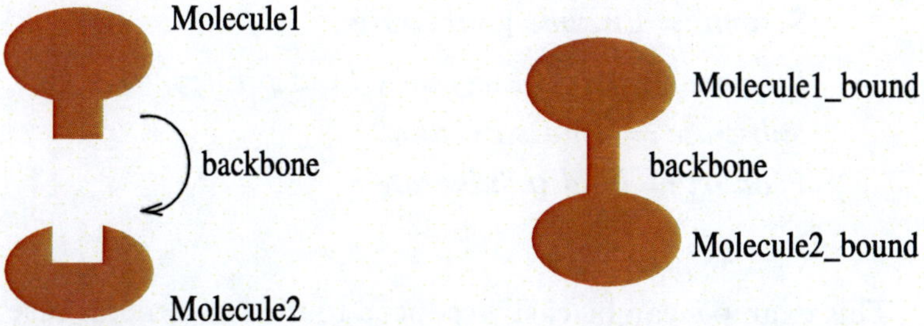

Figure 4.3 Visualization of the specification of "physical binding" reaction in π-calculus.

the shared private *backbone* channel can represents a spontaneous complex breakage. As a result the two processes return to the initial "free" state (*Molecule1* and *Molecule2*). If there are many copies of these processes, any two particular copies of Molecule1 and Molecule2 may communicate on the channel *bind*. However, the two resulting "bound" processes share a private channel, which is distinct from all other channels, and may allow only this particular pair to communicate with each other.

4.5.2.3 Enzymatic catalysis in π-calculus

The third example is a model for a single-substrate reversible enzymatic reaction with one product. In this type of reactions, a molecule of enzyme can interact either with a molecule of substrate or with a molecules of product. Both of these interactions leave the enzyme molecule unchanged, while the substrate is transformed into a product and the product is transformed back into the substrate. In the π-calculus model, the system is composed by two processes, that are *Enzyme* and *Substrate*. *Enzyme* includes a non deterministic choice (denoted by the symbol "+") between an interaction with *Substrate* on the *bind_s* channel and an interaction with Product on the *bind_p* channel. In the former a *Product* is released and in the latter a *Substrate*.

Published by Woodhead Publishing Limited, 2013

$$System := Enzyme \mid Substrate$$
$$Enzyme := \overline{bind\_s}.Enzyme + \overline{bind\_p}.Enzyme$$
$$Substrate := bind\_s.Product$$
$$Product := bind\_p.Substrate$$

This example can be easily reduced to the less general case of an single-substrate irreversible reaction with one product. In this case we modify the definition of the *Enzyme* to account for the irreversibility of the reaction in the following way:

$$Enzyme := \overline{bind\_s}.Enzyme$$

and the definition of the *Product* as

$$Product := 0$$

to indicate with the symbol "0" that the process *Product* will not have any further behavior i.e. it is the *deadlock* process).

4.5.2.4 Competitive inhibition in π-calculus

The fourth example we show in this introduction is the mechanism of competitive inhibition. It is a form of enzyme inhibition where binding of the inhibitor to the enzyme prevents binding of the substrate and vice versa. This can occur in two ways:

In classical competitive inhibition, the inhibitor binds to the same active site as the normal enzyme substrate, without undergoing a reaction. [3] The substrate molecule cannot enter the active site while the inhibitor is there, and the inhibitor cannot enter the site when the substrate is there. In this case, the maximum speed of the reaction is unchanged, because although the substrate's apparent affinity for the site is decreased, the substrate concentration will eventually increase to match the concentration of the inhibitor.

The π-calculus code for the classical competitive inhibition is given by the following statements.

Published by Woodhead Publishing Limited, 2013

$$System ::= Enzyme \mid Substrate \mid Inhibitor$$
$$Enzyme ::= (\nu\ s,\ i)\overline{bind\_s}\langle s\rangle.EX(s) + \overline{bind\_i}\langle i\rangle.\bar{i}.Enzyme$$
$$EX(release\_s) := \overline{release\_s}.Enzyme$$
$$Substrate ::= bind\_s(erel\_s).SX(erel\_s)$$
$$SX(rel\_s) ::= rel\_s.Substrate + rel\_ps.Product$$
$$Inhibitor ::= bind_i(rel\_i).rel\_i.Inhibitor$$

The system is composed by three processes in parallel *Enzyme*, *Substrate* and *Inhibitor*. *Enzyme* can communicate either with the substrate or the inhibitor via channels *bind_s* and *bind_i*, respectively. The process *Enzyme* sends *s* on channel *bind_s* or *i* on channel *bind_i*. If the communication on *bind_s* occurs, the result is the process *EX*(*s*) representing the bound state of the enzyme to substrate. *EX*(*s*) means that the process *EX* is equipped with (or parametrically defined by) the channel *s*, that, in this model, is the only channel on which it can communicate for detaching from the substrate. A communication of *EX* on its parameter channel realizes the change of the bound state of the enzyme (*EX*) into the enzyme in its free state (*Enzyme*). See Figs. 4.4 and 4.5.

The process substrate *Substrate* communicates with he process *Enzyme* on the shared channel *bind_s* and changes to *SX*(*erel_s*) representing the bound state of the substrate. *SX* is the counterpart of *EX*. The reversibility of the reaction $E + S \rightleftharpoons ES$ is represented by the non deterministic choice that defines the process *Substrate*. A communication on the channel parameter (*rel_s*) can result in an unchanged instance of *Substrate* or in an instance of *Product*.

The process *Inhibitor* communicates with *Enzyme* on channel *bind_i* and changes to a bound state represented by the process *rel_i.Inhibitor*. This process states for the bound state of the inhibitor with the enzyme. The counterpart of this complex is the process $\bar{i}$.*Enzyme*. After the communication on *bind_i*, *rel_i.Inhibitor* becomes *i.Inhibitor*, that by communicating with *Enzyme* on *i* return back to its free state.

Published by Woodhead Publishing Limited, 2013

4.5.3 The Beta binders formalism

The π-calculus efficiently handles the molecular and the biochemical aspects of a biomolecular network. However it provides only a limited solution for modeling biological compartments and site specific interactions.

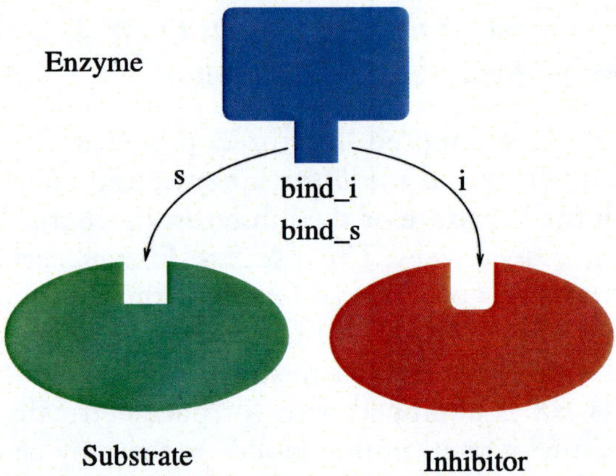

Figure 4.4 Competitive inhibition: substrate and inhibitor interact with enzyme in a mutually exclusive way.

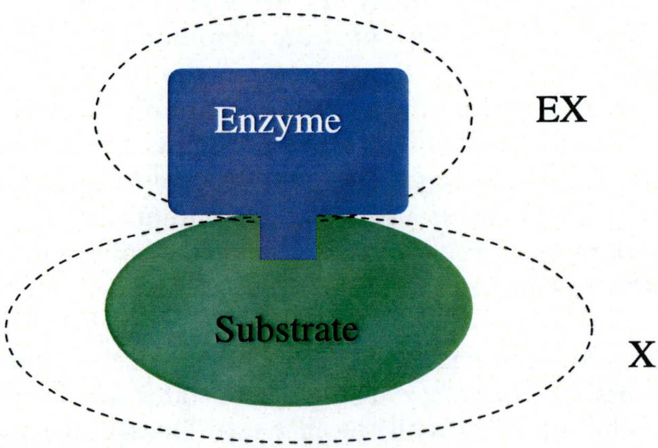

Figure 4.5 Pictorial representation of the bounded states of enzyme and substrate in the molecular complex enzyme-substrate.

Published by Woodhead Publishing Limited, 2013

More then 3 billion years ago, primitive replicating forms became enveloped in a lipid film, a biomolecular diffusion barrier that separated the living cell from its environment. Starting from the simple design of the membrane of these primitive beings, the cytoplasmatic membranes of all contemporary organisms have built a more elaborate design, with many, more-selective transport and interaction devices handling different jobs, often under separate physiological control. For instance, the compartments play an important role in the functional organization of the biomolecular systems, e.g. the cellular metabolism depends on compartmentalization and location of proteins and metabolites. Moreover, at the level of multicellular organism, the function of many systems depends on the location of different components in separate cells, histological compartments and even organs. Due to the presence of many highly specialized components, the membrane is a multifunctional surface studded with many interaction sites, each of which in turn can be seat of a multiplicity of interactions.

The multifuntionality expressed by the presence of many different proteins and molecules on a cell membrane is also a characteristic of proteins and bio-molecular complexes. For instance, the site-specific genetic recombination, that moves specialized nucleotide sequences between non homologous sites within a genome, is guided by specific recombination enzymes [3]. These enzymes recognize short specific nucleotides sequences present on one or both the recombining DNA molecules. The concepts of multifuncitonality and "chemical recognition" are strongly connected that at the formal level of the model specification need specific linguistic structures able to represent both of them.

Some calculi have been developed to model compartments and hierarchies of entities. They are BioAmbients [44] and Brane Calculi [7]. These calculi directly born from biological reasoning and are motivated by the ambient calculus, a process algebra for the specification of process location and movements through computational domains. Recently, the Beta binders formalism, largely inspired to the π-calculus,

Published by Woodhead Publishing Limited, 2013

has been introduced to represent not only the concept of biological compartments, but also those of site-specific interaction and multifunctionality.

The basic elements of the Beta binders formalism are the π-processes and the synchronous communication is the primitive form of interaction between parallel running processes. Then the formalism defines a special class of binders, the so-called beta binders, and uses them to model the π-processes encapsulated into *boxes* having interaction capability. The boxes are π-processes *prefixed* by beta binders and are named *bio-processes.* An elementary bio-process is written in the following formalism: $\beta(x,\Gamma)[P]$, where $\beta(x,\Gamma)$ is a *beta binder* and P is a *π-process.* A π-process is the following extension of the πprocess form given in (4.2)

$$P ::= \mathsf{nil} \mid \pi.P \mid P \mid P \mid \nu y P \mid !\pi.P \tag{4.6}$$

where

$$\pi ::= x(y) \mid \bar{x}z \mid \mathsf{hide}(x) \mid \mathsf{unhide}(x) \mid \mathsf{expose}(x,\Gamma) \tag{4.7}$$

Pi-processes behave just as π processes; the additional prefixes expose, hide and unhide are for manipulating beta binders. Prefixes $\mathsf{hide}(x)$ and $\mathsf{unhide}(x)$ make the elementary beta binder with subject x not available (hidden) and available (unhidden), respectively. When a binder is hidden it cannot be used in interaction. The prefix $\mathsf{expose}(x,\Gamma)$ adds to the box the elementary beta binder $\beta(x:\Gamma)$.

The pair $x:\Gamma$ denotes the site of interaction with which the box communicates with the external world; x is the *name* of the beta binder and Γ is a set of names (the *type* indicating the interaction capabilities at x). The name and the type of a beta binder specify in a very compact way the parametric view of the box interface.

Fig. 4.6 (A) depicts in graphic notation a system composed by two parallel bio-processes B_1 and B_2 defined as follows

$$B_1 ::= \beta(x,\Gamma)[x(y).P|\bar{x}z.P'_1] \text{and} \quad B_2 ::= \beta(u,\Delta)[\bar{u}w.P_2|P'_2]$$

Published by Woodhead Publishing Limited, 2013

A bio-process can evolve in four ways: (i) when the enclosed π-processes evolve independently from the external world and without affecting the configuration of the box interface (ii) when the bio-process communicate with other bio-processes of the system through its beta binders, provided that the types sets of the respective beta binders have a non-empty intersection, (iii) when the box interface undergoes a modification driven by the actions of the internal π-processes, and (iv) when the whole bio-process undergoes structural modifications due to the merging with other bio-processes or to the split in sub-boxes. The interface evolution of a bio-process can model the functional dependency of the interaction capabilities of a biological entity on its particular shape of folding, while the structural modification can model biological phenomena such as endocytosis and mitosis, for instance.

Fig. 4.6 (B) and (C) respectively show the case (i), named *intra reduction* and the case (ii), named *inter reduction*. In the first case, $\bar{x}z.P'_1$ can release z, by sending it on channel x, that can be caught by $x(y).P_1$ (by the input on channel x yet). At the end of the communication, the name y in P is substituted by z, and the system becomes $B'_1 \parallel B'_2$, where

$$B'_1 ::= \beta(x,\Gamma)[P_1\{z/y\} \mid P'_1]\text{and} \quad B'_2 ::= \beta(u,\Delta)[\bar{u}w.P_2 \mid P'_2]$$

In case (ii), if $\Gamma \cap \Delta \neq \emptyset$, the system in Fig. 4.6 (A) can evolve in the following way

$$B'_1 ::= \beta(x,\Gamma)[P_1\{w/y\} \mid P'_1]\text{and} \quad B'_2 ::= \beta(u,\Delta)[P_2 \mid P'_2]$$

This example shows how data (the name w) may flow from one biological entity (the bio-process B_2) to another (the bio-process B_1) through the appropriate site.

Moreover, in order to express the case (iii), the beta binders formalism introduces three new prefixes in the syntax of the π-processes: expose, hide and unhide. The prefix $expose(x,\Gamma).P$ calls for adding a new site of type Γ to the bio-process (Fig. 4.7 (A)). The prefix $hide(x,\Gamma).P$ and $unhide(x^h,\Gamma).P$ reads as a request to enable and, respectively, forbid the further interactions occurring through the x site

Published by Woodhead Publishing Limited, 2013

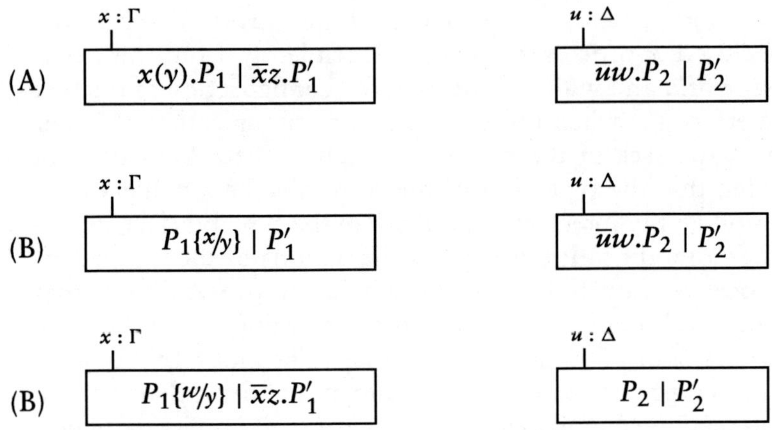

Figure 4.6 A system of two parallel bio-processes B_1 and B_2 (left and right box, respectively) in (A). bio-processes intra (B) and inter (C) reductions [41].

(Fig. 4.7 (B) and (C)). The superscript h over the names of a beta binder indicates that the communication through it is forbidden.

The structural modifications of the bio-processes are expressed by the axioms join (join two boxes together as in Fig. 4.8 (A)) and split (split a box in two as in Fig. 4.8 (B)). The merging of two boxes is modeled as the *absorption* of a bio-process by another one. After the merging of two boxes, the π-component of the absorbed box (P_2) is moved into the absorbing bio-process (P_1), and the engulfed material has no longer a proper beta binder to interact with the external world. Symmetrically to join, the rule split formalizes the splitting of a box in two parts each of them taking away a sub-component of the content of the original box. The rules join and split are parametric with regard to the functions f_{join} and f_{split} that define the binder set and the enclosed processes of the box resulting from the aggregation of two boxes and of the boxes resulting from the division of a box, respectively. For instance, consider the f_{join} and f_{split} instances as given in (4.8) and in (4.9).

Published by Woodhead Publishing Limited, 2013

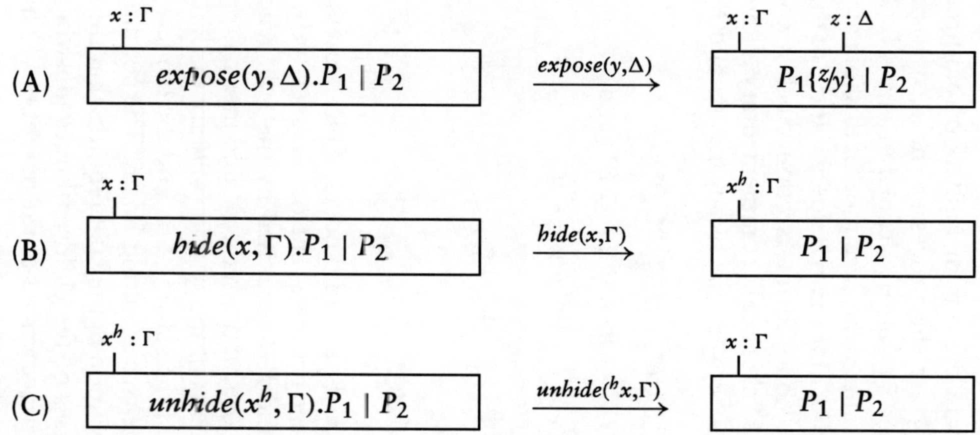

Figure 4.7 Graphical representation of the evolution of a bio-process due to expose (A), hide (B), and unhide reductions (C). The expose rule assumes that $z \notin \Delta$ and $z \neq x$.

$$
\begin{aligned}
&f_{join}(B_1^*, B_2^*, P_1^*, P_2^*) = \\
&\quad \text{if } (B_1^* = \beta(x:\Gamma)) \text{ and } B_2^* = \beta(y:\Delta) \text{ and } \Gamma \cap \Delta \neq \emptyset \\
&\quad \text{then}(B_1^*, (P_1 | P_2\{x/y\})) \\
&\quad \text{else}((B_1^*, P_1),\ (B_2^*, P_2))
\end{aligned}
\tag{4.8}
$$

where B_1^* and B_2^* are the sets of binders of bio-process B_1 and B_2, P_1^* and P_2^* are the processes enclosed respectively by the bio-processes B_1 and B_2 defined in Fig. 4.8 (A). The condition that guarantees the execution of this join algorithm requires that the types of the two binders x and y of the boxes B_1 and B_2 have a non-null intersection. Such a condition represents a sort of "chemical" complementary allowing the physical interaction between the two boxes B_1 and B_2.

$$
\begin{aligned}
&f_{split}(B^*, P^*, B_1^*, B_2^*, P_1^*, P_2^*) = \\
&\quad \text{if } (B_1^* \subseteq B^* \text{ and } B_2^* \subseteq B^*) \text{ and} \\
&\quad (P_1^* \subseteq P^* \text{ and } P_2^* \subseteq P^*) \\
&\quad \text{then} \\
&\quad ((B_1^*, P_1^*), (B_2^*, P_2^*)) \\
&\quad \text{else} \\
&\quad (B^*, p^*)
\end{aligned}
\tag{4.9}
$$

where B_1^* and B_2^* are the sets of binders of bio-process B_1 and B_2, P_1^* and P_2^* are the processes enclosed respectively by new bio-processes B_1 and B_2, generated by the split reaction of the bio-process whose binder set is B^* and internal processes are P^*. The arguments of the f_{split} in (4.9) are the binders set and the processes of the starting box and the binders set and the processes of the two resulting boxes. The condition that guarantees the execution of the split requires that the binders set and the processes of the children boxes must be subsets of binders set and processes set of the parent box. This avoid the unrealistic creation from nothing.

The pictorial representation of the execution of (4.8) and (4.9) is shown in Fig. 4.8.

Recently a stochastic extension of Beta binders has been developed [41]. As in the stochastic extension of

Published by Woodhead Publishing Limited, 2013

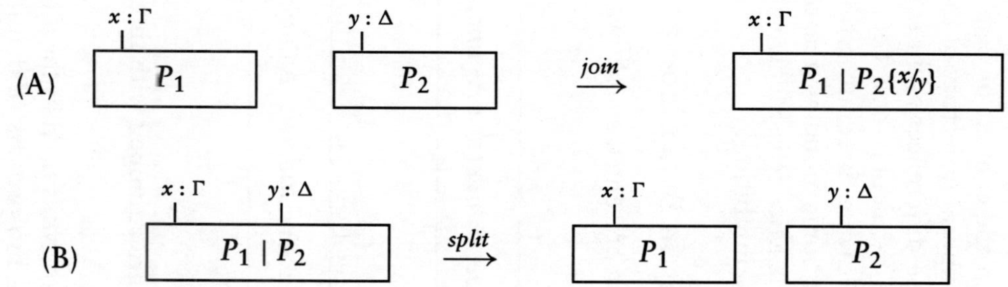

Figure 4.8 (A) The execution of the join reduction defined in (4.8). (B) The execution of the split reduction defined in (4.9). As far as join rule, note that, unlike BioAmbients, the Beta binders formalism forbids the nesting of boxes.

the π, also in stochastic Beta binders each action π is replaced by (π, r_π), where $\pi \in$ {intra, expose, hide, nhide, inter, join, split} and r_π is the parameter of an exponential distribution. The same race condition defined for the stochastic π can be used in the context of Beta binders formalism to compute the temporal trajectory of the system.

In order to introduce the differences and the analogies in the expressiveness of the π-calculus and Beta binder language we report in the following three subsections the models of specification of the ionic bonding reaction between sodium and chlorine, of the Michaelis-Menten catalysis, and of the classical competitive inhibition.

4.5.3.1 Ionic bonding in Beta binders

The system $Na|Cl$ can be represented as follows

$$x : \Gamma \qquad e : \Delta \qquad w : \Omega$$

$$!\overline{x}.hide(x : \Gamma).hide(e : \Delta).\overline{w}.unhide(e^h : \Delta).unhide(x^h : \Gamma)$$

Na

$$y : \Gamma \qquad (e')^h : \Delta \qquad z : \Omega$$

$$y.hide(y : \Gamma).unhide((e')^h : \Delta).z.hide(e' : \Delta).unhide(y^h : \Gamma)$$

Cl

Both of the bio-processes are equipped with three binders, as in the following.

- $\{x : \Gamma\}$ on the interface of Na is used in the inter communication with the bio-process Cl that involves the binder y having the same type Γ. This communication gives the starting signal to the series of reductions leading to the conversion from the system state $Na|Cl$ to the state $Na^+|Cl^-$.
- $e^h : \Delta$ represents the electron e in the valence shell of the sodium. $(e')^h : \Delta$ represents the vacancy of an electron in

Published by Woodhead Publishing Limited, 2013

the valence shell of the chlorine. Thus, the loss of an electron is modeled via a hide reduction, whereas the acquisition of an electron is modeled via an unhide reduction of the binder that specifies the electron.

- $\{w : \Delta\}$ on the interface of *Na* is used in the inter communication with the bio-process *Cl* that involves the binder *z* having the same type Ω. This communication abstracts the further interaction between *Na*+ and Cl^- that gives the starting signal to the backward reaction $Na^+|Cl^- \longrightarrow Na|Cl$.

The direct reaction that transforms the system $Na|Cl$ into $Na^+|Cl^-$ can be realized by the following sequence of reductions: $inter(x,y)$, $hide(y : \Gamma)$, $hide(x : \Gamma)$, , $hide(e : \Delta)$, $unhide(e' : \Delta)$, $inter(w,z)$. Moreover, in this model we assume that the reaction rates *r*'s satisfy the following equations

$$r_{inter(x,y)} < r_{hide(x:\Gamma)} \tag{4.10}$$

$$r_{inter(x,y)} < r_{unhide(y^h:\Gamma)} \tag{4.11}$$

$$r_{inter(x,y)} = r_{inter(w,z)} \tag{4.12}$$

$$r_{hide(e:\Delta)} = r_{unhide((e')^h:\Delta)} \tag{4.13}$$

$$r_{unhide(e^h:\Delta)} = r_{hide(e':\Delta)} \tag{4.14}$$

$$r_{unhide((x^h:\Gamma)^h)} = r_{unhide(y*h:\Gamma)} \tag{4.15}$$

For convenience, let introduce *P* and *Q* defined as follows and recall that $!P \equiv P|!P$.

$$\begin{aligned} P &\equiv !\overline{x}.hide(x:\Gamma).\overline{w}.unhide(e:\Delta).\overline{w} \\ &\quad .hide(e:\Delta).unhide(x^h:\Gamma) \\ Q &\equiv !y.hide(y:\Gamma).z.hide(e:\Delta) \\ &\quad .z.unhide(e^h:\Delta).unhide(y^h:\Gamma) \end{aligned}$$

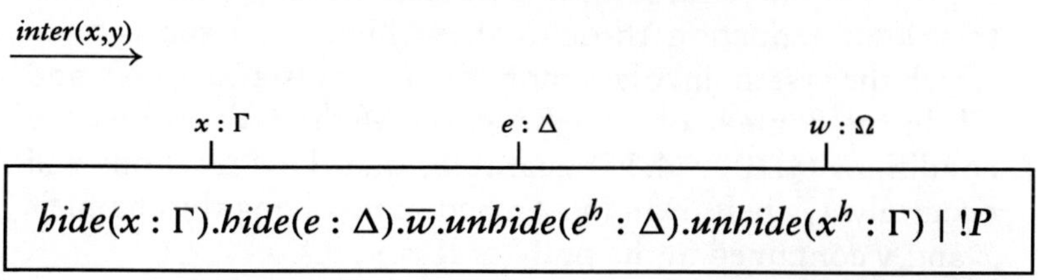

Na′

Published by Woodhead Publishing Limited, 2013

$y : \Gamma$ $\quad$ $(e')^h : \Delta$ $\quad$ $z : \Omega$

$$hide(y : \Gamma).unhide((e')^h : \Delta).z.hide(e' : \Delta).unhide(y^h : \Gamma) \mid !Q$$

Cl′

The consumption of the prefixes x and y after the inter reduction leaves disclosed the actions $hide(x : \Gamma)$ and $hide(y : \Gamma)$, that we introduced in this model only to avoid any further inter reduction through these binders in the case in which the system involves more than one instance of *Na* and *Cl*. In the context of First Reaction Method algorithm, the conditions (4.10) - (4.15) guarantee that these reactions will occur always before an inter reaction involving the channels x and y contained in the body of $!P$ and $!Q$.

Published by Woodhead Publishing Limited, 2013

$\xrightarrow{hide(y:\Gamma)}$

$x : \Gamma \qquad e : \Delta \qquad w : \Omega$

$$hide(x : \Gamma).hide(e : \Delta).\overline{w}.unhide(e^h : \Delta).unhide(x^h : \Gamma) \mid !P$$

Na′

$y^h : \Gamma \qquad (e')^h : \Delta \qquad z : \Omega$

$$z.unhide((e')^h : \Delta).z.hide(e' : \Delta).unhide(y^h : \Gamma) \mid !Q$$

Cl′

$\xrightarrow{hide(x:\Gamma)}$

$x^h : \Gamma \qquad e : \Delta \qquad w : \Omega$

$$hide(e : \Delta).\overline{w}.unhide(e^h : \Delta).unhide(x^h : \Gamma) \mid !P$$

Na′

$y^h : \Gamma \qquad (e')^h : \Delta \qquad z : \Omega$

$$unhide((e')^h : \Delta).z.hide(e' : \Delta).unhide(y^h : \Gamma) \mid !Q$$

Cl′

After the execution of $hide(x : \Gamma)$ and $hide(y : \Gamma)$, the reductions $hide(e : \Delta)$ and $unhide((e')^h : \Delta)$ model the passage of the valence electron *e* from *Na* to *Cl*, that become Na^+ and Cl^-.

Published by Woodhead Publishing Limited, 2013

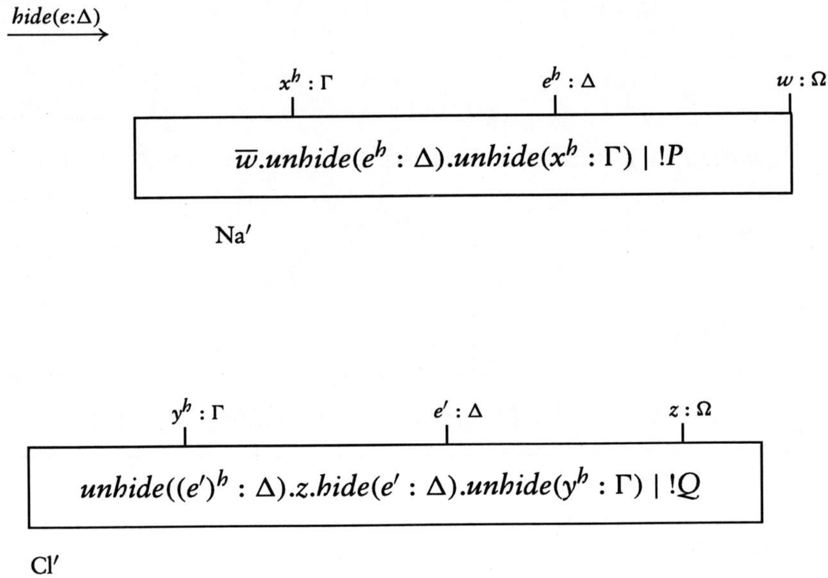

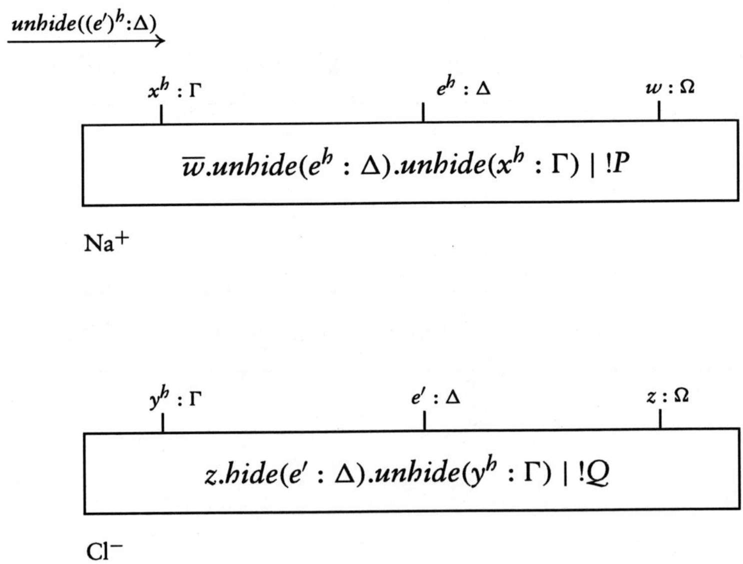

When the system is in the state $Na^+|Cl^-$, a reaction between the two ions, modeled as an inter reduction through the

Published by Woodhead Publishing Limited, 2013

channels w and z disclose the prefixes $unhide(e : \Delta)$ and $hide(e' : \Delta)$ whose execution represent the passage of the valence electron back from Cl to Na.

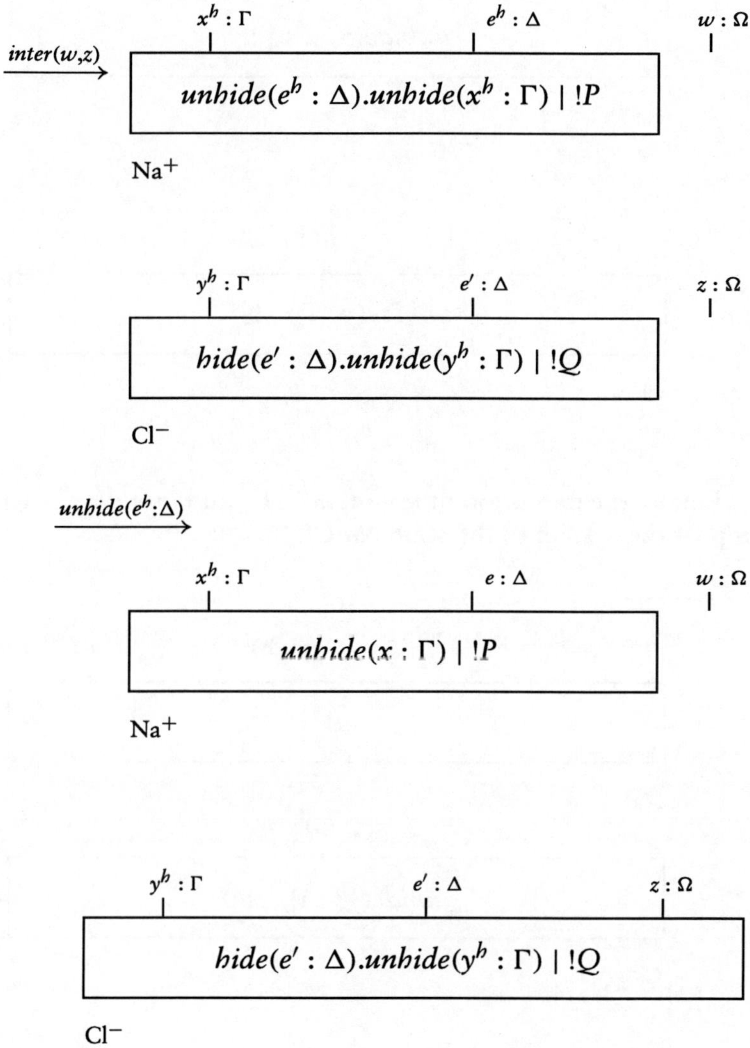

Published by Woodhead Publishing Limited, 2013

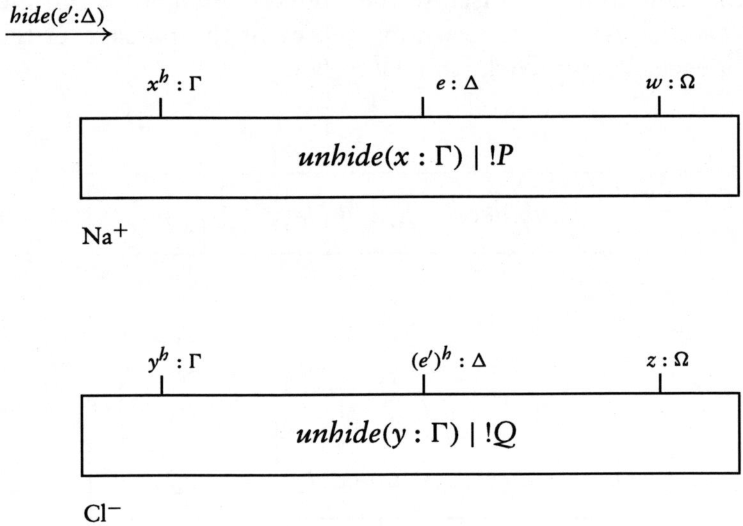

Finally the execution of $unhide(x^h : \Gamma)$ and $unhide(y^h : \Gamma)$ report the system to the state $Na|Cl$.

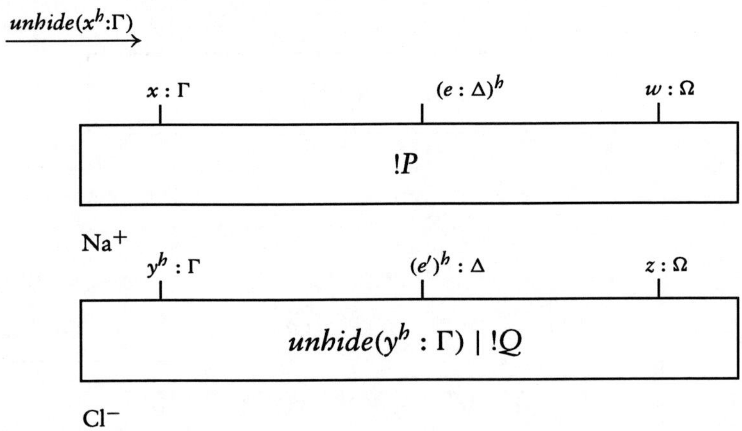

Published by Woodhead Publishing Limited, 2013

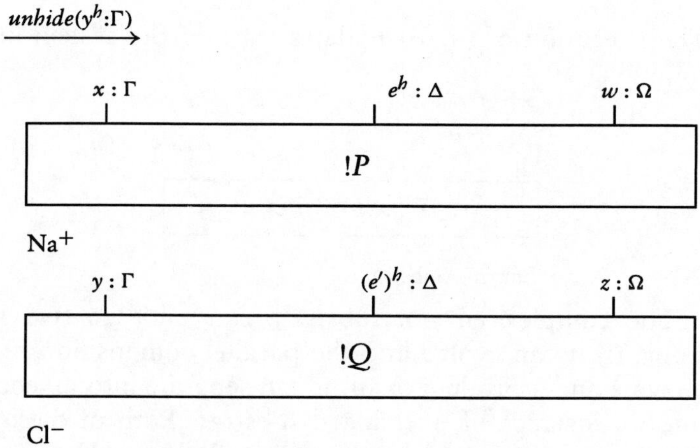

4.5.3.2 Enzymatic catalysis in Beta binders

The system is the parallel composition of two bio-processes: *Enzyme* and *Substrate*, as follows

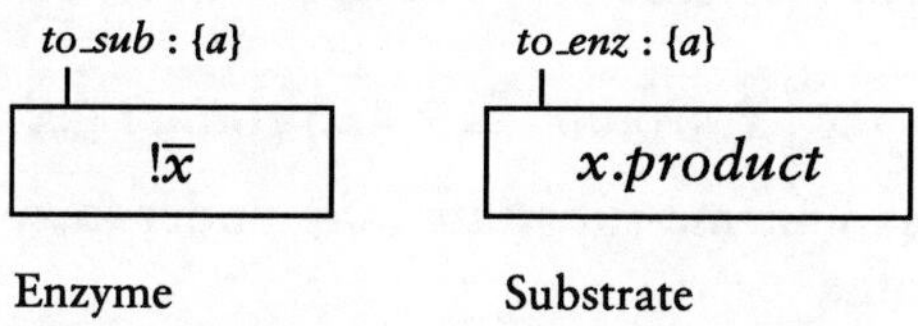

The bio-process *Enzyme* is equipped by the binder *to_sub* : $\{a\}$ and the bio-process *Substrate* is equipped by the binder *to_enz* : $\{a\}$. The type sets of these two binders has non-null intersection. This condition can be used to represent the structural complementary of the key-lock mechanism typical of the enzymatic reactions. Thus, it can be view as the necessary presupposition to allows the formation of the enzyme-substrate complex.

Let consider the function f_{join} given by

$$\begin{aligned}
&f_{join}(B_1^*, B_2^*, P_1^*, P_2^*) = \\
&\text{if } (B_1^* = \beta(x : \Gamma)) \text{ and } B_2^* = \beta(y : \Delta) \text{ and } \Gamma \cap \Delta \neq \emptyset \\
&\text{then}(\{(B_1^*)^h, (B_2^*)^h\}, (P_1 | P_2)) \\
&\text{else}((B_1^*, P_1), \ (B_2^*, P_2))
\end{aligned} \tag{4.16}$$

Published by Woodhead Publishing Limited, 2013

The execution of (4.16) updates state of the system in the following way

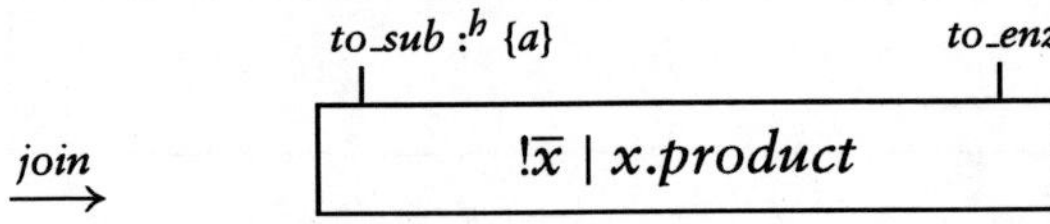

Enzyme-Substrate

The complex enzyme-substrate can undergo two reactions: (i) it can evolve into the parallel composition of and enzyme and a product or (ii) it can separate into an enzyme and an instance of unchanged substrate.Both of these two reactions can be modeled by split reductions. However, in order to give the correct result, the first split must follow an internal arrangement of the complex enzyme-substrate (intra reduction). This internal re-arrangement consists of an input/output communication on channel x, that consumes the channel and release the process *product*. 0. Since

$$!\overline{x} \mid x.product \equiv \overline{x} \mid x.product \mid !\overline{x}$$

The complex enzyme-substrate may undergo an intra reduction to become

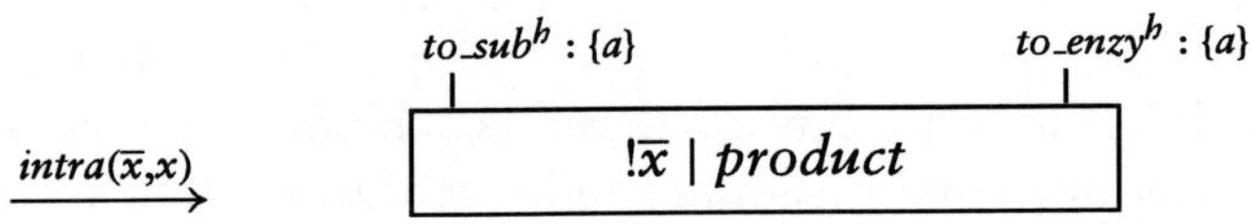

Enzyme-Substrate ′

Then a subsequent split produces the bio-processes *Enzyme* and *Product*

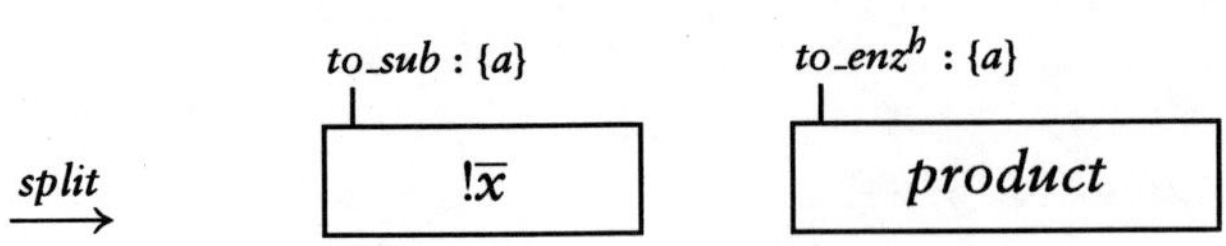

Enzyme Product

Alternatively, if the intra reduction is not "enabled", the bio-process *Enzyme-Substrate* will undergo a split reduction

Published by Woodhead Publishing Limited, 2013

that will reconstitute a bio-process *Enzyme* and a bio-process *Substrate*.

There are some remarkable observations to point out with regard to this example.

1. The first is the closure of the binders *to_enzy* and *to_sub* after the join reduction. The change of state from "unhidden" to "hidden" of the enzyme and substrate represent the "physical binding" between the two agents, analogously to what we made using the scope restriction of the channels *backbone* sent from *Molecule1* to *Molecule2* in the example of section 4.5.2.2 and of the channels *i* and *s* sent by the process *Enzyme* to the processes *Inhibitor* and *Substrate* in the example of section 4.5.2.4. All the binders of the bio-process *Enzyme-Substrate* are hidden to indicate that in this model, the complex cannot have further interactions with other bio-processes.
2. The second note is the "invisible" transition from the unhidden to hidden state of the binders and vice versa. We use the adjective "invisible", because we did not explicitly use the action hide and unhide int he declaration of π-processes inside a box to carry out these transformations. They are defined in the functions *f_join* and *f_split* and not in the internal body of the bio-processes to avoid the formation of spurious reaction compounds, that do not represent any biochemical entity and whose presence in the computation can slow down the simulation. Suppose for example to define the bio-process *Enzyme* in the following manner

$to\_sub : \{a\}$

$$!\overline{x}.hide(to\_sub : \{a\})$$

Enzyme′

and the bio-process *Substrate* as follows

Published by Woodhead Publishing Limited, 2013

$to\_enzy : \{a\}$

$$x.hide(to\_enzy : \{a\}).product$$

Substrate′

A join reduction with the substrate would result in the following bio-process

$to\_sub : \{a\}$ $to\_enzy : \{a\}$

$$!\overline{x}.hide(to\_sub : \{a\})|x.hide(to\_enzy : \{a\}).product$$

Enzyme-Substrate″

Thence, the intra communication on channel x results in

$to\_sub : \{a\}$ $to\_enzy : \{a\}$

$$hide(to\_sub : \{a\}) \mid !\overline{x}.hide(to\_sub : \{a\})|hide(to\_enzy : \{a\}).product$$

Enzyme-Substrate‴

The bio-process *Enzyme − Substrate‴* needs two hide reductions ($hide(to\_sub : \{a\})$ and $hide(to\_enzy : \{a\})$) to transform into the bio-process

$to\_enzy^h : \{a\}$ $to\_sub^h : \{a\}$

$$!\overline{x}.hide(to\_sub : \{a\}).product$$

Enzyme-Substrate′///

that represent the molecular complex enzyme-substrate that a split reduction can solve into a parallel composition of *Enzyme′* and a *Substrate′*. In terms of simulation and closeness to the reality of the interacting components, this explicit declaration of the hide actions not only generates a meaningless bio-process, but it also slow down the computation of the system evolution because it add two steps more to the creations of the evolutionary trajectory of the

Published by Woodhead Publishing Limited, 2013

system. Finally, most importantly, in the case in which many instances of *Enzyme* and *Substrate* are present in the system, the temporary "meaningless" complexes can interact with them to generate unwanted bio-process and meaningless evolutionary trajectories.

Note also that if we declare the unhide actions that transform the hidden binders back into their unhidden states to avoid the redefinition of the boxes interfaces in the f_{split} that de-complexes the enzyme-substrate box, the simulation is not only slowed down further on, but it also needs a very careful and precise definition of the rates of the disclosures of the binders, to avoid the above troubles. Namely, if we define the content of the box *Enzyme* as in the following

$to\_enzy^h : \{a\}$

$$!\overline{x}.hide(to\_enzy : \{a\}).unhide(to\_enzy : \{a\})$$

Enzyme″

and the content of the box *Substrate* as

$to\_enzy^h : \{a\}$

$$x.hide(to\_sub : \{a\}).unhide(to\_sub : \{a\})$$

Substrate″

The join of two boxes followed by the intra reduction on channel x generates this box

$to\_enzy : \{a\}$ $to\_sub : \{a\}$

$$\begin{aligned}&hide(to\_sub : \{a\}).unhide(to\_sub : \{a\})\\ &|hide(to\_enzy : \{a\}).unhide(to\_enzy : \{a\}).product\\ &|!\overline{x}.hide(to\_sub : \{a\}).unhide(to\_sub : \{a\})\end{aligned}$$

Enzyme-Substrate‴′

Published by Woodhead Publishing Limited, 2013

The execution of the action *hide*(*to_sub* : {*a*}) hides the binder *to_sub* : {*a*} and leaves unprefixed the next action *unhide*(*to_sub* : {*a*}, that can compete with *hide*(*to_enzy* : {*a*}) to occur. If in the competition to occur, the winner action is *unhide*(*to_sub* : {*a*} "wins" the state of the system will be updated in the following way

$to\_enzy : \{a\}$ $\quad$ $to\_sub : \{a\}$

$$hide(to\_enzy : \{a\}).unhide(to\_enzy : \{a\}).product$$
$$|!\overline{x}.hide(to\_sub : \{a\}).unhide(to\_sub : \{a\})$$

Enzyme-Substrate^{tmp}

The bio-process *Enzyme*–*Substrate*^{tmp} can follows at least two computations: 1. it can undergo the execution of the internal hide and unhide actions 2. or it can join with an other copy of *Sybstrate* and prime an intra reduction on x.

This problem can be avoided if the modeler assigns suitable values to the rates of the channels and using the First Reaction Method of the Gillespie stochastic simulation algorithm. The race condition of the smallest reaction time imposed by this methods guarantees that between two actions, the winner of the competition will be the fastest, or in analogous term, the one with the higher rate. This is the reason why in the prototype of simulator for Beta binders we present in this thesis, we have implemented a First Reaction Method-like algorithm for the selection of the next reaction.

Nevertheless, if we deal with a complex system made by many bio-processes whose internal processes themselves are complex in terms of number of their instances and their structural form, the result can be fairly hard to define for all involved channels the reaction rates that give the

Published by Woodhead Publishing Limited, 2013

desired temporal sequence of reductions. We think that the definition of a set of join and split functions templates implementing the control of the necessary premises in order that reacting bio-processes can perform complexation and complex breakage and the algorithm implementing the reaction mechanism, may simplify the specification of the model and reduce the quantity of information that the modeler has to fit in.

4.5.3.3 Competitive inhibition in Beta binders

The three actors *Enzyme Substrate* and *Inhibitor* are represented by the following three bio-processes

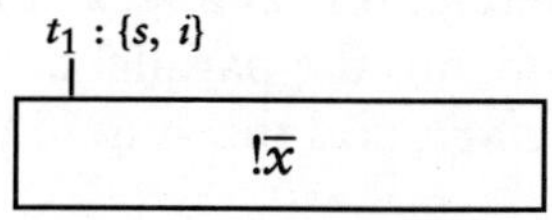

Enzyme

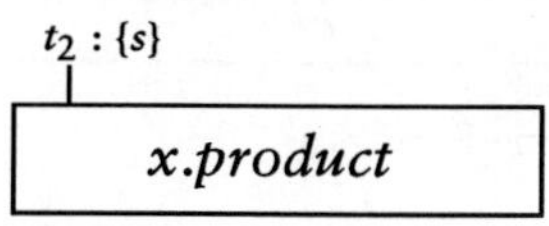

Substrate

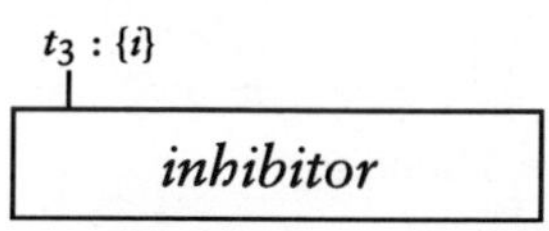

Inhibitor

The type set of the binder of the bio-process *Enzyme* has a non-null intersection both with the type set of *Substrate* and with the type set of *Inhibitor*. This specification is the analogous of the choice construct of the π-calculus. The type set of binder t_1 intersecting with the type sets of the binders t_2 and T_3 represents the potential interactions of the enzymes with its substrate and with its inhibitor.

A possible trajectory of the system is the following. First a the *Enzyme* form a complex *Enzyme – Substrate* that update

Published by Woodhead Publishing Limited, 2013

the system in the following manner

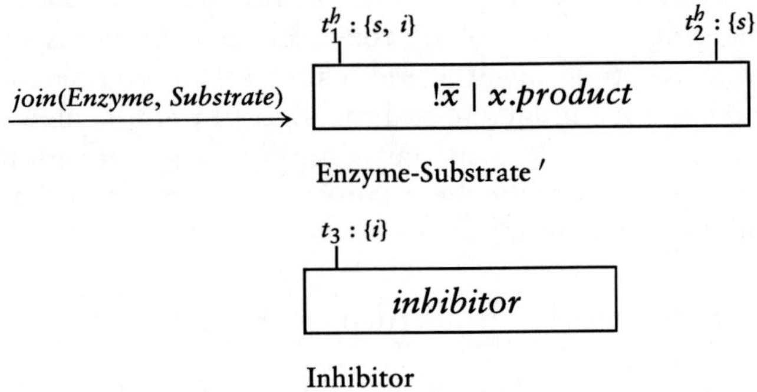

Then an intra reduction changes the processes inside *Enzyme – Substrate* into the parallel composition of process !*overlinex* and *product*, and the system becomes

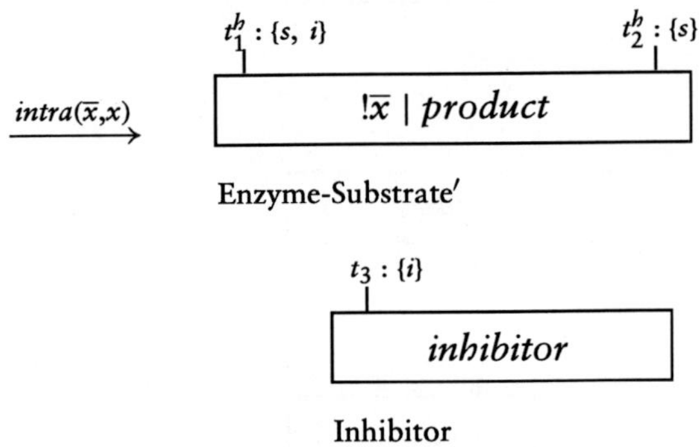

A split of the complex *Enzyme – Substrate′* generates the bio-process *Product* and thus updates the system in this way

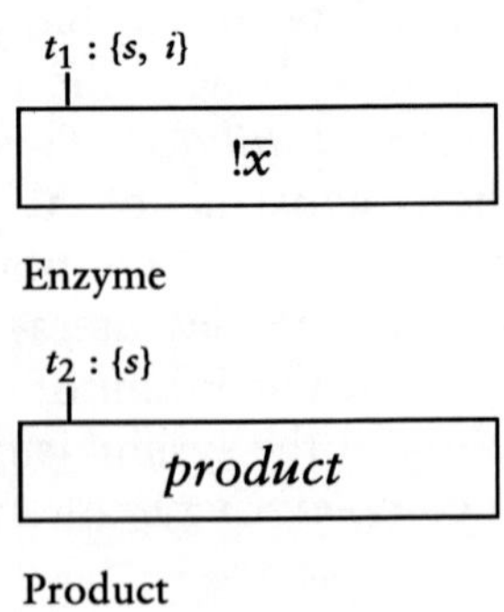

Published by Woodhead Publishing Limited, 2013

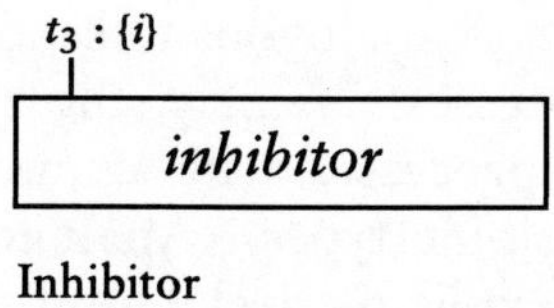

Another possible computation of the system could start from the join reduction between *Enzyme* and *Inhibitor*.

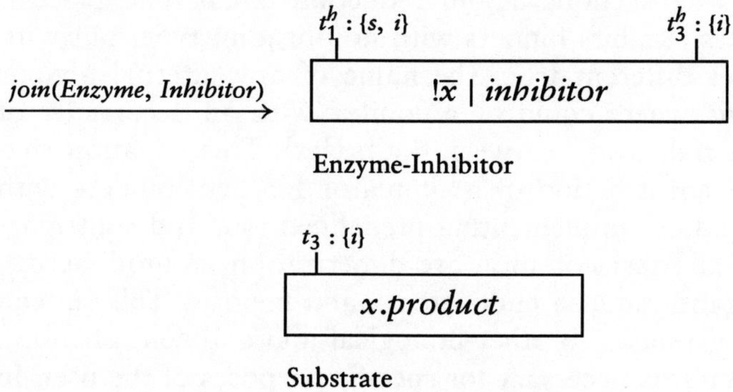

The resulting complex *Enzyme* – *Inhibitor* can then split again into *Enzyme* and *Inhibitor*

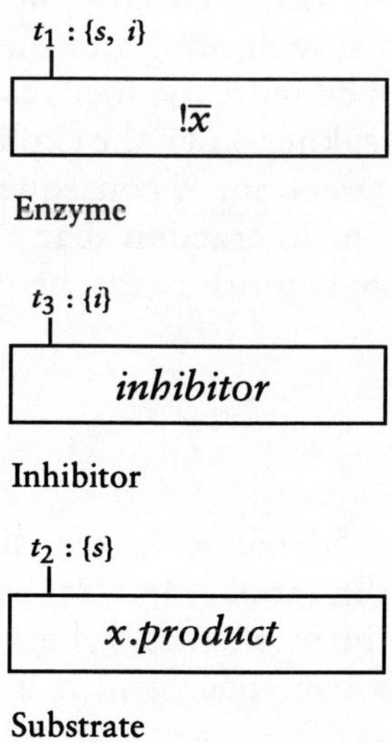

Published by Woodhead Publishing Limited, 2013

4.5.3.4 Some remarks

The reduction inter is analogous to sending-receiving reduction between processes. Namely, the rule models interaction between bio-processes with complementary sites (i.e. sites with non-disjoint types). Analogously, the rules join and split are non explicitly declared in the model (they only refer to the definition of the functions f_{join} and f_{split}. However, their occurrence verifies in dependence on the characteristics of the boxes. In the examples showed in the two previous subsections a join reduction could take place if the boxes exhibits binders with non-disjoint type, but with a subject different from the name of any internal actions (otherwise there could be a conflict with an occurrence of an inter reductions between the boxes). This situation suggest the construction of a simulator for Beta binders with a set modules implementing predefined join and split reactions. The most common are dimerization, homodimerization, ligand-induced endocytosis, and meiosis. This set can also be extended to other biological join and split phenomena when it is necessary for specific purposes of the user. In this way the task of the modeler is only that of accurately specifying each bio-processes involved in the system, in such a way that is exhibits the requested characteristics to satisfy the premises for the occurrence of the desired join and/or split reactions. Thus, with respect to the π, the Beta binders formalism shifts much more the focus on the specification of components, by breaking down the explicit user-driven coordination of the interaction. A consequence of this fact is a major promiscuity of interaction that, anyhow, in a framework of a FRM-like algorithm, can be partially triggered by the reaction rates.

Notes

1. With the term "dynamics", we simply mean "time-evolution". In this book the term is not used with the meaning it has in mechanics, where it is different from "kinetics" or "kinematics" and it is concerned with the

Published by Woodhead Publishing Limited, 2013

effects of forces on the motion of a particle or system of particles, especially of forces that do not originate within the system itself. On the contrary, in chemistry "dynamics" is synonymous of "kinetics", that is concerned with the rates of change in the concentration of reactants in a chemical reaction, and thus with the time-behavior of the system.

2. A lot of references about Finite Element Method can be found at http://www.solid.ikp.liu.se/fe/tit.html
3. In non-classical or allosteric competitive inhibition, the inhibitor binds away from the active site, creating a conformational change in the enzyme such that the substrate can no longer bind to it. Consequently, adding more substrate will not increase the reaction rate. Thus, the reaction rate cannot reach its maximum velocity.

References

1. S. C. J. L. Abkowitz and P. Guttorp. Evidence that hematopoiesis may be a stochastic process in vivo. *Nature Medicine*, 2(2):190–197, 1996.
2. C. Adami. What is complexity. *Bioessays*, 24:1085 – 1094, 2002.
3. B. Alberts, A. Johnson, J. Lewisa, M. Raff, K. roberts, and P. Walter. *Molecular biology of cell.* Garland Science, Taylor and Francis Group, 2002.
4. P. K. Banerjee. *The Boundary Element Methods in Engineering.* McGraw-Hill College, 1994.
5. S. Bornholdt. Modeling genetic networks and their evolution: a complex dynamical system perspective. *Biol. Chem.*, 382:1289–1299, 2001.
6. D. Bray. Reductionism for biochemists: how to survive the protein jungle. *Trends in biochemical science*, 22:325 – 326, 1997.
7. L. Cardelli. Membrane interactions. In *BioConcur '03 Workshop on Concurrent Models in Molecular Biology*, 2003.

Published by Woodhead Publishing Limited, 2013

8. C Chapon. Expression of malt, the regulator gene of the maltose region in escherichia coli, is limited both at transcription and translation. *EMBO J.*, 1(3):369–374, 1982.
9. E. H. Davidson et al. A genomic regulatory network for development. *Science*, 295(5560):1669 – 1678, 2002.
10. D. T. Gillespie. A general method for numerically simulating the stochastic time evolution of coupled chemical species. *J.Comp. Physics*, 22:403–434, 1976.
11. A. Gilman and A. Arkin. Genetic code: representations and dynamical models of genetic components and networks. *Annu. Rev. Genomics Hum. Genet.*, 3:341–369, 2002.
12. J. J. Han, N. Bertin, T. Hao, D. S. Goldberg, G. F. Berritz, D. Dupuy L. V. Zhang, A. J. M. Walhout, M. E. Cusick, F. P. Roth, and M. Vidal. Evidence for dynamically organized mdoularity in the yeast protein-protein interaction network. *Nature*, 430:88–92, 2004.
13. J. Hasty and J. J. Collins. Translating the noise. *Nature*, 31:1314, 2002.
14. J. Hasty and F. Issacs. Designer gene networks: toward fundamental cellular control. *CHAOS*, 11(1):207–220, 2001.
15. M. D. J. Hasty, F. Isaacs, and J. J. Collins. Designer gene networks: toward fundamental cellular control. *Chaos*, 11(1):207–220, 2001.
16. J. Hasty, J. Pradines, and J. J. Collins. Noise-based switches and amplifiers for gene expression. In *Proc. Natl. Acad. Sci. Usa*, pages 2075–2080, 2000. vol. 97.
17. P. Hänggi. Stochastic resonance in biology. *ChemPhysChem*, 3:285–290, 2002.
18. D. A. Hume. Probability in transcriptional regulation and its implications for leukocyte differentiation and inducible gene expression. *Blood*, 96(7):177–185, 2000. Available online at: http://www.bloodjournal.org/cgi/content/abstract/96/7/2323.
19. T. Ito, T. Chiba, R. Ozawa, M. Yoshida, M. Hatton, and Y. Sakaki. A comprehensive two-hybrid analysis to

Published by Woodhead Publishing Limited, 2013

explore the yeast proteins interactome. In *Proc. Natl. Acad. Sci.*, volume 98, pages 4569–4574, 2001.
20. P. Johnstone and R. Gulraiani. A new method for regularization parameter determination in the inverse problem of electrocardiography. *IEEE Trans. Biomed. Eng.*, 44(1):19–39, 1997.
21. T. Kadesh. Notch signaling: a dance of proteins changing partners. *Exp. Cell. Res.*, 260(1):1–8, 2000.
22. J. Kastner, J. Solomon, and S. Fraser. Modeling a nox gene network in silico using a stochastic simulation algorithm. *Developmental biology*, 246:122–131, 2002.
23. A. M. Kierzek, R. M. Mattheyses, and M. K. Simmons. Hybrid simulation of cellular behavior. *Bioinformatics*, 20:316–322, 2001.
24. H. Kitano. *Foundations of system biology*, chapter System biology: toward system-level understanding of biological systems. The MIT Press Cambridge, 2001.
25. H. Kitano. Computational system biology. *Nature*, 420, Nov. 2002.
26. K. Kohn. Molecular interaction maps as information organizers and simulation guide. *Chaos*, 11:84 - 97, 2001.
27. D. A. Lauffenburger. Cell signaling pathways as control modules: Complexity for simplicity? In *Proceedings of the National Academy of Science, USA,*, pages 97(10):5031–5033, 2000.
28. M. J. LaVoie and D. J. Selkoe. The notch ligands, jagged and delta, are sequentially processed by -secretase and presenilin/-secretase and release signaling fragments. *J. Biol. Chem.*, 278:34427–34437, 2003.
29. M. D. Levin. Noise in gene expression as the source of non-genetic individuality in the chemotactic response of escherichia coli. *FEBS Letters*, 2003.
30. M. Lynch and J. S. Conery. The origin of genome complexity. *Science*, 302(5649):1401–1404, Nov. 2003.
31. H. H. McAdams. Stochastic mechanism in gene expression. In *Proc. Natl. acad. Sci. USA*, volume 94(3), pages 814–819, 1997.

Published by Woodhead Publishing Limited, 2013

32. H. H. McAdams and A. Arkin. It's a noisy business! Genetic regulation at namolar scale. *Trends Genet.*, 15:65–69, 1999.
33. P. Mendes. Gepasi: a software package for modeling the dynamics, steady states and control of biochemical and other systems. *Comput. Appl. Biosci.*, 9:563–571, 1993.
34. M. D. Mesarović. System theory and biology - view of a theoretician. *System Theory and Biology, Springer-Verlag*, 1968.
35. R. Milner. *Communicating and mobile systems: the π-calculus*. Cambridge University Press, 1999.
36. D. Noble. The rise of computational biology. *Nat. Rev. Mol. Cell. Biol.*, 3:460–461, 2002.
37. M. A. Novak, M. C. Boerlijst, and J. Maynard Smith. Evolution of genetic redundancy. *Nature*, 388:167–171, 1997.
38. P. Nurse. Reductionism: the ends of understanding. *Nature*, 387:657, 1997.
39. J. Paulsson, O. G.Berg Dagger, and M. Ehrenberg. Stochastic focusing: Fluctuation-enhanced sensitivity of intracellular regulation. *PNAS*, 97(13):7148–7153, 2000.
40. C. Priami. Stochastic π-calculus. *The Computer Journal*, 6:578–589, 1995.
41. C. Priami and P. Quaglia. Beta binders for biological interactions. In *CMSB 2004*. LNBI, Springer, 2005.
42. F. Radtke and K. Raj. The role of notch in tumori geneis: oncogene or tumor suppressor? *Nat. Rev. Cancer*, 3(10):756–767, 2003.
43. S. Ramsey, D. Orrell, and H. Bolouri. Dizzy: stochastic simulation of large-scale genetic regulatory networks. *J. Bioinf. Comp. Biol.*, 3(2):415–436, 2005.
44. A. Regev, E. Panina, W. Silverman, L. Cardelli, and E. Shapiro. Bioambients: An abstraction for biological compartments. *TCS*, 325(1), 2004.
45. A. Regev and E. Shapiro. Cells as computations. *Nature*, 419:343, 2002.

Published by Woodhead Publishing Limited, 2013

46. A. Regev, W. Silverman, and E. Shapiro. Representation and simulation of biochemical processes using the π-calculus process algebra. In *Pacific Symposium on Biocomputing*, pages 459–470, 2001.
47. A. Ross. Transcription of individual genes in eukaryotic cells occurs randomly and infrequently. *Immunol. Cell Biol.*, 72:177–185, 1994.
48. E. Schrödinger and R. Penrose. *What is life: with mind and matter and autobiographical sketches*. Cambridge University Press, 1992.
49. B. Schwikowski, P. Uetz, and S. Fields. A network of protein-protein interactions in yeast. *Nature Biotech.*, 18:1257 – 1261, 2000.
50. S. Shaham. Apoptosis: a process with a beta(nac) for complexity. *Cell*, 114(6):659–661, Sep. 2003.
51. E. Six. The notch ligand delta1 is sequentially cleaved by and adam protease and -secretase. *PNAS*, 100:7638–7643, 2003.
52. T. F. Smith. Functional genomics - bioinformatics is ready for the challenge. *Trends in Genetics*, 14:291 – 293, 1998.
53. S. Till. A stochastic model of stem cell proliferation based on the growth of spleen colony-forming cells. In *Proc. Acad. Sci. USA*, volume 51, pages 29–36, 1964.
54. M. Tomita, K. Hashimoto, K. Takahashi, T. S. Shimizu, Y. Matsuzaki, F. Miyoshi, K. Saito, S. Tanida, K. Yugi, J. C. Venter, and C. A. Hutchison III. Software environment for whole cell simulation. *Comput. Appl. Biosci.*, 9:563–571, 1993.
55. M. C. Walters. Enhancers increase the probability but not the level of gene expression. In *Proc. Natl. Acad. Sci. USA*, pages 7125–7129, 1995.
56. J. D. Watson and F. H. C. Crick. Molecular structure of nucleic acid. a structure of deoxyribose nucleic acid. *Nature*, 4356:753, 1953.
57. O. Wolkenhauer and M. Mesarović. Feedback dynamics and cell function: why systems biology is called system biology. *Mo. BioSyst*, 1:14–16, 2005.

Published by Woodhead Publishing Limited, 2013

58. J. Yang and R. Lusk. Organismal complexity, protein complexity and gene duplicability. In *Proc. Natl. Acad.. Sci. USA*, volume 302(15), pages 661–665, Dec 2003.
59. H. Zhu, S. Huang, and P. Dhar. The next step in system biology: simulating the temporospatial dynamics of molecular networks. *Bioessays*, 26:68–72, 2004.
60. M. Zurita and C. Merino. The transcriptional complexity of the fifth complex. *Trends Genet.*, 19(10):578–584, Oct. 2003.

Published by Woodhead Publishing Limited, 2013

5

The structure of biochemical models

Abstract Modeling and simulation of biochemical reactions is of great interest in the context of system biology. The central dogma of this re-emerging area states that it is system dynamics and organizing principles of complex biological phenomena that give rise to functioning and function of cells. Cell functions, such as growth, division, differentiation and apoptosis are temporal processes, that can be understood if they are treated as dynamic systems. System biology focuses on an understanding of functional activity from a system-wide perspective and, consequently, it is defined by two hey questions: (i) how do the components within a cell interact, so as to bring about its structure and functioning? (ii) How do cells interact, so as to develop and maintain higher levels of organization and functions? In recent years, wet-lab biologists embraced mathematical modeling and simulation as two essential means toward answering the above questions. The credo of dynamics system theory is that the behavior of a biological system is given by the temporal evolution of its state. Our understanding of the time behavior of a biological system can be measured by the extent to which a simulation mimics the real behavior of that system. Deviations of a simulation indicate either limitations or errors in our knowledge.

Published by Woodhead Publishing Limited, 2013

The aim of this chapter is to summarize and review the main conceptual frameworks in which models of biochemical networks can be developed. In particular, we review the stochastic molecular modelling approaches, by reporting the principal conceptualizations suggested by A. A. Markov, P. Langevin, A. Fokker, M. Planck, D. T. Gillespie, N. G. van Kampfen, and recently by D. Wilkinson, O, Wolkenhauer, P. S. Jöberg and Lecca.

Keywords: deterministic processes, stochastic process, state space, differential equations, simulation algorithms.

5.1 Classification of biological processes and mathematical formalism

We can distinguish four fields of application of mathematical models to biology: 1. population dynamics; 2. cell and molecular biology; 3. physiological systems; 4. spatial modeling.

Different formalisms are usually applied to describe the dynamics of these different fields. In general the mathematical structure of a model of a physical phenomenon depends on the nature of the determination, of the time, and of the space state. The determination of a model can be deterministic or stochastic, or also hybrid deterministic and stochastic. The time course can be continuous of discrete, and the state space can also be continuous of discrete. The combination of the these characteristics give rise to different mathematical approaches to the modeling the dynamics of the phenomenon. Here following we list some of the most common mathematical formalism and approaches to specify the dynamics of a system with respect to the four categories listed above.

1. **Deterministic processes** (Newtonian dynamical systems). A fixed mapping between an initial state and a final state. Starting from an initial condition and moving forward in

Published by Woodhead Publishing Limited, 2013

time, a deterministic process will always generate the same trajectory and no two trajectories cross in state space.

- *Ordinary differential equations* (Continuous time. Continuous state space. No spatial derivatives.)
- *Partial differential equations* (Continuous time. Continuous state space. Spatial derivatives.)
- *Maps* (Discrete time. Continuous state space)

2. **Stochastic processes** (random dynamical systems) A random mapping between an initial state and a final state, making the state of the system a random variable with a corresponding probability distribution.

- *Jump Markov process* – Master equation (Continuous time with no memory of past events. Discrete state space. Waiting times between events discretely occur and are exponentially distributed.)
- *Continuous Markov process* – stochastic differential equations or a Fokker-Planck equation (Continuous time. Continuous state space. Events occur continuously according to a random Wiener process.)
- *Non-Markovian processes* – Generalized master equation (Continuous time with memory of past events. Discrete state space. Waiting times of events (or transitions between states) discretely occur and have a generalized probability distribution.)
- *Stochastic simulation algorithms*: Gillespie exact simulation and StochSim

3. **Hybrid stochastic/deterministic systems (metabolic and signaling pathways)**

- *Gillespie τ-leap algorithm* – Differential equations for the simulation of fats reactions and Gillespie algorithm for the exact simulation of slow reactions.

Published by Woodhead Publishing Limited, 2013

Table 5.1 Classes of biological phenomena and most used formalisms to describe them.

Population dynamics	Deterministic processes Ordinary differential equations.
Cell and molecular biology	Stochastic processes: Jump Markov processes and continuous Markov processes.
Physiological systems Spatial modeling (epidemiology)	Deterministic processes Deterministic processes. Partial differential equations.

With regard to modeling the chemistry of intracellular dynamics, the two most popular frameworks are the deterministic modeling and the stochastic modeling. The deterministic modeling is based on the construction of a set of rate equations to describe the reactions in the biochemical pathways of interest. These rate equations are ordinary differential equations with concentrations of chemical species as variables. In general, given the complexity of biological pathways we have to deal with non-linear differential equations. Deterministic simulations produce the time course of the concentrations by solving the differential equations.

In its most known aspect stochastic modeling involves the formation of a set of chemical master equations with probabilities as variables [32]. Stochastic simulation produces counts of molecules of some chemical species as realizations of random variables drawn from the probability distribution described by the master equations.

Which framework is appropriate for a given biological system is not only a question of what biological phenomena are investigated but also influenced by assumptions one makes to simplify the analysis. For instance, the scale, and thus the level of granularity at which a phenomenon is investigated may be parameters to choose a deterministic of a stochastic approach [26].

Published by Woodhead Publishing Limited, 2013

In this paper we firstly review the deterministic approach to chemical kinetics, then we examine thoroughly and critically discuss the main concepts of stochastic chemical kinetics and highlight the necessary re-formulations to adapt them to the biological simulation context.

5.2 Spatially Homogeneous Models

The most part of the mathematical models of chemical reactions is based on the assumption of spatial homogeneity. This means that in these models diffusion and other transport processes can be neglected. Thence, from the formal point of view chemical reaction is handled as a temporal process and a network of chemical interactions is considered a dynamical system.

A dynamic system is an ordered pair: $(\mathbb{A}, \phi)$, where $\mathbb{A}$ is the state space, and $\phi : \mathbb{T} \times \mathbb{A} \to \mathbb{A}$ is a function which assigns to an arbitrary point $x_0 \in \mathbb{A}$ the point $x \in \mathbb{A}$, that characterizes the state at the time t, assuming that the system was in x_0 at $t = 0$. A fundamental property of ϕ is the validity of the identity

$$\phi((t+s), x_0) = \phi(s, \phi(t, x_0)). \tag{5.1}$$

The motion of a dynamics system is the one variable function

$$\phi_{x_0} : T \to A \tag{5.2}$$

$$\phi_{x_0} \equiv \phi(\cdot, x_0) \tag{5.3}$$

where $\mathbb{T} \subset \mathbb{R}$ and $\mathbb{A} \subset \mathbb{R}^M$, $(M \in \mathbb{N})$, or $\mathbb{A}$ consists of random variables taking their values from $\mathbb{R}^M$. For every $t \in T$ $\phi(t, \cdot) : \mathbb{A} \to \mathbb{A}$ is an automorphism.

The process, or equivalently the chemical reaction, to be described can be classified either by the properties of the *process-time*, or by the structure of the *state space*, or by the nature of *determination*.

Published by Woodhead Publishing Limited, 2013

5.2.1 Properties of process-time

The time can be chosen as continuous ($T \subset \mathbb{R}$) or a discrete ($T \subset \mathbb{Z}$) variable. Both the continuous and the discrete time models presents advantages, disadvantages, arguments in favor and arguments in disfavor. The arguments generally adopted for choosing a continuous time variable are:

1. Calculation with continuous time models have greater tradition. Continuous models have the advantage over discrete time models in that they are more amenable to algebraic manipulation, although they are slightly harder to implement on a computer.
2. Most physical processes are inherently continuous in time. In particular, some physico-chemical quantities can be transduced continuously. Thus, the parameters in the models are strongly correlated with the physical properties of the systems; something that is very appealing to an engineer. Moreover, as cost of computation becomes cheaper, today's data acquisition equipment can provide nearly continuous-time measurements. Fast sampled data can be more naturally dealt with using continuous-time models than discrete-time models.

Arguments for selecting a discrete time variable are the following:

1. Time is really discrete. The idea that time has no objective existence but depends on events led some scientists to abandon the assumption that it is a continuous variable. Moreover, we perceive temporal intervals of finite duration rather than durationless instants; and the researcher prefers to assume that the nature has properties that can be verified.
2. The notion of "immediate next time" can be easily interpreted, and this is not so easy in the case of continuous time.
3. The experimentalists measure at discrete points only.

5.2.2 Properties of state-space

The state space can be chosen either continuous or discrete. To emphasize the existence of elementary particles of a population as in reaction kinetics a discrete state space formalism is preferred.

The notion of *state* was derived from the theory of mechanics and of thermodynamics and generalized by mathematical system theory. The quantities of a model can be classified into two categories: state variables and constitutive quantities. State variables are functions such that their values specify the state of the system. The constitutive quantities are functions of the state, in the sense that their value is univocally determined once the state of the system has been assigned. Thus, a constitutive quantity Ω can be expressed as follows

$$\Omega(t) = \omega(g(t), t) \tag{5.4}$$

where g denotes the state of the system and $\omega : \mathbb{A} \times T \rightarrow \mathbb{R}^r$ is the constitutive functional[1] mapping the state into a constitutive quantity $r \in \mathbb{N}$. The case $r = 1$ means that the value of the constitutive quantity is a scalar.

As we already introduced in the previous section, the state of an M-component chemical system is described by a vector:

$$\vec{x} : \mathbb{T} \rightarrow \mathbb{R}^M, \ t \mapsto \vec{x}(t) \in \mathbb{R}^M \tag{5.5}$$

In this section we also said that a state is described by function. The two statements are not in contradiction, namely a finite-dimensional vector can also be interpreted as a function: $\mathbb{R}^M$ can be considered as an abbreviation

$$\mathbb{R}^M := \mathbb{R}^{\{1,2,\cdots,M\}} = \{f; f : \{1, 2, \cdots, M\} \rightarrow \mathbb{R}\}.$$

The state of the system with continuously changing components is described at a fixed point of time by a (not necessarily scalar valued) function $f : \mathbb{R}^M \rightarrow \mathbb{R} * m$. The state of the system is $\tilde{n}$, where $\tilde{n} : \mathbb{T} \rightarrow (\mathbb{R}^m)^{\mathbb{R}^M}$, or it is an element of the set

$$[(\mathbb{R}^m)^{\mathbb{R}^M}]^{\mathbb{T}} = \{f; f : \mathbb{T} \rightarrow (\mathbb{R}^m)^{\mathbb{R}^M}$$

Published by Woodhead Publishing Limited, 2013

According to the convectional treatment of pure homogeneous reaction kinetics the state is a finite-dimensional vector and the only constitutive quantities are the reaction rates.

The theory of thermodynamics adopts the concept of 'particles with memory'. According to this concept, the constitutive quantities depend on the history of the independent variables, and not only on their present value. This means that it is not definite that the instantaneous value of state variables (i.e. state) completely determines the state.

Let introduce the *site* function $h : \mathbb{T} \to \mathbb{R}^M$. Since the state is determined by earlier values of the site function, therefore the state g is interpreted as

$$g : \mathbb{T} \to G, \quad t \mapsto h'$$

where h' is known as *history* function defined as

$$h'(s) = h(t - s), \quad s > 0.$$

Knowing the history, the state can be set up

$$\mathscr{H}(h, \cdot) = g,$$

i.e.

$$\mathscr{H}(h, t) = g(t) = \mathscr{H}(h') = h'$$

$\mathscr{H}$ is a mapping assigning a function to a function and to a number. If we assume that the history of the site does not influence the state, then the constitutive functional reduces to a function. Furthermore, if we also assume the invertibility of this function then the differences between the state variables and constitutive quantities are not significant. These two assumptions are tacitly adopted in the classical theories of the thermodynamics.

The stochastic version of a memory-free deterministic process is a Markov process (more precisely, a first-order Markov process).

Published by Woodhead Publishing Limited, 2013

5.2.3 Nature of determination

An $(\mathbb{A}, \phi)$ dynamic system is deterministic if knowing the state of the system at one time means that the system is uniquely specified for all $t \in \mathbb{T}$.

When the state of the system can be assigned to a set of values with a certain probability distribution, the future behavior of the system can be determined stochastically. Discrete time, discrete state space (first order) Markov processes (i.e. Markov *chain*) are defined by the formula

$$\begin{aligned}\mathscr{P}(\xi_{t+1} = a|\xi_0 = a, \xi_1 = a_1, \ldots, \xi_t = a_t) \\ = \mathscr{P}(\xi_{t+1} = a|\xi_t = a_t).\end{aligned} \tag{5.6}$$

where the set $\{\xi_t|t = 0, 1, 2, \ldots\}$ is a discrete time stochastic process.

Knowing the total history of the process we can extrapolate its future behavior with the same probability as if we knew only the actual current state. Put another way, a Markov process is a stochastic process which possesses the property that the future behavior depends only on the current state of the system. Thus, given information about the current state of the system, information about the past behavior of the system is no help in predicting the time-evolution of the process. The behavior of the chain is therefore determined by $\mathscr{P}(\xi_{t+1} = a|\xi_t = a_t)$, and thus it depends on a and t. However, if there is no t dependence, so that

$$\mathscr{P}(\xi_s = x|\xi_t = y) = \mathscr{P}_{xy}(s - t),$$

i.e. the transition probabilities are stationary, the Markov chain is said to be *time homogeneous*. In this case the law of evolution of the system does not depend explicitly on time and consequently, the time origin can be defined arbitrarily. Deterministic dynamics systems generated by ordinary differential equations

$$\frac{dx(t)}{dt} = f(x(t))$$

can be associated with the time homogeneous Markov processes.

Published by Woodhead Publishing Limited, 2013

Markov processes are particularly amenable for both theoretical and computational analysis and the dynamic behavior of biochemical networks can be effectively modeled by a Markov chain. Moreover, a Markovian description can be introduced by generalizing deterministic systems modeled by ordinary differential equations, since the stochastic version of a deterministic process without 'after-effect' is a Markov process.

However, the Markov character of the chemical process represented by the state vector has not been derived from microscopic models of the chemical dynamics. Therefore Markovicity is not more and not less than a plausible assumption.

5.2.4 XYZ models

At least eight different kinetic models can be defined, depending on the specification of time (X), state space (Y) and nature of determination (Z). As was explained earlier, time can be discrete (D) or continuous (C), the state space can be also discrete (D) or continuous (C), and the nature of determination can be deterministic (D) or stochastic (S).

Mass-action type kinetic differential equations (Chapter 1) can be identified with the CCD model, while the more often used stochastic model is the CDS model. DCD models have achieved a significance in the last decade in connection with chaotic phenomena. There are at least two distinct methods of relating DCD models to CCD models. The first is the discretization of time. An autonomous differential equation

$$\frac{dx}{dt} = f(x(t), t), \quad x_{t=0} = x_0$$

can be transformed as

$$x(t + h) = x(t) + f(x(t), t)h + o(h)$$

The second method can be applied if the differential equation

Published by Woodhead Publishing Limited, 2013

has a periodic solution. take a hyperplane of dimension $n-1$ traverse to the curve $t \longrightarrow x(t)$ through x_0. A map

$$F : U \longrightarrow \mathbb{R}^{n-1}$$

is induced by associating with t_0 the nearest intersection of the trajectory (with initial condition $x_{t=0} = x_0$) with the given hyperplane. If the first such intersection occurs at x_1, we define $F(x_0) \equiv x_1$. Since the form of F is independent of the index of the series and also of the coordinates, we can specify

$$x_{n+1} = F(x_n).$$

Thus a difference equation has been obtained from a differential system.

5.3 Variants of the SSA for non-Markovian and non-homogeneous processes

The structures of certain biological and biochemical models are not conducive to the exact SSA described in Chapter 3, owing to qualities such as time-dependence of rate constants or non-homogeneous spatial distributions of molecular populations. Simulation of such models requires specialized versions of the SSA.

5.3.1 *Time-dependent extension of First Reaction Method*

The First Reaction Method may be extended to the case of time-depending rates and changes in volume. Lu and coworkers [22] developed one of the first approaches which we reproduce here. This re-formulation has been adapted to be incorporated in the framework of stochastic π-calculus

Published by Woodhead Publishing Limited, 2013

and its implementation has been succesfully applied to a sample simulation in biology: the passive glucose cellular transport [19, 18].

Let us suppose that the volume $V_s(t)$ contains a mixture of chemical species, X_i $(i = 1,\dots,N)$ which may interact through the reaction channels R_μ, $\mu = 1,\dots,M$. Let suppose furthermore that a subset of these channels is characterized by the time-dependent propensities

$$a_s(t) = a'_s/V(t), \;\; s = 1,\dots,S \tag{5.7}$$

and an other sub-set is characterized by the time-dependent propensities

$$a_q(t) = a'_q/V(t), \;\; q = S+1,\dots,M \tag{5.8}$$

where a'_s and (a'_q) are the time-independent propensities, that have to be computed according to the type of reaction (see Chapters 2 and 3).

Following the Gillespie approach, let introduce these probabilities:

1. $P(\tau,\mu|Y,t)d\tau$: probability that, given the state $Y = (X_1,\dots,X_N)$ at time t, the next reaction will occur in the infinitesimal time interval $(t+\tau, t+\tau+d\tau)$, at it will be reaction R_μ
2. $a_\mu(t)dt$: probability that, given the state $Y = (X_1,\dots,X_N)$ at time t, reaction R_μ will occur within the interval $(t, t+dt)$.

$P(\tau,\mu|Y,t)d\tau$ is computed as a product of the probabilities that no reaction will occur within $(t, t+\tau)$ times the probability that R_μ will occur within the subsequent interval $(t+\tau, t+\tau+d\tau)$

$$P(\tau,\mu|Y,t)d\tau = P_0(\tau|Y,t)\cdot a_\mu(\tau+t)d\tau \tag{5.9}$$

where, summing over all reaction channels $\mu = 1,\dots,M$ and splitting the sum in the two terms over s and q

Published by Woodhead Publishing Limited, 2013

$$P_0(\tau + d\tau | Y, t) = P_0(\tau | Y, t)\left[1 - d\tau \sum_{s=1}^{S} a_s(t+\tau) - d\tau \sum_{q=S+1}^{M} a_q(t+\tau)\right]$$

With the initial condition $P_0(\tau = 0 | Y, t) = 1$, the solution of this differential equation is

$$P_0(\tau | Y, t) = \exp\left[-\sum_s \int_t^{t+\tau} a_s(t+\tau')d\tau' - \sum_q \int_t^{t+\tau} a_q(t+\tau')d\tau'\right] \quad (5.10)$$

Now, by combining Eq. (5.9) with the Eq. (5.10), we obtain

$$P(\tau, \mu|, Y, y) = a_\mu(t+\tau) \times \exp\left[-\sum_s \int_t^{t+\tau} a_s(t+\tau') - \sum_q \int_t^{t+\tau} a_q(t+\tau')d\tau'\right] \quad (5.11)$$

By introducing two functions $f_s(\tau)$ and $f_q(\tau)$ describing the variation of volume in time, the time-dependence of the volumes can be described by these expressions:

$$V_s(t+\tau) = V_s(t)f_s(\tau) \quad \text{and} \quad V_q(t+\tau) = V_q(t)f_q(\tau).$$

Consequently, the propensities are

$$a_s(t+\tau) = a_s(t)/f_s(\tau) \quad \text{and} \quad a_q(t+\tau) = a_q(t)/f_q(\tau).$$

Substituting these expressions in Eq. (5.11), and introducing, for convenience

$$A_s \equiv \sum_s a_s(t) \quad \text{and} \quad A_q \equiv \sum_q a_q(t)$$

$$F_s(\tau) \equiv \int_t^{t+\tau} \frac{1}{f_s(\tau')}d\tau' \quad \text{and} \quad F_q(\tau) = \int_t^{t+\tau} \frac{1}{f_q(\tau')}d\tau'$$

so that Eq. (5.11) can be re-written as

Published by Woodhead Publishing Limited, 2013

$$P(\tau,\mu|Y,t)=\begin{cases}\frac{a_s(t)}{f_s(\tau)}\cdot\exp\left[-A_sF_s(\tau)-A_qF_q(\tau)\right]\\ \frac{a_q(t)}{f_q(\tau)}\cdot\exp\left[-A_sF_s(\tau)-A_qF_q(\tau)\right]\end{cases}\tag{5.12}$$

Finally, the probability of any reaction occurring between time t and the time $t+T$, is obtained by integrating Eq. (5.12) over time and summing over all channels:

$$\int_0^T\sum_\mu P(\tau,\mu|Y,t)d\tau=\begin{cases}\int_0^T\sum_{s'=1}^{S}\frac{a_{s'}(t)}{f_s(\tau)}\cdot\exp\left[-A_sF_s(\tau)-A_qF_q(\tau)\right]d\tau\\ \int_0^T\sum_{q'=S+1}^{M}\frac{a_{q'}(t)}{f_q(\tau)}\cdot\exp\left[-A_sF_s(\tau)-A_qF_q(\tau)\right]d\tau\end{cases}\tag{5.13}$$

Generalizing, in systems where the physical reaction space is divided into n sub-spaces whose volumes change in time, the probability density function of reaction is split into n exponential terms multiplied by the ratio between reaction propensity and volume of the subspace. The volume of each subspace can follow a different temporal behavior. Consequently a different reaction probability and a different expression of reaction time are obtained for each sub-regions of the space.

The effect of temperature changes on the probability density function can be simulated by expressing the time dependence of the propensity of a reaction μ as $a_\mu(t+\tau)=a_\mu(t)\cdot T(\tau)$, where $T(\tau)=\exp(1/(a+b\tau))$ models the variation of the propensity function following the Arrhenius formula (for instance, in Lecca [18] $a=37°C$ and $b=1\ °C/min$).

5.3.2 Spatio-temporal algorithms

Previous sections cover the stochastic algorithms for modeling biological pathways with no spatial information. However, the real biological world consists of components which interact in a three dimensional space. Within a cell compartment, the intracellular material is not distributed homogeneously in space and molecular localization plays an important role, e. g. diffusion of ions and molecules across membranes and propagation of an action potential along a nerve fiber's axon. Thus, basic assumption of spatial homogeneity

Published by Woodhead Publishing Limited, 2013

and large concentration diffusion is no longer valid in realistic biological systems [4]. In this context, stochastic spatio-temporal simulation of biological system is required.

The enhancement on the performance of Gillespie Algorithms has made the spatio-temporal simulation tractable. Stundzia and Lumsden [31], and Elf et al. [4], extended the Gillespie Algorithms to model intracellular diffusion. They formalized the reaction-diffusion master equation and the diffusion probability density functions. The entire volume of a model was divided into multiple subvolumes and by treating diffusion processes as chemical reactions, the Gillespie Algorithm was applied without much modification. Stundzia has showcased the application of the algorithm on calcium wave propagation within living cells and has observed regional fluctuations and spatial correlations in the small particles limit. However, this approach requires detailed knowledge about the diffusion processes to be available, in order to estimate the probability density function for diffusion. Furthermore, the algorithms have only been applied to small systems with finite number of molecular species but require large amount of computational power.

Shimizu in [30] also extended the Stochsim algorithm to include spatial effects of the system. In his approach, spatial information was added to the attributes of each molecular species and a simple two dimensional lattice was formed to enable interaction between neighboring nodes. The algorithm was applied to study the action of a complex of signaling proteins associated with the chemotactic receptors of coliform bacteria. He showed that the interactions among receptors could contribute to high sensitivity and wide dynamic range in the bacterial chemotaxis pathway.

Another way of simulating stochastic diffusion is to directly approximate the Brownian movements of the individual molecules (MCell [2]). In this case, the motion and direction of the molecules are determined by using random numbers during the simulation. Similarly, collisions with potential binding sites and surfaces are detected and handled by using only random numbers with a computed binding probability. MCell is capable of treating stochastic and a

Published by Woodhead Publishing Limited, 2013

3-dimensional biological model that involves a discrete number of molecules. Though MCell incorporates 3D spatial partitioning and parallel computing to increase algorithmic efficiency, the simulation is limited to the microphysiological processes such as synaptic transmission due to high computational requirement.

Apart from the enhancements on various algorithms, the simulation of a spatio-stochastic biological system is still a challenging problem. To address it the author recently proposed a new mathematical treatment of diffusion that can be incorporated in a stochastic algorithm simulating the dynamics of a reaction-diffusion system is presented [20, 21]. The movement of a molecule A from a region i to a region j of the space is represented as a first order reaction $A_i \xrightarrow{k} A_j$, where the rate constant k depends on the diffusion coefficient. The diffusion coefficients are modeled as function of the local concentration of the solutes, their intrinsic viscosities, their frictional coefficients and the temperature of the system. The stochastic time evolution of the system is given by the occurrence of diffusion events and chemical reaction events. At each time step an event (reaction or diffusion) is selected from a probability distribution of waiting times determined by the intrinsic reaction kinetics and diffusion dynamics. To demonstrate the method the simulation results of the reaction-diffusion system of chaperone-assisted protein folding in cytoplasm are shown.

5.3.3 The Langevin equation

While internal fluctuations are self-generated in the system, and they can occur in closed and open systems as well, external fluctuations are determined by the environment of the system. We have seen that a characteristic property of internal fluctuations is that they scale with the system size and tend to vanish in the thermodynamics limit. External noise has a crucial role in the formation of ordered biological structures. External noise-induced ordering was introduced to model the ontogenetic development and plastic behavior of certain neural structures [5]. Moreover, it was demonstrated that noise can support the transition of a system from

Published by Woodhead Publishing Limited, 2013

a stable state to another stable state. Since stochastic models might exhibit qualitatively different behavior than their deterministic counterpart, external noise can support transitions to states which are not available (or even do not exist) in a deterministic framework [15].

In the case of extrinsic stochasticity, the stochasticity is introduced by incorporating multiplicative or additive stochastic terms into the governing reaction equations. These terms, normally viewed as random perturbations to the deterministic system, are also known as stochastic differential equations. The general equation is:

$$\frac{dx}{dt} = f(x) + \xi_x(t) \tag{5.14}$$

The definition of the additional term ξ_x differs according to the formalism adopted. In Langevin Equations [12], ξ_x is represented by Eq. (5.15) Other studies [14] adopt a different definition where $\xi_i(t)$ is a rapidly fluctuating term with zero mean $(\{\xi_i(t)\} = 0)$. The statistics of $\xi(t)$ are such that $(\{\xi_i(t)\xi_i(t')\} = 0) = D\delta_{ij}(t - t')$ to maintain independence of random fluctuations between different species (D is proportional to the strength of the fluctuation).

$$\xi_x(t) = \sum_{j=1}^{M} V_{ij}\sqrt{\alpha_j}X(t)N_j(t) \tag{5.15}$$

where V_{ij} is the change in number of molecules of species i brought by one reaction j and N_j are statistically independent normal random variables with mean 0 and variance 1.

The way in which Langevin introduced fluctuations into the equation of molecular population level evolution does not carry over nonlinear systems. This section briefly sketch the difficulties to which such a generalization leads. External noise denotes fluctuations created in an otherwise deterministic system by the application of a random force, whose stochastic properties are supposed to be known. Internal noise is due to the fact that the system itself consists of discrete particles. It is inherent in the mechanism by which the

Published by Woodhead Publishing Limited, 2013

state of the system evolves and cannot be divorced from its evolution equation. A Brownian particle, with its surrounding fluid is a closed physical system with internal noise. Langevin, however, treated, the particle as a mechanical system subject to the force exerted by the fluid. This force he subdivided in a deterministic damped force and a random force, which he treated as external, i.e. its properties as a function of time were supposed to be known. For the physical pictures, these properties will not be altered if an additional force on the particle is introduced.

In more recent years, however, Eq. (5.14) has been used also in modeling the evolution of biochemical systems, although the noise source in a chemical reacting network is internal and no physical basis is available for a separation into a mechanical part and a random term with known properties. The strategy used in the application of the Langevin equation in modeling the evolution of a system of chemical reacting particles is the following. Suppose a system whose evolution is described phenomenologically by a deterministic differential equation

$$\frac{dx}{dt} = f(x) \tag{5.16}$$

where x stands for a finite set of macroscopic variables, but for simplicity in the present discussion we take the case that x is a single variable. Let suppose to know that for some reason there must also be fluctuations about this macroscopic values. Therefore, we supplement (5.16) with a Langevin term

$$\frac{dx}{dt} = f(x) + L(t) \tag{5.17}$$

Note now, that on averaging (5.17) one does not find that $\langle x \rangle$ obeys to the phenomenological equation (5.16), rather than

$$\partial_t \langle x \rangle = \langle f(x) \rangle = f(\langle x \rangle) + \frac{1}{2} \langle (x - \langle x \rangle)^2 \rangle \partial_t^2 (\langle x \rangle) + \ldots$$

It follows that $\langle x \rangle$ does not obey any differential equation at all. This reveals the basic flaw in the application of the

Published by Woodhead Publishing Limited, 2013

Langevin approach to the internal noise of systems whose phenomenological law is nonlinear. The phenomenological equation (5.16) holds only in the approximation in which fluctuations are neglected. That implies that $f(x)$ is determined phenomenologically with an inherent margin of uncertainty of the order of fluctuations. If we deduce a certain form of $f(x)$ from a theory or experiment in which fluctuations are ignored there is no justification for postulating that $f(x)$ is to be used in (5.17). There may be a mismatch between both of the same size as the fluctuations; that would not show up in macroscopic results, but cannot of course be neglected in the equation of the fluctuations themselves.

5.3.4 Hybrid algorithms

Biological systems are stiff by nature in the sense that processes with very different time scales are coupled. Some molecues are quickly synthesized and degenrated (typically metabolites) and take a long time to run over (typically macromolecules). Some biochemical reactions involve a chain of many steps, while other reactions just involve a single association of dissociation event. This difference in time scales can be exploited by assuming quasi-equilibrium and usign the equilibrium constant to eliminate from the model some components, and thus to reduce its complexity.

Stochastic algorithms suffer from the same "stiffness" problems as that of deterministic algorithms. In order to capture the fast dynamics of the system, entire simulation is slowed down significantly. Hence, the basic idea of hybrid algorithms aims to exploit the advantages of other algorithms to offset the disadvantages of the stochastic algorithms.

Several attempts have been made to illustrate the relevance and feasibility of hybrid algorithms. Bundschuh et al [28], Haseltine and Rawlings [13], and Puchalka and Kierzek [27], have used a similar approach to integrate ODE/Langevin with Gillespie algorithms. In both cases, the modeler has to identify methods and criteria to partition the system into fast

Published by Woodhead Publishing Limited, 2013

dynamics and slow dynamics sub-systems. The fast dynamics subsystem can be handled by either ODE or Langevin Equations while the slow dynamics subsystem can be handled by Gillespie algorithms. In addition, numerical treatment such as the "slow variables" in [28], and the "probability of no reaction" in Haseltine and Rawlings [13], is required to maintain accuracy of the solutions. The algorithms show promising results and the results are consistent with those from Gillespie algorithms. Haseltine and Rawlings in [13], showed the applicability of hybrid algorithms by simulating the effect of stochasticity to the bi-modality of an intracellular viral infection model using the algorithm. Kiehl et al. [17], also tested the algorithms on the λ phage model.

The relevance of hybrid algorithms has been pointed out in several papers (Alur et al. [1]; Matsuno et al. [24]; Bockmayr and Courtois [3]). Bockmayr and Courtois used hybrid constraint programming methods to model an alternative splicing regulation model. This implementation is very useful under circumstances where detailed knowledge about the model is unavailable. Meanwhile, Alur et al. used CHARON, a formal description language of hybrid system which combines ODE with "mode switching" mechanism to model the quorum sensing phenomenon in *Vibrio fischeri*, a marine bacterium that involves the Lux regulon. A Hybrid Petri Net [24] approach has been employed to model a hybrid system using ODEs and discrete events. This method has been used to model the growth pathway control of λ phage.

Hybrid algorithms aim to close the gap between macroscopic and mesoscopic scales of the system. In particular, the relevance of hybrid modeling has been proved necessary to capture the behavior of a real biological system. Moreover, hybrid algorithms have substantially cut down the computational cost of large scale modeling and simulation. One major drawback here is that by introducing additional numerical treatment to the algorithms, more parameters have to be defined and the accuracy of the solutions is dependent on the accuracy of parameters. Mostly, the

Published by Woodhead Publishing Limited, 2013

simulations result in solutions of highly tuned parameters. Although these hybrid approaches show significant improvements in the computational cost, there are still lots of computational issues to be resolved before it can be applied to a realistic problem. Some of the issues are:

- accuracy of results,
- consistency of system parameters between different levels of abstraction,
- highly non-linear system,
- methodology to separate the systems into different subsystems, dynamic switching between different mathematical formalisms.

We end this chapter by noticing that the biochemical approach to understand biological processes is essentially one of simulation. A biochemist typically prepares a cell-free extract that can mediate a well-described physiological process. Once the extract is fractioned to purify the components that catalyze individual reactions, the physiological process in reconstructed *in vitro*. The validity of this approach is measured by how closely the *in vitro* reconstructed process matches physiological observations.

Similarly, the validity of a model in its conceptual framework is measured by how closely its simulation matches physiological observations. Unfortunately, often controlled experiments cannot be performed on the system to validate our model (for example, how can the model be validated if only a single historical dataset exists?). The validation becomes difficult also when the model is stochastic, i.e. it has random elements. However, whatever the nature of the model is, in general validation ensures that the model meets its intended requirements in terms of the methods employed and the results obtained. The ultimate goal of model validation is to make the model useful in the sense that the model addresses the right problem, provides accurate information about the system being modeled, and makes the model actually used [23, 29].

Published by Woodhead Publishing Limited, 2013

Notes

1. The functional assigns a number to a function. Here the term refers to every mapping having the function as argument.

References

1. R. Alur, C. Belta, F. Ivancic, V. Kumar, M. Mintz, G. Pappas, H. Rubin, and J. Schug. Hybrid modeling and simulation of biomolecular networks. In *Hybrid System. Computation and Control, 4th International Workshop, HSCC, Rome Italy*, 2001.
2. T. M. Bartol and J. R. Stiles. *M-cell, http://www.MCell.cnl.salk.edu*, 2002.
3. A. Bockmayr and A. Courtois. Using hydrid concurrent programming to model dynamics biological systems. In *18th International Conference on Logic Programming, ICLP02*, pages 85–99. Springer, LNCS 2401, July 2002.
4. J. Elf, A. Doncic, and M. Eherenberg. Mesoscopic reaction-diffusion in intracellular signaling. In *Proceedings of SPIE 5110*, pages 114–124, 2003.
5. P. Erdi and G. Barna. Self-organisation in neural systems. some illustrations. *Lecture Notes in Bioinformatics*, 71, 1993.
6. B. Ermentrout. *Simulating, analyzing, and animating dynamical systems: a guide to XPPAUT for researchers and students.* 1st edition. SIAM New York, 2002.
7. M. Gibson and J. Bruck. Efficient exact stochastic simulation of chemical systems with many species and many channels. *J. Phys. Chem. A*, 104, 2000.
8. D. T. Gillespie. A general method for numerically simulating the stochastic time evolution of coupled chemical species. *J.Comp. Physics*, 22:403–434, 1976.
9. D. T. Gillespie. Exact stochastic simulation of coupled chemical reactions. *The J. of Physical Chemistry*, 81(25), 1977.

Published by Woodhead Publishing Limited, 2013

10. D. T. Gillespie. *Markov Processes*. Academic Press, 1992.
11. D. T. GIllespie. A rigorous derivation of the chemical master equation. *Physica A*, 188:404–425, 1992.
12. D. T. Gillespie. Approximate accelerated stochastic simulation of chemically reacting systems. *J. Chem. Phys.*, 115:1716–1733, 2001.
13. E. L. Haseltine and J. B. Rawlings. Approximate simulation of coupled fast and slow reactions for stochastic chemical kinetics. *J. Chem. Phys.*, 117:6959–6969, 2002.
14. J. Hasty and F. Issacs. Designer gene networks: toward fundamental cellular control. *CHAOS*, 11(1):207–220, 2001.
15. W. Horsthemke and L. Hanson. Non equilibrium chemical instabilities in continuous flow stirred tank reactors: the effect of stirring. *J. Chem. Phys.*, 81, 1993.
16. P. S. Jöberg. Numerical solution of the Fokker-Planck approximation of the chemical master equation. Master's thesis, Dept. of Information Technology, Uppsala University, 2005.
17. T. R. Kiehl, R. M. Mattheyses, and M. K. Simmons. Hybrid simulation of cellular behavior. *Bioinformatics*, 20:316–322, 2004.
18. P. Lecca. Simulating the cellular passive transport of glucose using a time-dependent extension of gillespie algorithm for stochastic π-calculus. *Int. Journal of Data Mining and Bioinformatics*, 1(4), 2006.
19. P. Lecca. A time-dependent extension of gillespie algorithm for biochemical stochastic π-calculus. In *SAC ACM '06*, pages 137–144, 2006.
20. P. Lecca and L. Demattè. Stochastic simulation of reaction diffusion systems. *In. Journal of Medical and Biological Engineering*, 1(4):211–231, 2008.
21. P. Lecca, L. Demattè, and C. Priami. Modeling and simulating reaction-diffusion systems with state-dependent diffusion coefficients. In *Int. Conference on Bioinformatics and Biomedicine 2008*, volume 34, page 361.

Published by Woodhead Publishing Limited, 2013

World Academy of Science, Engineering and Technology, 2008.

22. T. Lu., L. Tsimring D. Volfson, and J. Hasty. Cellular growth and division in the gillespie algorithm. *Syst. Biol.*, 1, 2004.
23. C. M. maclan. Model verification and validation. In *Workshop on Threat Anticipation: Social Science Methods and Models, The University of Chicago and Argonne National Laboratory.*
24. H. Matsuno, A. Doj, M. Nagasaki, and S. Miyano. Hybrid petri net representations of gene regulatory networks. In *Pac. Symp. Biocomput.*, pages (5) –333–349, 2000.
25. D. A. McQuarrie. Stochastic approach to chemical kinetics. *J. Appl. Prob.*, 4:413–478, 1967.
26. W. Kolch O. Wolkenhauer, M. Ullah and K. Cho. Modelling and simulation of intracellular dynamics: Choosing an appropriate framework. *IEEE Transaction on Nano-Bioscience Special Issue molecular and subcellular system biology*, 2004.
27. J. Puchalka and A. M. Kierzek. Binding the gap between stochastic and deterministic regimes in the kinetic simulations of the biochemical reaction networks. *Biophys. J.*, 86:1357–1372, 2004.
28. F. Hayot R. Bundschuh and C. Javaprakash. Fluctuations of slow variables in generic netwroks. *Biophys. J.*, 84:1606–1615, 2003.
29. R. G. Sargent. Simulation model verification and validation. In *Proceedings of the 23rd conference on Winter simulation*, pages 37–47. IEEE Computer Society Washington, DC, USA.
30. T. S. Shimizu. *The spatial organisation of cell signaling pathways - a computer based study*. PhD thesis, University of Cambridge, UK, 2002.
31. A. B. Stundzia and C. J. Lumsden. Stochastic simulation of coupled reaction-diffusion processes. *J. Comput. Phys.*, 127:196–207, 1996.

Published by Woodhead Publishing Limited, 2013

32. N. G. van Kampfen. *Stochastic Processes in Physics and Chemistry*. Elsevier Amsterdam, 1992.
33. D. J. Wilkinson. *Stochastic Modeling for System Biology*. Chapman & Hall, 2006.

Published by Woodhead Publishing Limited, 2013

6

Reaction-diffusion systems

Abstract Reaction-diffusion systems are mathematical models that describe how the concentrations of substances distributed in space change under the influence of local chemical reactions, and diffusion which causes the substances to spread out in space. The classical representation of a reaction-diffusion system is given by semi-linear parabolic partial differential equations, whose solution predicts how diffusion causes the concentration field to change with time. This change is proportional to the diffusion coefficient. If the solute moves in a homogeneous system in thermal equilibrium, the diffusion coefficients are constants that do not depend on the local concentration of solvent and solute. However, in non-homogeneous and structured media the assumption of constant intracellular diffusion coefficient is not necessarily valid, and, consequently, the diffusion coefficient is a function of the local concentration of solvent and solutes. In this paper we propose a stochastic model of reaction-diffusion systems, in which the diffusion coefficients are function of the local concentration, viscosity and frictional forces. We then describe the software tool *Redi* (REaction-DIffusion simulator) which we have developed in order to implement this model into a Gillespie-like stochastic simulation algorithm. Finally, we show the ability of our model implemented in the Redi tool to reproduce the observed gradient of the bicoid protein in the Drosophila Melanogaster embryo. With Redi, we were able to simulate with an accuracy of 1%

Published by Woodhead Publishing Limited, 2013

the experimental spatio-temporal dynamics of the bicoid protein, as recorded in time-lapse experiments obtained by direct measurements of transgenic bicoid-enhanced green fluorescent protein.

Keywords: reaction-diffusion systems, state-dependent diffusion, CoSBiLab Redi.

6.1 Introduction

As the name indicates, reaction-diffusion models consist of two components. The first is a set of biochemical reactions which produce, transform or remove chemical species. The second component is a mathematical description of the diffusion process. At molecular level, diffusion is due to the motion of the molecules in a medium. If solutions of different concentrations are brought into contact with each other, the solute molecules tend to flow from regions of higher concentration to regions of lower concentration, and there is ultimately an equalization of concentration. The driving force leading to diffusion is the Gibbs energy difference between regions of different concentration.

The great majority of mesoscopic reaction-diffusion models in intracellular kinetics is usually performed on the premise that diffusion is so fast that all concentrations are maintained homogeneous in space. However, recent experimental data on intracellular diffusion constants, indicate that this supposition is not necessarily valid even for small prokaryotic cells [1]. If the system is composed by a sufficiently large number of molecules, the concentration, i.e. the number of molecules per unit volume, becomes a continuum and differentiable variable of space and time. In this limit a reaction diffusion system can be modeled by using differential equations. In an unstructured solvent, ideally behaving solutes (i.e. solutes for which solute-solute interaction

Published by Woodhead Publishing Limited, 2013

are negligible) obey Fick's law of diffusion. However in biological system even for purely diffusive transport phenomena the classical Fick's diffusion is at best a first approximation [2, 3]. Spatial effects are present in many biological systems, so that the spatially homogeneous assumption will not always hold. Examples of spatial effects include mRNA movement within the cytoplasm [4], Ash 1 mRNA localization in budding yeast [5], morphogen gradients across egg-polarity genes in Drosophyla oocyte [5], and the synapse-specificity of long-term facilitation in Aplysia [7]. The intracellular medium is not a homogeneous mixture of chemical species, but a highly structured environment partitioned into compartments in which the distribution of the biomolecules could be non-homogeneous. The description of diffusion processes in this environment has to start from a model of the diffusion coefficient containing its dependency on the local concentrations of the solutes and solvent.

In order to tackle about this problem, P. Lecca et al. [6] presented a new model of diffusion coefficient for a non-homogeneous non-well-stirred reaction-diffusion system. In this model the diffusion coefficient explicitly depends on the local concentration of solute, frictional coefficient and temperature. In turn, the rate of diffusion of the biochemical species are expressed in terms of this concentration-dependent diffusion coefficients. In this study purely diffusive transport phenomena of non-charged particles, and, in particular, the case in which the diffusion is driven by a chemical potential gradient in x direction only (the generalization to the three-dimensional case poses no problems) are considered. The derivation, introduced in this work, consists of five main steps: 1. calculation of the *local* virtual force F per molecules as the spatial derivative of the chemical potential 2. calculation of the particles mean drift velocity in terms of F and local frictional coefficient f; 3. estimation of the flux J as the product of the mean drift velocity and the local concentration; 4. definition of diffusion coefficients as function of local activity and frictional coefficients and concentration, and 5. calculation of diffusion rates as the negative first spatial derivative of the flux J. The determination of the

Published by Woodhead Publishing Limited, 2013

activity coefficients has required the estimation of the second virial coefficient, that is calculated by using the Lennard-Jones potential to describe the inter-molecular interactions. The frictional coefficient is assumed to be linearly dependent on the local concentration of solute.

The diffusion events are modeled as reaction events and the spatial domain of the reaction chamber is divided into cubic subvolumes of size l, that from now on will be called indifferently *cells*, *meshes* or *boxes*. The movement of a molecule A from box i to box j is represented by the reaction $A_i \xrightarrow{k} A_j$, where A_i denotes the molecule A in the box i and A_j denotes the molecule A in the box j. The reaction-diffusion system is thus modeled as a purely reaction system in which the diffusion events are first order reactions whose rate coefficients ks are expressed in terms of state-dependent diffusion coefficients.

The space domain of the system is divided into a number N_s of subvolumes. The time evolution of the system is computed by a Gillespie-like algorithm [8] that at each simulation step selects in each subvolume the fastest reaction, compares the velocities of the N_s selected reactions and finally executes the reaction that is by far the fastest. To make the Gillespie approach applicable in each subvolumes, the size of the mesh has to be chosen sufficiently small so that the homogeneity and well-stirred assumption on the distribution of the molecules inside are good approximations, and sufficiently large to have a number of reaction events significantly greater than one.

The chapter is organized as follows: Section 6.2 illustrates the mathematical model of the diffusion as a time dependent process. In the subsection of this section the new model of diffusion coefficient depending on the state variables of the system, the models of virial coefficient, intrinsic viscosity and frictional coefficient are described. In Section 6.3 the method to estimate the suitable size of the subvolumes in which the entire reaction space has to be subdivided is explained. Section 6.4 describes the algorithm implementing the simulation of the model of reaction-diffusion systems. Sections 6.5

Published by Woodhead Publishing Limited, 2013

and 6.6 show the results obtained by applying the new proposed algorithm respectively on chaperone-assisted protein folding and to the diffusion of bicoid gradient in Drosophila melanogaster to investigate the influence of spatial effect on these processes.

6.2 A generalization of the Fick's law

The Gibbs energy difference between regions of different concentration, i.e. the gradient of the chemical potential μ, causes diffusive transport of molecules. Let consider a solution containing N different solutes. The chemical potential μ_i of any particular chemical species i is defined as the partial derivative of the Gibbs energy G with respect to the concentration of the species i, with temperature and pressure held constant. Species are in equilibrium if their chemical potentials are equal.

$$\mu_i \equiv \frac{\partial G}{\partial c_i} = \mu_i^0 + RT \ln a_i \tag{6.1}$$

where c_i is the concentration of the species i, μ_i^0 is the standard chemical potential of the species i (i.e. the Gibbs energy of 1 mol of i at a pressure of 1 bar), $R = 8.314\ \mathrm{J} \cdot \mathrm{K}^{-1} \cdot \mathrm{mol}^{-1}$ is the ideal gas constant, and T the absolute temperature. The quantity a_i is called *chemical activity* of component i, and it is given by

$$a_i = \frac{\gamma_i c_i}{c^0} \tag{6.2}$$

where γ_i is the *activity coefficient*, c^0 being a reference concentration, which, for example, could be set equal to the initial concentration. The activity coefficients express a deviation of a solution from the ideal thermodynamic behavior and in general they may depend on the concentration of all the solutes in the system. For an ideal solution, the limit of γ_i which is recovered experimentally at high dilutions is $\gamma_i = 1$. If the concentration of species i varies from point to point

Published by Woodhead Publishing Limited, 2013

in space, then so does the chemical potential. For simplicity, here the case in which there is only a chemical potential gradient in the x direction only is taken into account. Chemical potential is the free energy per mole of substance, free energy is the negative of the work W which a system can perform, and work is connected to force F acting on the molecules by $dW = Fdx$. Therefore an inhomogeneous chemical potential is related to a virtual force per molecule of

$$F_i = -\frac{1}{N_A}\frac{d\mu_i}{dx} = -\frac{k_B T c^0}{\gamma_i c_i}\sum_j \frac{\partial a_i}{\partial c_j}\frac{\partial c_j}{\partial x} \tag{6.3}$$

where $N_A = 6.022 \times 10^{23}$ mol^{-1} is the Avogadro's number, $k_B = 1.381 \times 10^{-23}$ J $\cdot$ K^{-1} is the Boltzmann's constant, and the sum is taken over all species in the system other than the solvent. This force is balanced by the drag force experienced by the solute ($F_{drag,i}$) as it moves through the solvent. Drag forces are proportional to the speed. If the speed of the solute is not too high in such a way that the solvent does not exhibit turbulence, the drag force can be written as follows

$$F_{drag,i} = f_i v_i \tag{6.4}$$

where $f_i \propto c_i$ is the frictional coefficient, and v_i is the mean drift speed.

Moreover, if the solvent is not turbulent, the *flux*, defined as the number of moles of solute which pass through a small surface per unit time per unit area, can be approximated as in the following

$$J_i = c_i v_i \tag{6.5}$$

i.e. the number of molecules per unit volume multiplied by the linear distance travelled per unit time.

Since the virtual force on the solute is balanced by the drag force (i.e. $F_{drag,i} = -F_i$), the following expression for the mean drift velocity is obtained

$$v_i = \frac{F_i}{f_i}$$

Published by Woodhead Publishing Limited, 2013

so that Eq. (6.5) becomes

$$J_i = -\frac{k_B T}{\gamma_i f_i} \sum_j \frac{\partial a_i}{\partial c_j} \frac{\partial c_j}{\partial x} \equiv - \sum_j D_{ij} \frac{\partial c_j}{\partial x} \tag{6.6}$$

where

$$D_{ij} = \frac{k_B T c^0}{\gamma_i f_i} \frac{\partial a_i}{\partial c_j} \tag{6.7}$$

are the diffusion coefficients. The Eq. (6.7) states that, in general, the flux of one species depends on the gradients of all the others, and not only on its own gradient. However, here it is supposed that the chemical activity a_i depends only weakly on the concentrations of the other solutes, i.e. it is assumed that $D_{ij} \approx 0$ for $i \neq j$ and the Fick's laws still holds. Let D_i denote D_{ii}. It is still generally the case that D_i depends on c_i in sufficiently concentrated solutions since γ_i (and thus a_i) has a non trivial dependence on c_i [9]. It is only in one very special case, namely that of an ideal solution with $\gamma_i = 1$, where the diffusion coefficient, $D_i = k_B T / f_i$, is constant. In order to find an analytic expression of the diffusion coefficients D_i in terms of the concentration c_i, let us consider that the rate of change of concentration of the substance i due to diffusion is given by

$$\mathscr{D}_i = -\frac{\partial J_i}{\partial x} \tag{6.8}$$

Substituting Eq. (6.7) into Eq. (6.6), and then substituting the obtained expression for J_i into Eq. (6.8), gives

$$\mathscr{D}_i = -\frac{\partial}{\partial x}\left(- D_i(c_i) \frac{\partial c_i}{\partial x} \right) \tag{6.9}$$

so that

$$\begin{aligned} \mathscr{D}_i &= \left(\frac{\partial D_i(c_i)}{\partial x} \right) \frac{\partial c_i}{\partial x} + D_i(c_i) \frac{\partial^2 c_i}{\partial x^2} = \\ &= \frac{\partial D_i(c_i)}{\partial c_j} \frac{\partial c_j}{\partial x} \frac{\partial c_i}{\partial x} + D_i(c_i) \frac{\partial^2 c_i}{\partial x^2} \end{aligned} \tag{6.10}$$

Published by Woodhead Publishing Limited, 2013

Let $c_{i,k}$ denote the concentration of a substance i at coordinate x_k, and $l = x_k - x_{k-1}$ the distance between adjacent mesh points. The derivative of c_i with respect to x calculate in $x_{k-\frac{1}{2}}$ is

$$\left.\frac{\partial c_i}{\partial x}\right|_{x_{k-\frac{1}{2}}} \approx \frac{c_{i,k} - c_{i,k-1}}{l} \tag{6.11}$$

By using Eq. (6.11) into Eq. (6.6) the diffusive flux of species i midway between the mesh points $J_{i,k-\frac{1}{2}}$ is obtained:

$$J_{i,k-\frac{1}{2}} = -D_{i,k-\frac{1}{2}} \frac{c_{i,k} - c_{i,k-1}}{l} \tag{6.12}$$

where $D_{i,k-\frac{1}{2}}$ is the diffusion coefficient midway between the mesh points.

The rate of diffusion of substance i at the mesh point k is

$$\mathscr{D}_{ik} = -\frac{J_{i,k+\frac{1}{2}} - J_{i,k-\frac{1}{2}}}{l}$$

and thence

$$\mathscr{D}_{ik} = \frac{D_{i,k-\frac{1}{2}}}{l^2}(c_{i,k-1} - c_{i,k}) - \frac{D_{i,k+\frac{1}{2}}}{l^2}(c_{i,k+1} - c_{i,k}) \tag{6.13}$$

To determine completely the right-hand side of Eq. (6.13) is now necessary to find an expression for the activity coefficient γ_i and the frictional coefficient f_i, contained in the formula (6.7) for the diffusion coefficient. In fact, by substituting Eq. (6.2) into Eq. (6.7) an expression of the diffusion coefficient in terms of activity coefficients γ_i is obtained

$$D_{ii} = \frac{k_B T}{f_i}\left(1 + \frac{c_i}{\gamma_i}\frac{\partial \gamma_i}{\partial c_i}\right) \tag{6.14}$$

Let focus now on the calculation of the activity coefficients, while a way to estimate the frictional coefficients will be presented in Section 6.2.1. By using the subscript '1' to denote the solvent and '2' to denote the solute, it can be written that

Published by Woodhead Publishing Limited, 2013

$$\mu_2 = \mu_2^0 + RT\ln\left(\frac{\gamma_2 c_2}{c^0}\right) \quad (6.15)$$

where γ_2 is the activity coefficient of the solute and c_2 is the concentration of the solute. Differentiating with respect to c_2 gives

$$\frac{\partial \mu_2}{\partial c_2} = RT\left(\frac{1}{c_2} + \frac{1}{\gamma_2}\frac{\partial \gamma_2}{\partial c_2}\right) \quad (6.16)$$

The chemical potential of the solvent is related to the osmotic pressure (Π) by

$$\mu_1 = \mu_1^0 - \Pi V_1 \quad (6.17)$$

where V_1 is the partial molar volume of the solvent and μ_1^0 its standard chemical potential. Assuming V_1 to be constant and differentiating μ_1 with respect to c_2 yield

$$\frac{\partial \mu_1}{\partial c_2} = -V_1\frac{\partial \Pi}{\partial c_2} \quad (6.18)$$

Now, from the Gibbs-Duhem relation [10], the derivative of the chemical potential of the solute with respect to the solute concentration is

$$\frac{\partial \mu_2}{\partial c_2} = -\frac{M(1 - c_2\overline{v})}{V_1 c_2}\frac{\partial \mu_1}{\partial c_2} = \frac{M(1 - c_2\overline{v})}{c_2}\frac{\partial \Pi}{\partial c_2} \quad (6.19)$$

where M is molecular weight of the solute and $\overline{v}$ is the partial molar volume of the solute divided by its molecular weight. The concentration dependence of osmotic pressure is usually written as

$$\frac{\Pi}{c_2} = \frac{RT}{M}\left[1 + BMc_2 + O(c_2^2)\right] \quad (6.20)$$

where B is the second virial coefficient (see Section 6.2.2), and thence the derivative of Π with respect to the solute concentration is

Published by Woodhead Publishing Limited, 2013

$$\frac{\partial \Pi}{\partial c_2} = \frac{RT}{M} + 2RTBc_2 + O(c_2^2) \tag{6.21}$$

Introducing Eq. (6.21) into Eq. (6.19) gives

$$\frac{\partial \mu_2}{\partial c_2} = RT(1 - c_2\overline{v})\left(\frac{1}{c_2} + 2BM\right) \tag{6.22}$$

From Eq. (6.16) and Eq. (6.22) it can be obtained that

$$\frac{1}{\gamma_2}\frac{\partial \gamma_2}{\partial c_2} = \frac{1}{c_2}\Big[(1 - c_2\overline{v})(1 + 2BMc_2) - 1\Big]$$

so that

$$\int_1^{\gamma_2'} \frac{d\gamma_2}{\gamma_2} = \int_{c^0}^{c_2'} \frac{1}{c_2}\Big[(1 - c_2\overline{v})(1 + 2BMc_2) - 1\Big]dc_2$$

On the grounds that $c_2\overline{v} \ll 1$ [11], solving the integral yields

$$\gamma_2' = \exp[2BM(c_2' - c^0)] \tag{6.23}$$

The molecular weight $M_{i,k}$ of the species i in the mesh k can be expressed as the ratio between the mass $m_{i,k}$ of the species i in that mesh and the Avogadro's number $M_{i,k} = m_{i,k}/N_A$. If p_i is the mass of a molecule of species i and $c_{i,k}l$ is the number of molecules of species i in the mesh k, then the molecular weight of the solute of species i in the mesh k is given by

$$M_{i,k} = \frac{p_i\ l}{N_A} c_{i,k} \tag{6.24}$$

Substituting this expression in Eq. (6.23) gives for the activity coefficient of the solute of species i in the mesh k ($\gamma_{i,k}$), the following equation

$$\gamma_{i,k} = \exp\left(2B\,\frac{p_i\,l}{N_A}\,c_{i,k}^2\right) \tag{6.25}$$

Published by Woodhead Publishing Limited, 2013

6.2.1 Intrinsic viscosity and frictional coefficient

The diffusion coefficient depends on the ease with which the solute molecules can move. It is a measure of how readily a solute molecule can push aside its neighboring molecules of solvent. An important aspect of the theory of diffusion is how the magnitude of the frictional coefficient f_i of a solute of species i and, hence, of the diffusion coefficient D_i, depends on the properties of the solute and solvent molecules. Examination of well-established experimental data shows that diffusion coefficients tend to decrease as the molecular size of the solute increases. The reason is that a larger solute molecule has to push aside more solvent molecules during its progress and will therefore move slowly than a smaller molecule. A precise theory of the frictional coefficients for the diffusion phenomena in biological context cannot be simply derived from the elementary assumptions and model of the kinetic theory of gases and liquids. Stokes's theory considers a simple situation in which the solute molecules are so much larger than the solvent molecules that the latter can be regarded as a continuum (i.e. not having molecular character). For such a system Stokes deduced that the frictional coefficient of the solute molecules is $f_i = 6\pi r_i^{(H)}\eta$, where $r_i^{(H)}$ is the hydrodynamical radius of the molecule and η is the viscosity of the solvent. For proteins diffusing in the cytosol, estimating the frictional coefficient through the Stokes's law is hard, for several reasons. First of all, the assumption of very large spherical molecules in a continuous solvent is not a realistic approximation for proteins moving through the cytosol: proteins may be not spherical and the solvent is not a continuum. Furthermore, in the protein-protein interaction, in the cytosol, water molecules should be included explicitly, thus complicating the estimation of the hydrodynamical radius. Finally, the viscosity of the solvent η within the cellular environment cannot be approximated either as the viscosity of liquid or the viscosity of gas. In both cases, the theory predicts a strong dependence on the temperature of the system, that has not been found in

Published by Woodhead Publishing Limited, 2013

the cell system, where the most significant factor in determining the behavior of frictional coefficient is the concentration of solute molecules. To model the effects of non-ideality on the friction coefficient it is assumed that it linearly depends on the concentration of the solute as in sedimentation processes [12]. The equation (6.26) give the frictional coefficient $f_{i,k}$ of species i at mesh k. In this equation k_f is an empirical constant, whose value can be derived from the knowledge of the ratio $R = k_f/[\eta]$.

$$f_{i,k} = k_f c_{i,k} \tag{6.26}$$

Accordingly to the Mark-Houwink equation [10], $[\eta] = kM^{\alpha}$ is the intrinsic viscosity coefficient, α is related to the shape of the molecules of the solvent, and M is the molecular weight of the solute. If the molecules are spherical, the intrinsic viscosity is independent of the size of the molecules, so that $\alpha = 0$. All globular proteins, regardless of their size, have essentially the same $[\eta]$. If a protein is elongated, its molecules are more effective in increasing the viscosity and $[\eta]$ is larger. Values of 1.3 or higher are frequently obtained for molecules that exist in solution as extended chains. Long-chain molecules that are coiled in solution give intermediate values of α, frequently in the range from 0.6 to 0.75 [13]. For globular macromolecule, R has a value in the range of 1.4 - 1.7, with lower values for more asymmetric particles [14].

Although Eq. (6.26) is a simplified linear model of the frictional forces, it works quite well in many case studies and can be easily extended to treat more complex frictional effects. At the moment of writing the authors are developing a new linear model of frictional forces including the effects of macromolecular crowding on the protein diffusion. The study of macromolecular crowding effects on protein properties has a long history and currently is re-drawing the attention of the scientific community (see for example [15, 16, 17, 18, 19]) due to current interests on protein aggregation as a potential

Published by Woodhead Publishing Limited, 2013

cause for neurodegenerative diseases. The cellular environment is a crowded solution. Namely, the cellular environments are packed with other biomolecules and this crowdedness may affect the stability and aggregation rates of proteins inside cells [18, 19]. Unlike in typical biochemical experiments in which the proteins of interest are purified and diluted, the living cell is crowded with a wide variety of other proteins and macromolecules which generally occupy 20-30% of the total cell volume. This percentage is called *excluded volume*. The effects imposed by the excluded volume, that is caused by the volume excluded by the "inert" macromolecules, are called *macromolecular crowding* effects and those macromolecules are called *crowding agents*.

6.2.2 Calculated second virial coefficient

The mechanical statistical definition of the second virial coefficient is given by the following expression

$$B = -2\pi N_A \int_0^\infty r^2 \exp\Big[-\frac{u(r)}{k_B T}\Big]dr \qquad (6.27)$$

where $u(r)$, which is given in Eq. (6.28), is the interaction free energy between two molecules, r is the intermolecular center-center distance, k_B is the Boltzman constant, and T the temperature. In this work, it is assumed that $u(r)$ is the Lennard-Jones pair (12,6)-potential (Eq. 6.28), that captures the attractive nature of the Van der Waals interactions and the very short-range Born repulsion due to the overlap of the electron clouds.

$$u(r) = 4\Big[\Big(\frac{1}{r}\Big)^{12} - \Big(\frac{1}{r}\Big)^{6}\Big] \qquad (6.28)$$

By expanding the term $\exp\left(\frac{4}{k_B T}\frac{1}{r^6}\right)$ into an infinite series, the Eq. (6.27) becomes

$$B = -2\pi N_A \sum_{j=0}^{\infty} \frac{1}{j!}(T^*)^j \int_0^\infty r^{2-6j} \exp\Big[-T^*\frac{1}{r^2}\Big]dr$$

Published by Woodhead Publishing Limited, 2013

where $T^* \equiv 4/(k_B T)$ and thus

$$B = -\frac{\pi N_A}{6} \sum_{j=0}^{\infty} \frac{1}{!j} 4^j (k_B T)^{-\frac{1}{4}+\frac{1}{2}j}\, \Gamma\Big(-\frac{1}{4} + \frac{1}{2} j\Big) \qquad (6.29)$$

The estimate of B is given by truncating the infinite series of Γ functions to $j = 4$, since, results not shown here prove that taking into account the additional terms, obtained for $j > 4$, does not significantly influence the simulation results.

6.3 The optimal size of the system's subvolumes

The reaction chamber volume V is divided into subvolumes of volume Δ and side length l, on the basis of the kinetic and dynamical properties of the diffusion particles. The subvolumes has been chosen sufficiently small, so that the probability distributions of the reactants can be treated as uniform inside each subvolume. This means that the rate by which two molecules in a subvolume react does not depend on their initial locations.

Let consider diffusion as a time dependent process, in which some distribution of concentration is established at some moment, and then allowed to disperse without replenishment. The Fick's law and its analogues for the transport of other physical properties relate to the flux under the influence of a constant gradient. They therefore describe time-independent processes. They refer, for example, to the flow of particles along a constant concentration gradient which is sustained by injecting particles in one region, and drawing them off in another. From the second Fick's law, the mean distance through which particle of solute has spread after time t is

$$l_f = 2\sqrt{\frac{D\,t}{\pi}} \qquad (6.30)$$

where D is the diffusion coefficient of the particle.

Published by Woodhead Publishing Limited, 2013

Let t_e be the the mean free time with respect to non-reactive (elastic) collisions and t_r the mean free time with respect to reactive collisions. The net distance covered by the particle during its lifetime is

$$L = 2\sqrt{\frac{D\,t_r}{\pi}} = 2\sqrt{\frac{\pi l_f^2 t_r}{4t_e\pi}} = l_f\sqrt{\frac{t_r}{t_e}} \tag{6.31}$$

In order to have a homogeneous mixing inside boxes, the length l of the box side has to fulfill the following inequality.

$$l \ll L \tag{6.32}$$

If this inequality is fulfilled, the particles in each box obeys the Einstein formula for the probability of fluctuations around the steady state [42]. Note also that the rate by which two molecules in a subvolume react does not depend on their initial location if the inequality (6.32) is fulfilled.

In terms of the diffusion coefficient D, Eq. (6.31) and (6.32) can be re-written as

$$l \ll 2\sqrt{Dt_r} \tag{6.33}$$

Now, in order to estimate the upper bound of l the diffusion coefficient D and the reaction time t_r have to be determined. The diffusion coefficient differs from species to species, and, in general, depends on the local concentration of solute. Since the local concentration of solute changes in time as consequence of the occurrence of the chemical reaction events and the diffusion events themselves, this would entail a dynamical change of l through the Eq. (6.33). This could make the algorithm of simulation more complex, so that, it is profitable to fix the value of l at the initialization time at

$$l \approx \sqrt{\langle D\rangle t_r} \tag{6.34}$$

where

$$\langle D\rangle = \frac{1}{R_{diff} + R_{react}} \sum_{i=1}^{M^{(diff)}} D_i^0 \tag{6.35}$$

Published by Woodhead Publishing Limited, 2013

and D_i^0 is the diffusion coefficient of the species i-*th* at time $t = 0$, and M^{diff} is the number of species that diffuse. In the next section the model of the diffusion coefficient as function of local concentration and the waiting time of reaction t_r is explained.

6.3.1 The waiting time of reaction

Let R_i be the i-*th* reaction channel expressed as

$$R_i : l_{i1} S_{p(i,1)} + l_{i2} S_{p(i,2)} + \cdots + l_{iL_i} S_{p(i,L_i)} \xrightarrow{r_i} \ldots$$

where l_{ij} is the stoichiometric coefficient of reactant $S_{p(i,j)}$, $p(i,j)$ is the index that selects the species S that participate to R_i, L_i is the number of reactants in R_i, and r_i is the rate constant. If the fundamental hypothesis of stochastic chemical kinetics [8] holds within a box, both diffusion and reaction events waiting times are distributed according to a negative exponential distribution, so that a typical time step has size

$$t_r \approx \frac{1}{R}\left(\sum_{\nu=1}^{R} a_\nu\right)^{-1} = \frac{1}{R}\left(\sum_{i=1}^{R_{diff}} a_i^{(diff)} + \sum_{i=1}^{R_{react}} a_i^{(react)}\right) \quad (6.36)$$

where R is the number of events. It is given by $R = R_{diff} + R_{react}$, where R_{diff} is the number of diffusions and R_{react} is the number of reaction events [20]. The diffusion and reaction propensities are given by the following expressions, respectively

$$a_i^{(diff)} = r_i^{(diff)} \frac{\prod_{j=1}^{M_i^{(diff)}} (\#S_{p(i,j)})^{l_{ij}}}{\prod_{j=1}^{L_i^{(diff)}} l_{ij}!} \quad (6.37)$$

$$a_i^{(react)} = r_i^{(react)} \frac{\prod_{j=1}^{M_i^{(react)}} (\#S_{p(i,j)})^{l_{ij}}}{\prod_{j=1}^{L_i^{(react)}} l_{ij}!} \quad (6.38)$$

Published by Woodhead Publishing Limited, 2013

where $M_i^{(diff)}$ and $M_i^{(react)}$ are the number of chemical species that diffuse and the number of those the undergo to reactions, respectively. In general $M \neq M_i^{(diff)} + M_i^{(react)}$, since some species are involved both in diffusions and reactions. In Eq. (6.37), $r_i^{(diff)}$ is the kinetic rate associated to the jumps between neighboring subvolumes, whereas in Eq. (6.38), $r_i^{(react)}$ is the stochastic rate constants of the i-*th* reaction.

From Eq. (6.13), the rate coefficient of the first order reaction representing a diffusion is recognized to be as follows.

$$r_i^{(diff)} = \frac{D_{ii}}{l^2} \tag{6.39}$$

6.4 The algorithm and data structure

In this section the new stochastic simulation algorithm developed by the authors is illustrated. It incorporates into a Gillespie-like approach the spatial effects of diffusive phenomena accordingly to the diffusion model presented in the previous sections.

For the reader's convenience, a brief description of the Gillespie Direct and First Reaction methods is here reported. Let suppose that in the system there are R reactions and M chemical species. at any instant of time the system is decribed by the state vector $\vec{X}(t) = \{X_1(t), \ldots, X_M(t)\}$ Gillespie's algorithm asks two questions:

1. Which reaction occurs next?
2. When does it occur?

Both of these questions must be answered probabilistically by specifying the probability density $P(\mu, \tau)$ that the next reaction is μ and it occurs at time τ. It can be shown [8] that

$$P(\mu, \tau) = a_\mu \exp\Big(-\tau \sum_{j=1}^{R} a_j\Big) d\tau \tag{6.40}$$

Published by Woodhead Publishing Limited, 2013

This equation leads directly to the answers of the two afore mentioned questions. First, what is the probability distribution for reactions? Integrating $P(\mu,\tau))$ over all τ from 0 to ∞ results in

$$\Pr(\text{Reaction} = \mu) = \frac{a_\mu}{\sum_{j=1}^{R} a_j a_j} \tag{6.41}$$

where a_j the propensity of reaction j as in Eqs. (6.37) and (6.38).

Second, what is the probability distribution for times? Summing $P(\mu,\tau)$ over all τ results in

$$P(\tau)d\tau = \Big(\sum_{j=1}^{R} a_j a_j\Big) \exp\Big(-\tau \sum_{j=1}^{R} a_j\Big) d\tau \tag{6.42}$$

These two distributions lead to Gillespie's direct algorithm:

1. Set initial numbers of molecules in $\vec{X}(t)$, set $t \leftarrow 0$, and the absolute simulation time T.
2. Calculate the propensity function, a_μ, for all j, $j = 1,\ldots,R$.
3. Choose j according to the distribution in Eq. (6.41).
4. Choose τ according to an exponential with parameter $\sum_{j=1}^{R} a_j$ (as in Eq. (6.42)).
5. Change the number of molecules to reflect execution of reaction μ. Set $t \leftarrow t+\tau$.
6. Go to Step 2 and repeat the procedure until $t < T$.

The algorithm is direct in the sense that it generates μ and τ directly. Gillespie also developed the First Reaction Method (FRM) which generates a putative time τ_j for each reaction to occur - a time the reaction would occur if no other reaction occurred first - then lets μ be the reaction whose putative time is first, and lets τ be the putative time τ_j. Formally, the algorithm for the First Reaction Method is as follows:

1. Set initial numbers of molecules in $\vec{X}(t)$, set $t \leftarrow 0$, and the absolute simulation time T.

Published by Woodhead Publishing Limited, 2013

2. Calculate the propensity function, a_{μ}, for all j, $j = 1, \ldots, R$..
3. For each μ, generate a putative time, τ_j, according to an exponential distribution with parameter a_j.
4. Let μ be the reaction whose putative time, τ_j, is least. 5. Let τ be τ_j.
5. Change the number of molecules to reflect execution of reaction μ. Set $t \leftarrow t + \tau$.
6. Go to Step 2 and repeat the procedure until $t \leq T$.

At first glance, these two algorithms may seem very different, but they are provably equivalent [8] that is, the probability distributions used to choose μ and τ are the same. With regard to the complexity of the procedure, this algorithm uses R random numbers per iteration (where R is the number of reactions), takes time proportional to r to update the a's, and takes time proportional to R to identify the smallest τ_j.

The design of the algorithm is inspired to the one proposed by Elf et al. [21] in the so-called *Next sub-volume method*. This method selects the next reaction and the time at which it will occur by using the Gillespie First Reaction method [8]. Each cell and the corresponding reaction time and reaction type is stored in a global priority queue that is sorted with increasing writing reaction time. From this queue at each time step, the fastest reaction (i.e. the reaction with the smallest waiting time) is picked and executed. Once the reaction has been executed the state of the cell, as well as the state of the neighboring cells that eventually have been affected by the occurrence of this reaction are updated. This approach is efficient as it does not update the state of all the cells, but only the one of the cells in which the occurrence of a reaction has produced changes in the inner amount of molecules. However, the method is centralised and sequential and does not scale to very large systems. Moreover, it cannot be easily adapted to turn parallel or distributed computing procedures to profit. Since the number of reactions involved in the system could be of the order of millions, the

Published by Woodhead Publishing Limited, 2013

property of scalability is required to make large simulations feasible. The algorithm proposed by the authors overcomes the scalability's limitations of the Next subvolume method by renouncing to the use of a global priority queue.

For each cell a set of *dependency relations* with neighbor cells is drawn; in a cell an event (reaction of diffusion) can be executed only if it is quicker than the diffusion events of the neighbor cells, since the diffusion events in and out of the cell could change the reactant concentrations, and, consequently the reaction propensities and the waiting times of the events in the neighbor cells. The algorithm has still the same average computational complexity of Elf's methods. Nevertheless, by removing the global priority queue and introducing a dependency relations graph, the algorithm gains the scalability property. The new algorithm consists of the following steps.

1. Set initial numbers of molecules in $\vec{X}(t)$, set $t \leftarrow 0$, and the absolute simulation time T. Divide the reaction chamber volume V into boxes of size l as in Eq. (6.34).
2. In each cell, calculate the time and the type of the next event with the FRM are and store them in a private priority queue, ordered with increasing waiting time.
3. Each cell "communicates" with its neighbors, in a hierarchical way on the basis of the dependency relations, to decide which one holds the event with the smallest waiting time, say τ_S. that will be executed next. Execute the event and update the state of the cell and the one of the neighbor cells, in the case in which the event is a diffusion, are updated.
4. Update the time variable: $t \leftarrow t + \tau_S$.
5. Go to Step 2 and repeat the steps until $t \leq T$.

6.4.1 *The prototype* Redi

Redi is a simulator of the stochastic kinetics of a reaction-diffusion system, that implements the model described in

Published by Woodhead Publishing Limited, 2013

the previous sections. *Redi* is a command line tool, that is invoked by typing the command

```
redi.exe [params]
```

Models for the reaction-diffusion system are given in a simple input language in a text input file Named entities that appear in the system are declared in the first section of the input file. In addition, it is possible to specify a base rate for entities that undergo diffusion:

```
var Y : rate 10e-6;
var Yp : rate 10e-6;
var Z2 : weight 24.0;
var M;
var MYp;
```

In this example, Y and Yp undergo diffusion events. Their basal diffusion rate is fixed, and it is specified after the *rate* keyword. Z2 can diffuse as well, but its basal diffusion rate is not fixed, but it is computed by Eq. (6.39 .

The algorithm takes as input the molecular weight of the protein to compute its chemical activity (see Eqs. (6.24) and (6.25)); the weight is expressed in kDA after the *weight* keyword. M and MYp ado not undergo diffusion event, so they have no rate or weight associated to them.

Reactions can be specified in a very intuitive format:

```
Yp + Z2 -> Y + Z2  [1.90E-03];
M + Yp <- -> MYp  [5.93E-03, 20];
```

In the first line, a bimolecular reaction is specified. The rate constant is given in scientific notation between square brackets. The second reaction is a shorthand to specify reversible reactions. In this case, it is necessary to specify a pair of rates between the square brackets for the forward and for the backward reaction, respectively.

In the input file the reaction list has to be followed by the specification of the number of molecules and spatial location of each species on the grid. For example:

Published by Woodhead Publishing Limited, 2013

```
Y [0, 0, 0, 500];
Y [0, 1, 0, 500];
Y [1, 0, 0, 500];
Y [1, 1, 0, 500];
```

After the species name, the first three number within the square brackets indicate the x, y, and z coordinates, and the fourth number is the amount of and an integer number. Although this input format may be verbose, it allows for great control and precision. We are currently studying better ways of inputting both data (location and amount of particles) and geometry (that is, for now, fixed and with a regular polyhedric shape).

The Redi package includes a number of modules providing different kinds of output, e. g. text files, writing data files to be used with professional visualization packages, visualizing in a simple 3D window concentrations in space, etc.) as well as a simple interface to write custom output modules (see an example of output visualization in Fig. 6.1).

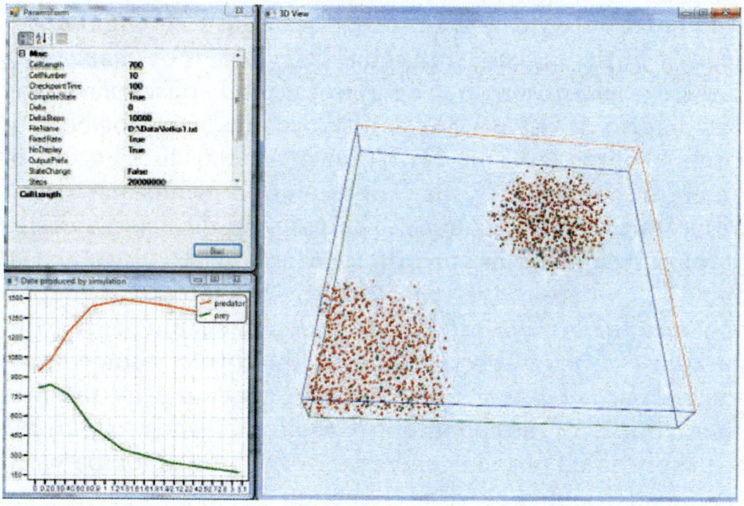

Figure 6.1 Redi's screenshots of output's visualization (http://www.cosbi.eu).

Published by Woodhead Publishing Limited, 2013

6.5 Case study 1: chaperone-assisted folding

Although a protein chain can fold in its correct conformation without outside help, protein folding in a living cell is often assisted by special proteins called *molecular chaperones*. These proteins bind to partly folded polypeptide chains and help them progress along the most energetically favorable folding pathways. Chaperones are vital in the crowded conditions of the cytoplasm, since they prevent the temporarily exposed hydrophobic regions in newly synthesized protein chains from associating with each other to form proteins aggregates.

In the healthy cells, if a protein does not assume the correct 3D shape, or a cellular stress induces a right-folded protein to assume a wrong folding, the chaperones re-shape it correctly. In the case in which the protein is not correctly refolded, and the ubitiquitin-proteasome system, designed to its digestion, does not correctly work, as in many neurodegenerative disorders, the faulty proteins accumulate and damage the cell.

Protein folding, chaperone binding, and misfolded protein accumulation - all of these processes take place inhomogeneously in the space. The spatial distribution of chaperones in the cytoplams may not be uniform, and consequently the distribution of correct and faulty proteins may not be uniform. In turn, the time evolution of spatial distribution of chaperones may affect the time evolution of the spatial distribution of faulty proteins. The reaction-diffusion systems of the case study consisting of the four reactions is shown in Table 6.1, where *chaperone* represents the molecular chaperone, *nascent_protein* presents the protein chain release from the ribosome, *right_protein* denotes the correctly folded protein, *misfolded_1* is a faulty protein generated by the first interaction with the chaperone (Reaction 2), and *misfolded_2* is the misfolded protein generated by the interaction between *misfolded_1* and *chaperone* (Reaction 4).

Published by Woodhead Publishing Limited, 2013

Accordingly to the measurements reported in [17] the following values of diffusion coefficients have been used to simulate the system : $D^0_{protein} = D^0_{right\_protein} = D^0_{misfolded\_1} = D^0_{misfolded\_2} = 10\ \mu m^2 \text{sec}^{-1}$, and $D^0_{chaperone} = 1\ \mu m^2 \text{sec}^{-1}$. As simulation space, a square grid $9 \times 9 \mu m^2$, thus consisting of 81 cells (each cell has size $l = 1$ nm) is considered. A 2D diffusion model is simulated and a spatially homogeneous distribution of *nascent_protein* and an initial null concentration of *right_protein* in every cell are assumed. The density (expressed in number of molecules per μm^3) and the spatial distribution of *chaperone*, *misfolded_1*, and *misfolded_2* in the first instants of simulation are shown in the first plots (at time $t \approx 10^{-5}$ sec), in Fig. 6.2 (A), Fig. 6.2 (B), Fig. 6.2 (C), and Fig. 6.2 (D) respectively.

At time $t = 1.1054 \times 10^{-5}$ sec - immediately after the begging of the simulation, the correctly folded proteins are located in the regions where the concentrations of chaperones is high (see Fig. 6.2 (A) and (B)). The misfolded proteins produced by Reaction 2 and Reaction 4 in the first instants of the simulation are close to the chaperones (Fig. 6.2 (C) and (B), Fig. 8.2 (C) and (D), and Fig. 6.4 (C) and (D)). at time $t = 0.000483$ sec, the chaperones and the correctly folded proteins start to leave their initial positions to migrate toward the central area of the system (Fig. 6.5 (A)). The concentration of misfolded proteins (of type 1 and 2) increses and their distributions spread in the space (Fig. 6.5 (C) and (D)). From $t = 0.003211$ sec to $t = 0.005080$ sec, the concentration of the chaperones is non-null in all the simulation space with a peak in the right upper corner (Fig. 6.8 (A) and Fig. 6.9 (A)). The distribution of right folded proteins is similar (Fig. 6.8 (B) and 6.9 9B). The concentration of misofled proteins produced by Reaction 2 is almost null in all the space except along the borders (Fig. 6.8 (C) and Fig. 6.9 (C)). Nevertheless, the concentration of misfolded proteins produced by Reaction 4 is significantly different from zero and fairly homogeneous (Fig 6.8 (D) and Fig. 6.9 (D)). At $t = 0.007749$ sec, the chaperones shift to the upper order of

Published by Woodhead Publishing Limited, 2013

Table 6.1 Chaperone-assisted protein folding. Reaction 1 describes the folding of the nascent protein into a correctly working protein (right_protein). Reaction 2 describes the uncorrect folding of the nascent protein into a misfolded protein (misfolded_1). Reaction 3 describes the interaction between the chaperone and the misfolded protein, that, consequently, is transformed into a correctly folded protein. Finally, reaction 4 describes the interaction between the chaperone and the misfolded proteins, that is not converted into a correctly working protein.

Reaction 1: $nascent\_protein + chaperone \xrightarrow{100\mu M^{-1}sec^{-1}} chaperone + right\_protein$
Reaction 2: $nascent\_protein + chaperone \xrightarrow{100\mu M^{-1}sec^{-1}} chaperone + misfolded\_1$
Reaction 3: $misfolded\_1 + chaperone \xrightarrow{100\mu M^{-1}sec^{-1}} chaperone + rprotein$
Reaction 4: $misfolded\_1 + chaperone \xrightarrow{100\mu M^{-1}sec^{-1}} chaperone + misfolded\_2$

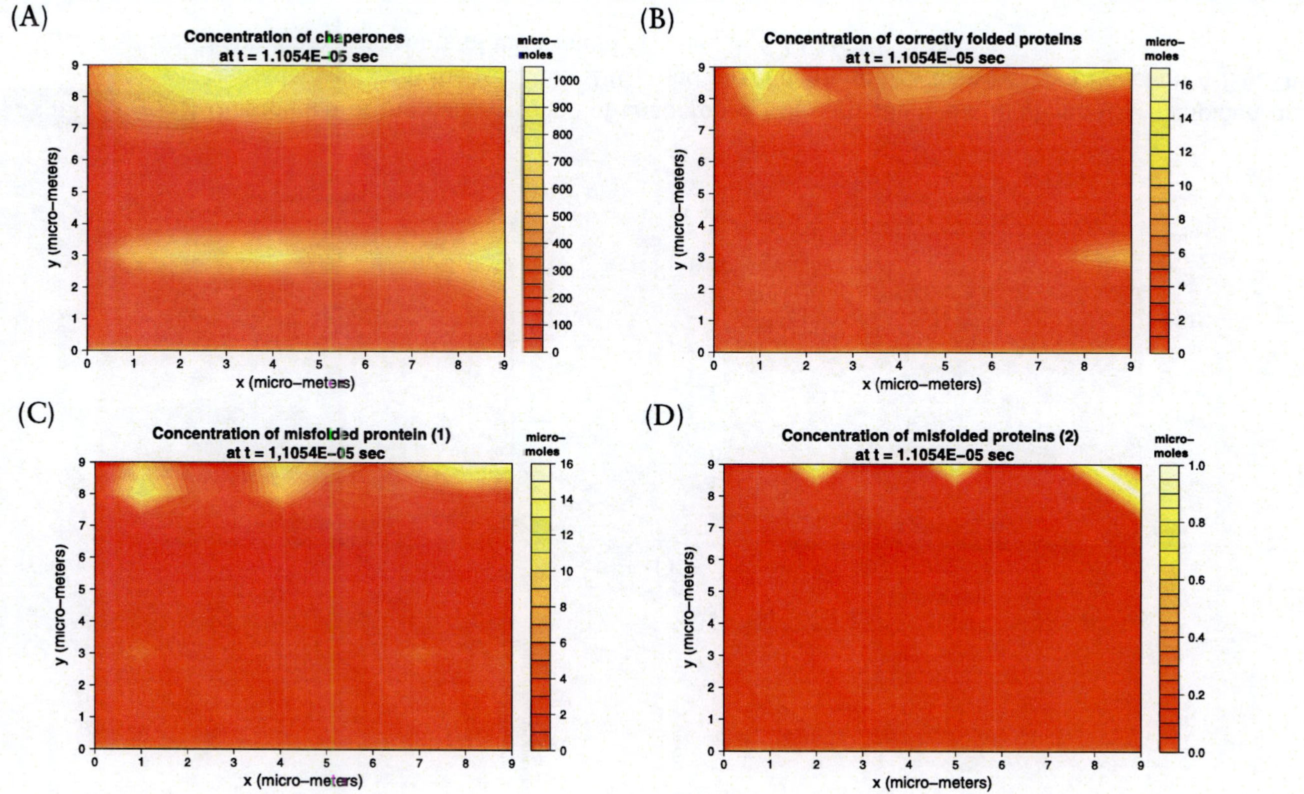

Figure 6.2 Distribution of the concentration of chaperones (A), correctly folded proteins (B), misfolded proteins deriving from the Reaction 2 (C), and misfolded proteins deriving from Reaction 4 (D). The figures are snapshots of the system at time $t = 1.1054 \times 10^{-5}$ sec.

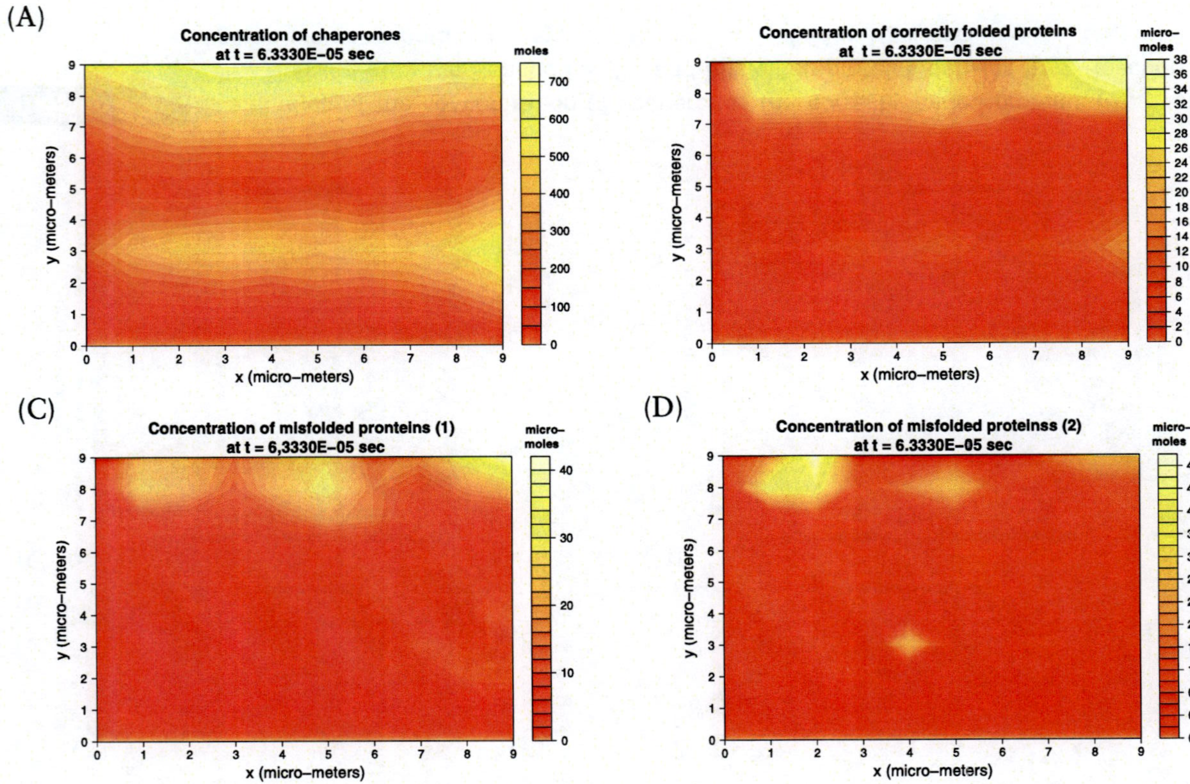

Figure 6.3 Distribution of the concentration of chaperones (A), correctly folded proteins (B), misfolded proteins deriving from the Reaction 2 (C), and misfolded proteins deriving from Reaction 4 (D). The figures are snapshots of the system at time $t = 6.333 \times 10^{-05}$ sec.

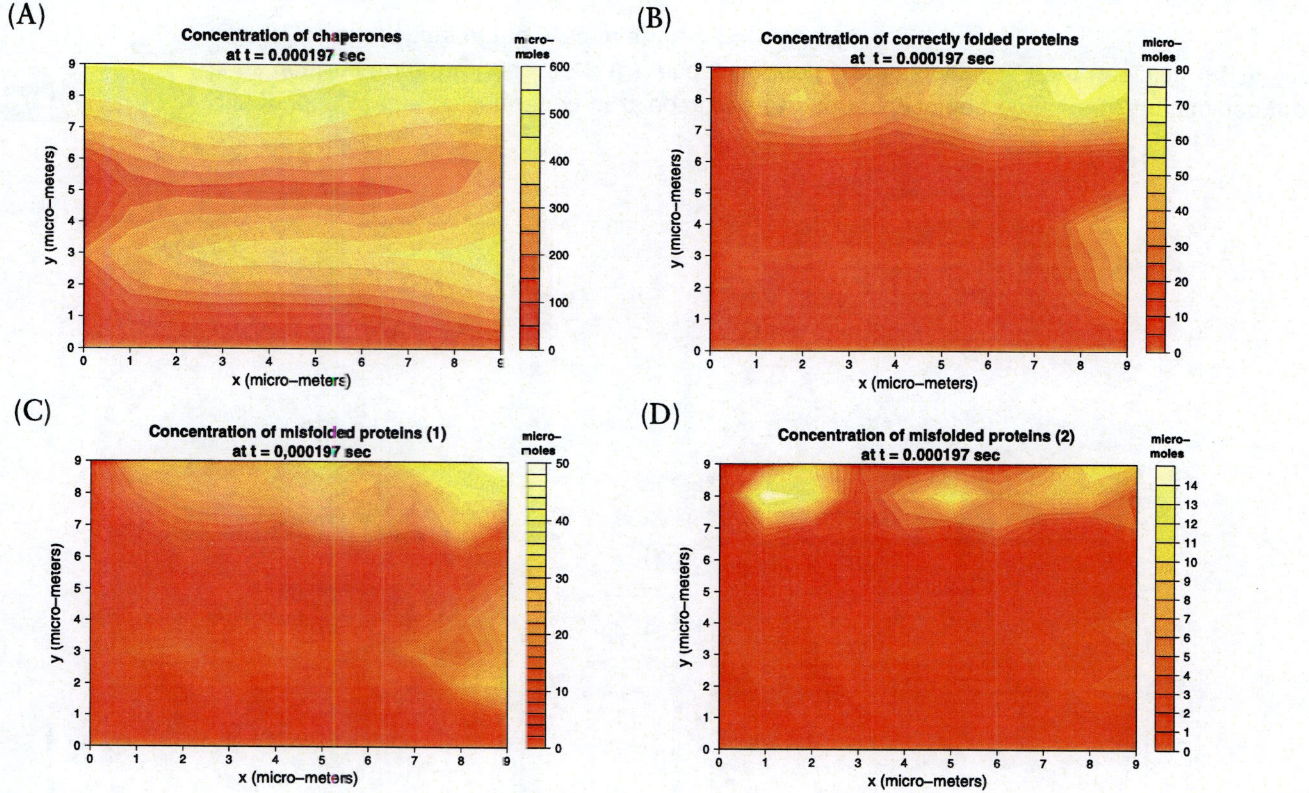

Figure 6.4 Distribution of the concentration of chaperones (A), correctly folded proteins (B), misfolded proteins deriving from the Reaction 2 (C), and misfolded proteins deriving from Reaction 4 (D). The figures are snapshots of the system at time $t = 0.000197$ sec.

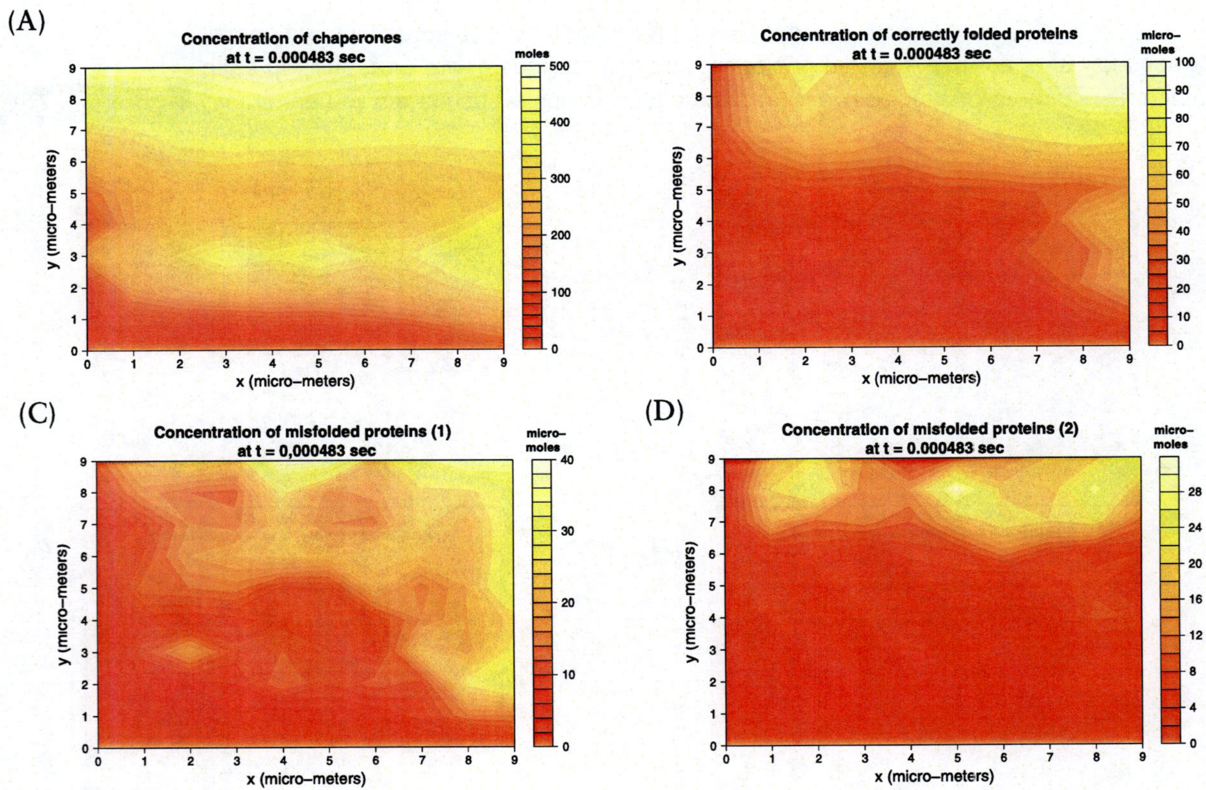

Figure 6.5 Distribution of the concentration of chaperones (A), correctly folded proteins (B), misfolded proteins deriving from the Reaction 2 (C), and misfolded proteins deriving from Reaction 4 (D). The figures are snapshots of the system at time $t = 0.000483$ sec.

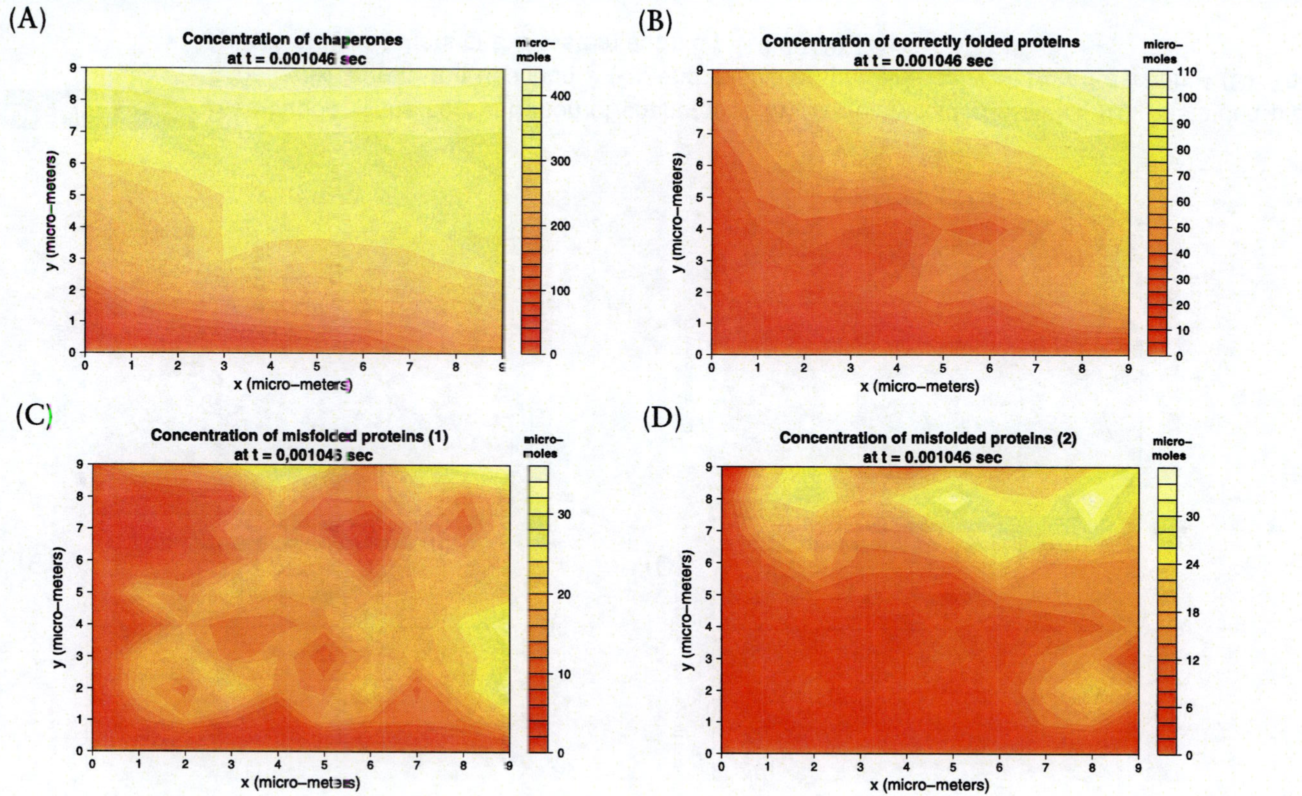

Figure 6.6 Distribution of the concentration of chaperones (A), correctly folded proteins (B), misfolded proteins deriving from the Reaction 2 (C), and misfolded proteins deriving from Reaction 4 (D). The figures are snapshots of the system at time $t = 0.001046$ sec.

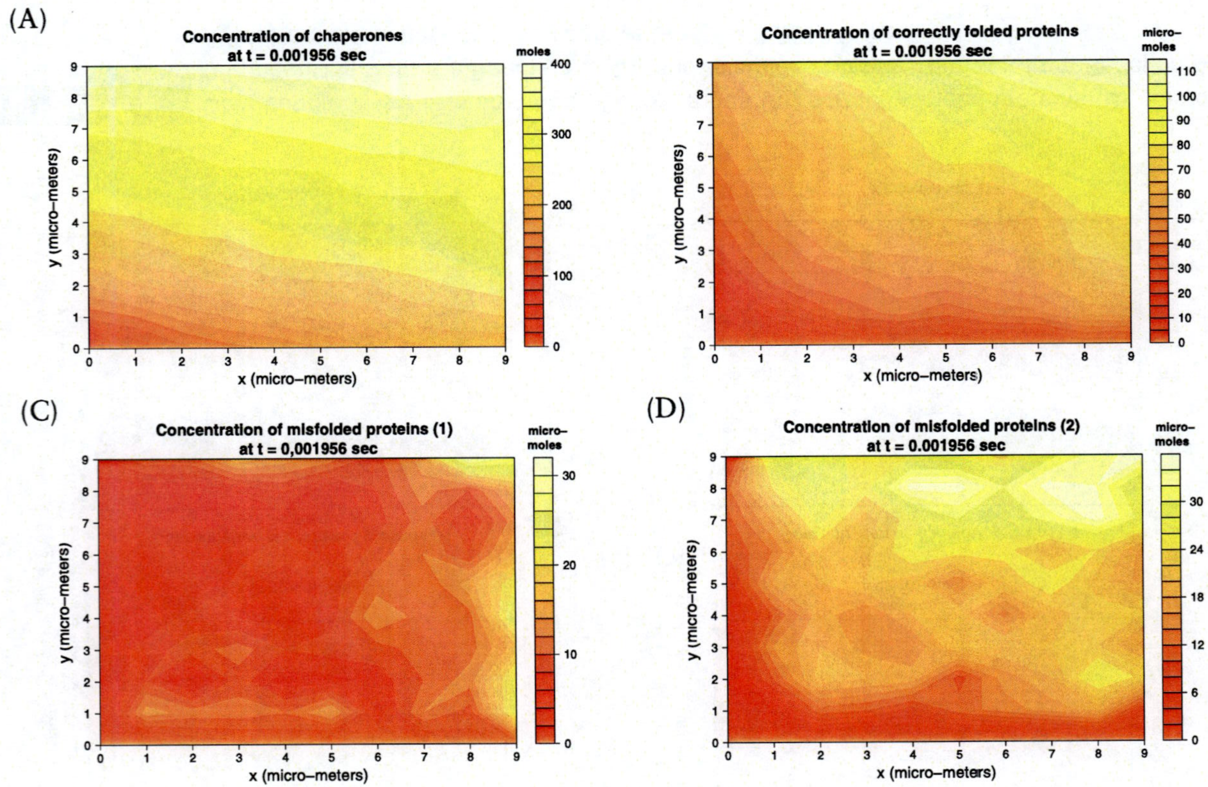

Figure 6.7 Distribution of the concentration of chaperones (A), correctly folded proteins (B), misfolded proteins deriving from the Reaction 2 (C), and misfolded proteins deriving from Reaction 4 (D). The figures are snapshots of the system at time $t = 0.001956$ sec.

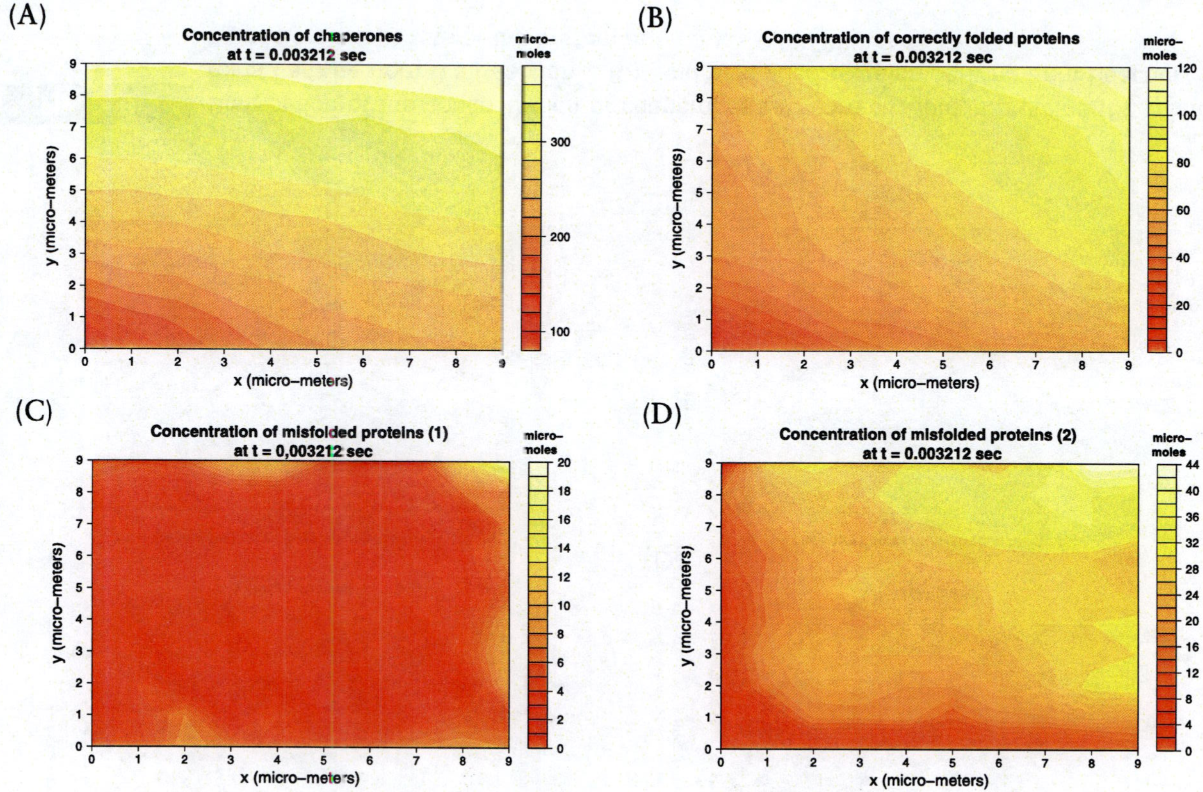

Figure 6.8 Distribution of the concentration of chaperones (A), correctly folded proteins (B), misfolded proteins deriving from the Reaction 2 (C), and misfolded proteins deriving from Reaction 4 (D). The figures are snapshots of the system at time $t = 0.003212$ sec.

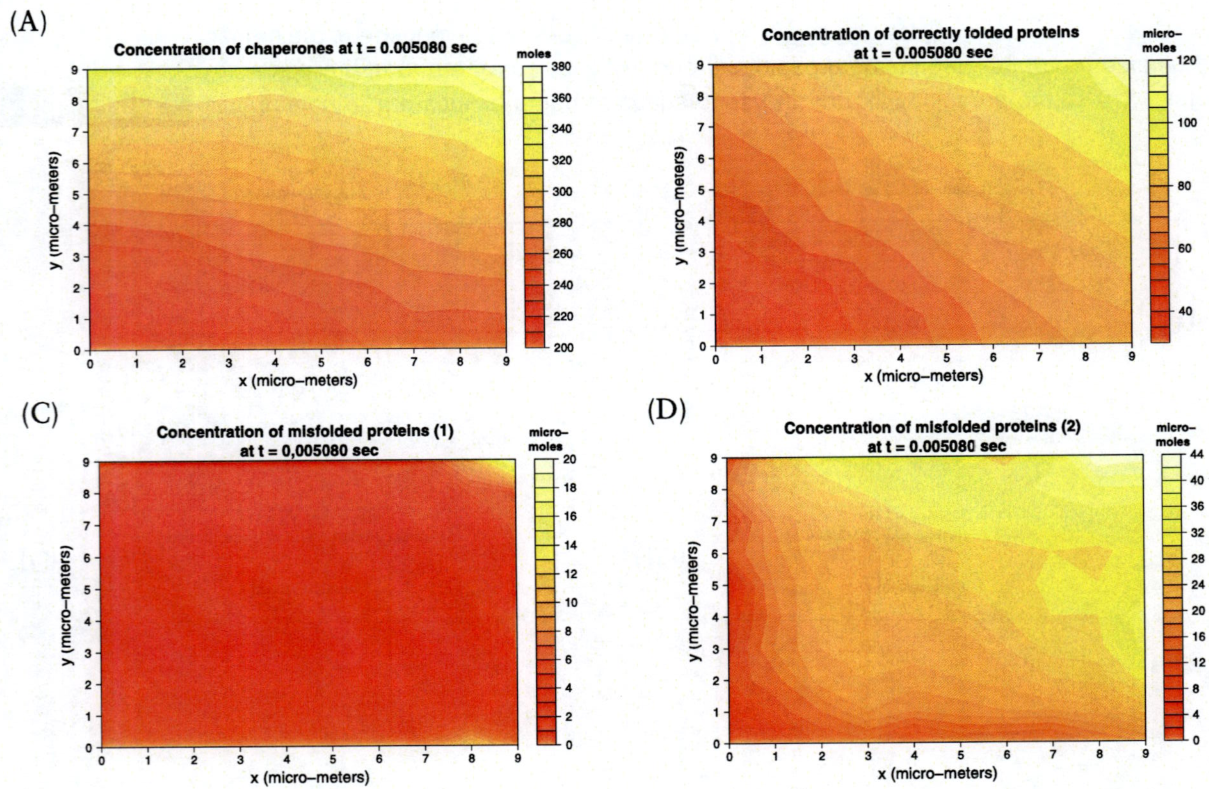

Figure 6.9 Distribution of the concentration of chaperones (A), correctly folded proteins (B), misfolded proteins deriving from the Reaction 2 (C), and misfolded proteins deriving from Reaction 4 (D). The figures are snapshots of the system at time $t = 0.005080$ sec.

the simulation space (Fig. 6.10 (A)); the correctly folded proteins concentration has a maximum in the right upper corner (Fig. 6.10 (B)); the concentration of misfolded proteins by Reaction 2 is almost everywhere except that on the borders, whereas the distribution of misfolded produced in Reaction 4 is almost everywhere null, but it has a peak in the right upper corner (Fig. 6.10 (D)). Finally, at time $t = 0.014273$ sec is non-null over all the space. It increases linearly from the upper border (Fig. 6.11 (A)). The concentration of correctly folded proteins increases from the lower left corner to the right upper corner (Fig. 6.11 (B)). Unlike the distribution of misfolded proteins deriving from Reaction 2, the distribution of misfolded proteins deriving from reaction 4 is different from zero everywhere (Fig. 6.11 (C) and increases from the left lower corner to the right upper corner (Fig. 6.11 (D)).

6.5.1 Spatial correlation between chaperones and proteins

The spatial correlation between the proteins and chaperones has been monitored in terms of the quantity $C_{p,c}$, which is defined by

$$C_{p,c} = \frac{\langle(\Phi_p - \langle\Phi_p\rangle)(\Phi_c - \langle\Phi_c\rangle)}{\langle\Phi_p\rangle\langle\Phi_c\rangle} \tag{6.43}$$

where $\Phi_p = \Phi_p(x, y, z)$ and $\Phi_c = \Phi_c(x, y, z)$ are function of spatial coordinates and denote the concentrations of nascent proteins and chaperones, respectively. The symbol $\langle\cdot\rangle$ denotes the mean value of "·". The subscript p ranges over the following species *right_protein*, *misfolded_1*, and *misfolded_2*, whereas the subscript c denotes *chaperone*. The positive value of $C_{p,c}$ means that the species p and c on average tend to be close each other in space.

Published by Woodhead Publishing Limited, 2013

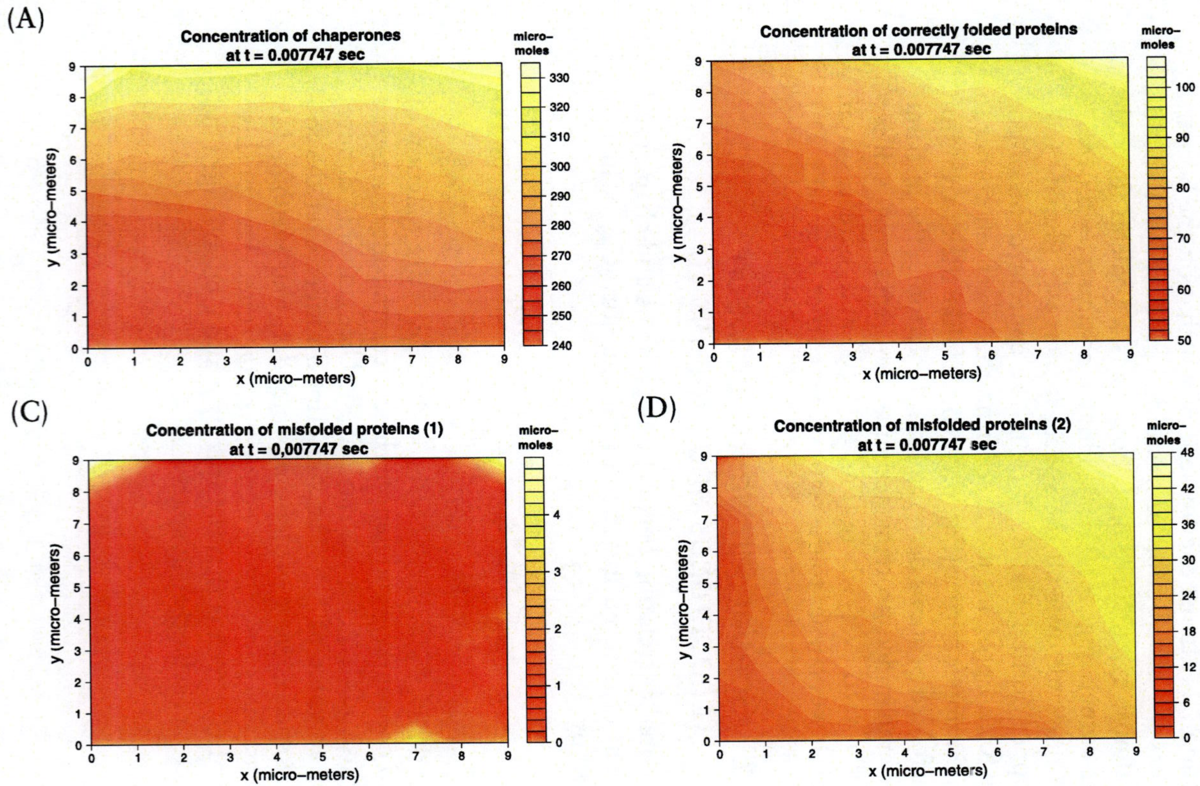

Figure 6.10 Distribution of the concentration of chaperones (A), correctly folded proteins (B), misfolded proteins deriving from the Reaction 2 (C), and misfolded proteins deriving from Reaction 4 (D). The figures are snapshots of the system at time $t = 0.007747$ sec.

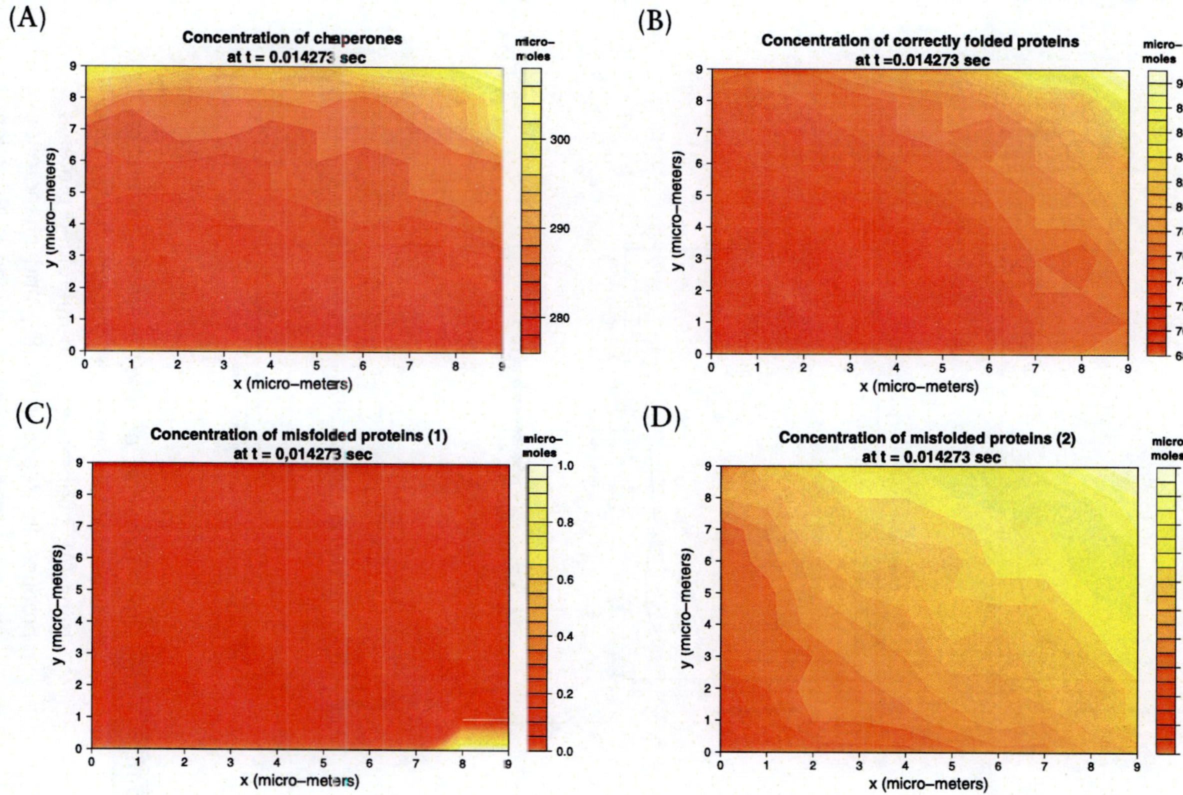

Figure 6.11 Distribution of the concentration of chaperones (A), correctly folded proteins (B), misfolded proteins deriving from the Reaction 2 (C), and misfolded proteins deriving from Reaction 4 (D). The figures are snapshots of the system at time $t = 0.014273$ sec.

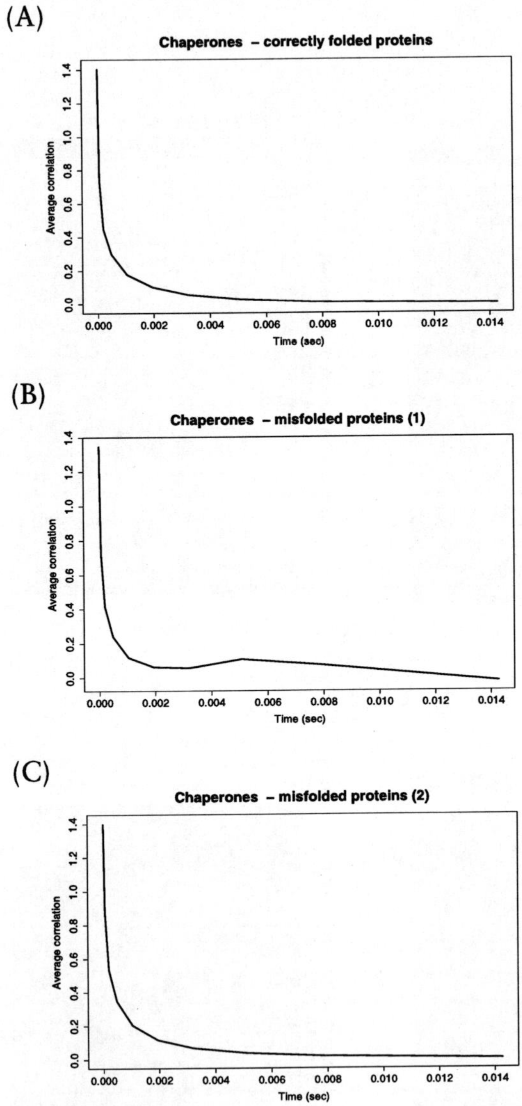

Figure 6.12 Time behavior of the average correlation between chaperones and correctly folded proteins (A), chaperones and misfolded proteins produce in Reaction 2 (B), and chaperones and misfolded proteins produced in Reaction 4 (C).

Published by Woodhead Publishing Limited, 2013

The average correlation between chaperones and correctly folded proteins, chaperones abd misfolded proteins derived from Reaction 2 and chaperones and misfolded proteins derived from reaction 4 decrease with increasing time (Fig. 6.12 (A), (B), and (C), respectively). The distribution of the intensity of these correlations in the simulation space is shown in Figs. 6.14 - Figs. 6.23. These results show that, at the beginning of the simulation, both the correctly folded and the misfolded proteins are likely to appear near the chaperones, that is they are released by the chaperones, and then they diffuse away from them, as it was obtained also in [17]. The figures 6.13 (A), (B), and (C) show that the total concentrations of correctly proteins, misfolded proteins (1) and (2), respectively, have a time behavior symmetric to the time behavior of their average correlations with the concentration of chaperones. In fact, the maximum of the correlation between chaperones and both correctly and misfolded proteins correspond to the onset of increase in protein concentration. The figures 6.12 (B) and 6.13 (B) show that the concentration of misfolded proteins produced in reaction 2 reaches the maximum when their correlation with the chaperones has a minimum. This behavior is due to the fact that the misfolded proteins of type 1 are released from the chaperones and then they quickly diffuse away from them. The chaperones also diffuse away from their initial positions but less quickly, so that they reach later the misfolded proteins of type 1. Once the chaperones reached the misfolded proteins, the occurrence of Reaction 4 causes the decreasing of the concentration of misfolded protein of type 1.*

6.5.2 *Validity of the model*

In this model describing the effects of an irregular distribution of chaperones on the kinetics of the chaperone-assisted protein folding, the internal structure and mechanism of the chaperone, as well as the size and the internal dynamics of the protein folding are not treated. No external source of energy is exerted upon the system in the present simulations: the diffusive transport is caused by spatial differences

Published by Woodhead Publishing Limited, 2013

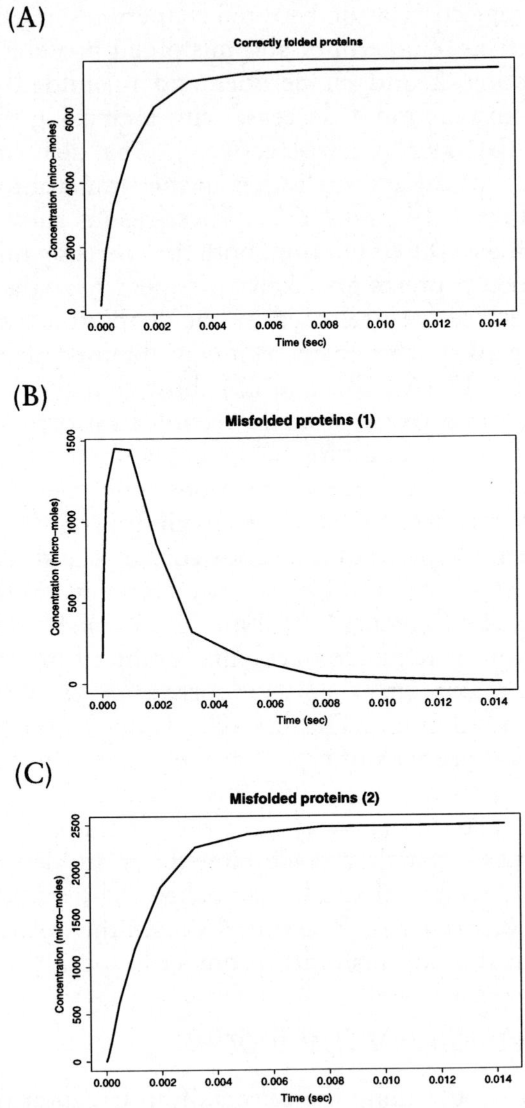

Figure 6.13 Time behavior of the total concentration of correctly folded proteins (A), misfolded proteins produce in Reaction 2 (B), and misfolded proteins produced in Reaction 4 (C).

Published by Woodhead Publishing Limited, 2013

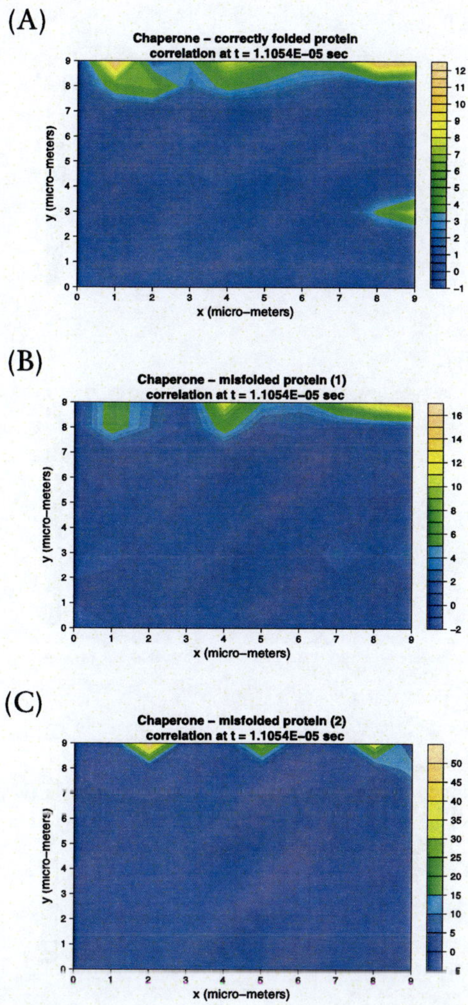

Figure 6.14 Matrices of correlation (Eq. (6.43)) between chaperones and correctly folded proteins concentrations (A), chaperones and misfolded proteins concentrations deriving from the Reaction 2 (B), and chaperones and misfolded proteins concentration deriving from Reaction 4 (C). The figures are snapshots of the system at time $t = 1.1054 \times 10^{-05}$ sec.

Published by Woodhead Publishing Limited, 2013

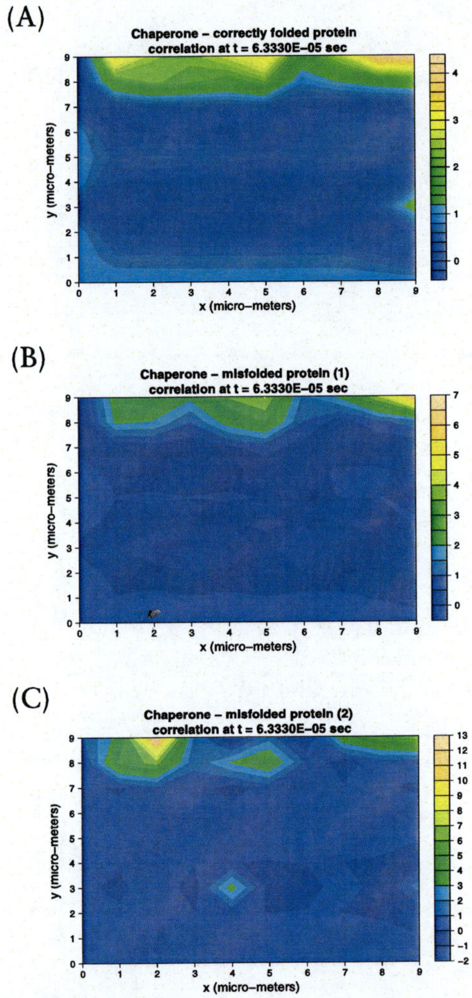

Figure 6.15 Matrices of correlation (Eq. (6.43)) between chaperones and correctly folded proteins concentrations (A), chaperones and misfolded proteins concentrations deriving from the Reaction 2 (B), and chaperones and misfolded proteins concentration deriving from Reaction 4 (C). The figures are snapshots of the system at time $t = 6.333 \times 10^{-05}$ sec.

Published by Woodhead Publishing Limited, 2013

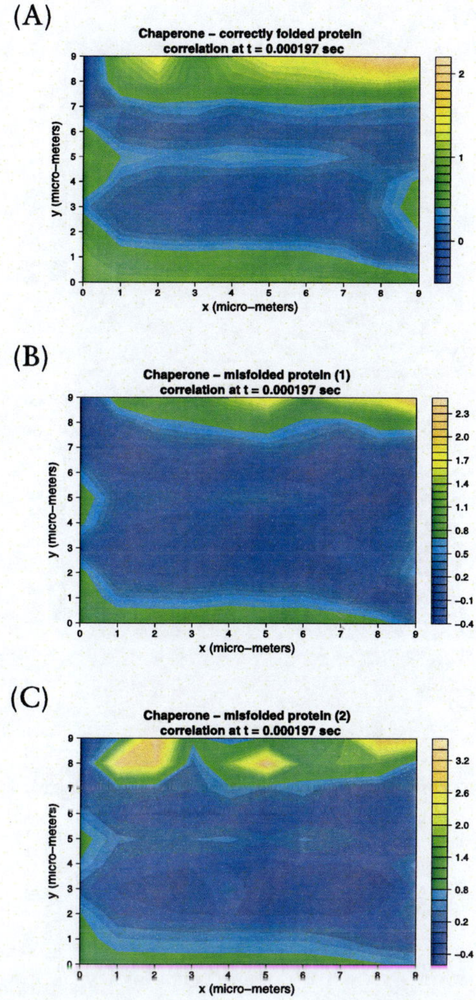

Figure 6.16 Matrices of correlation (Eq. (6.43)) between chaperones and correctly folded proteins concentrations (A), chaperones and misfolded proteins concentrations deriving from the Reaction 2 (B), and chaperones and misfolded proteins concentration deriving from Reaction 4 (C). The figures are snapshots of the system at time $t = 0.000197$ sec.

Published by Woodhead Publishing Limited, 2013

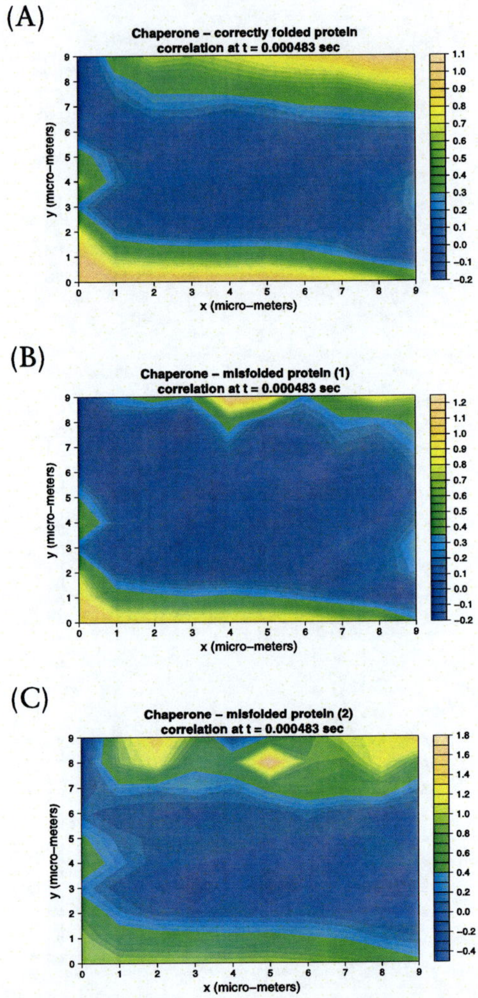

Figure 6.17 Matrices of correlation (Eq. (6.43)) between chaperones and correctly folded proteins concentrations (A), chaperones and misfolded proteins concentrations deriving from the Reaction 2 (B), and chaperones and misfolded proteins concentration deriving from Reaction 4 (C). The figures are snapshots of the system at time $t = 0.000483 \times 10^{-05}$ sec.

Published by Woodhead Publishing Limited, 2013

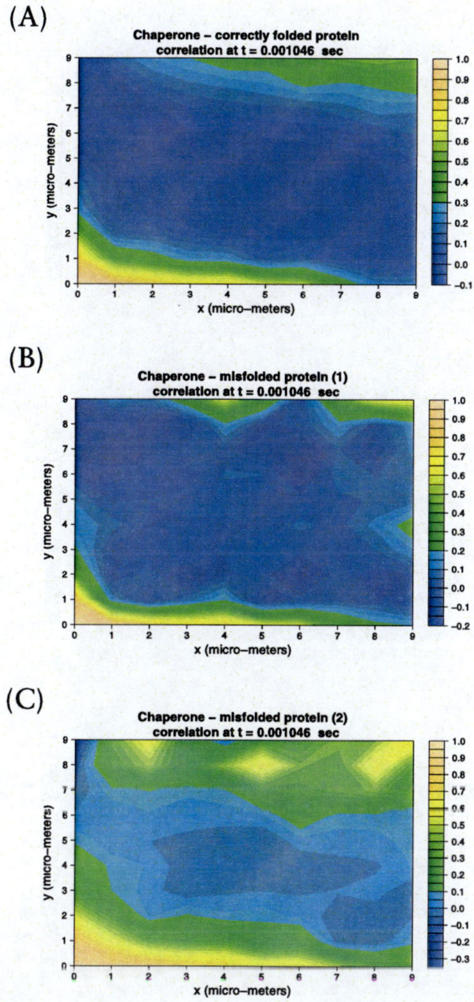

Figure 6.18 Matrices of correlation (Eq. (6.43)) between chaperones and correctly folded proteins concentrations (A), chaperones and misfolded proteins concentrations deriving from the Reaction 2 (B), and chaperones and misfolded proteins concentration deriving from Reaction 4 (C). The figures are snapshots of the system at time $t = 0.001046$ sec.

Published by Woodhead Publishing Limited, 2013

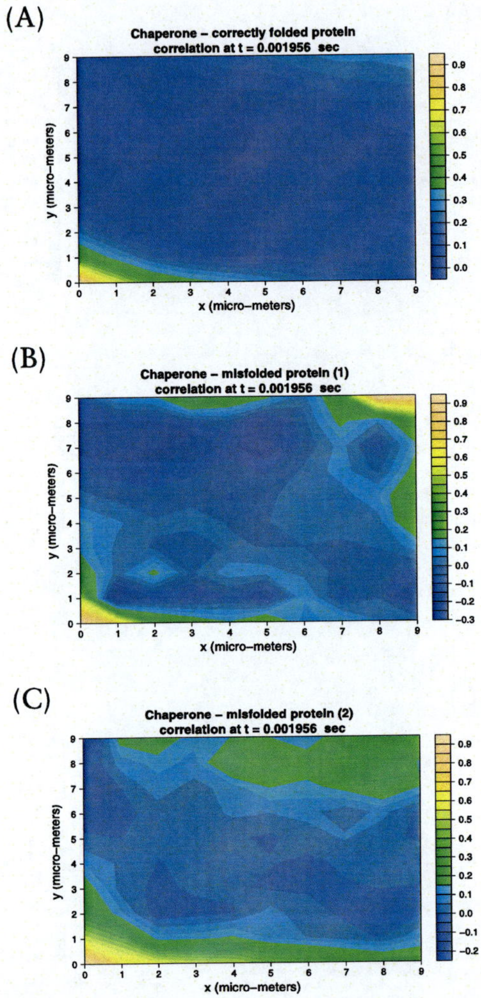

Figure 6.19 Matrices of correlation (Eq. (6.43)) between chaperones and correctly folded proteins concentrations (A), chaperones and misfolded proteins concentrations deriving from the Reaction 2 (B), and chaperones and misfolded proteins concentration deriving from Reaction 4 (C). The figures are snapshots of the system at time $t = 0.001956$ sec.

Published by Woodhead Publishing Limited, 2013

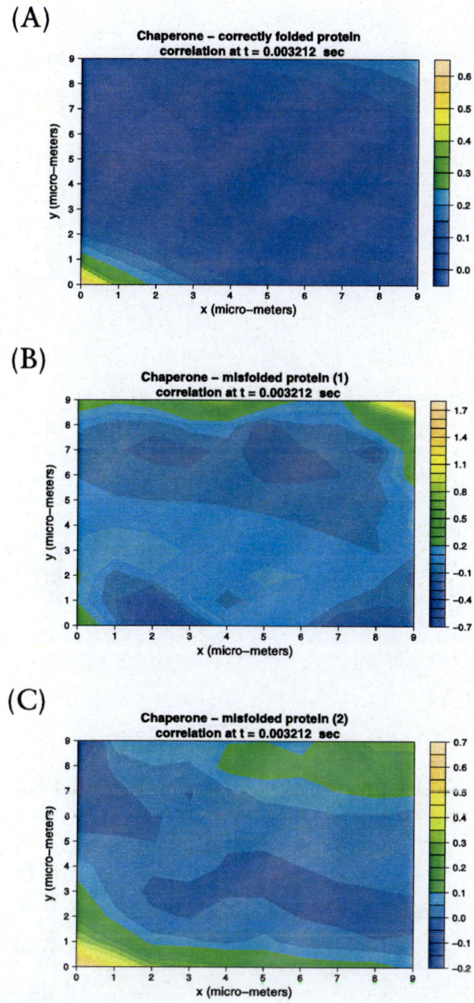

Figure 6.20 Matrices of correlation (Eq. (6.43)) between chaperones and correctly folded proteins concentrations (A), chaperones and misfolded proteins concentrations deriving from the Reaction 2 (B), and chaperones and misfolded proteins concentration deriving from Reaction 4 (C). The figures are snapshots of the system at time $t = 0.003212$ sec.

Published by Woodhead Publishing Limited, 2013

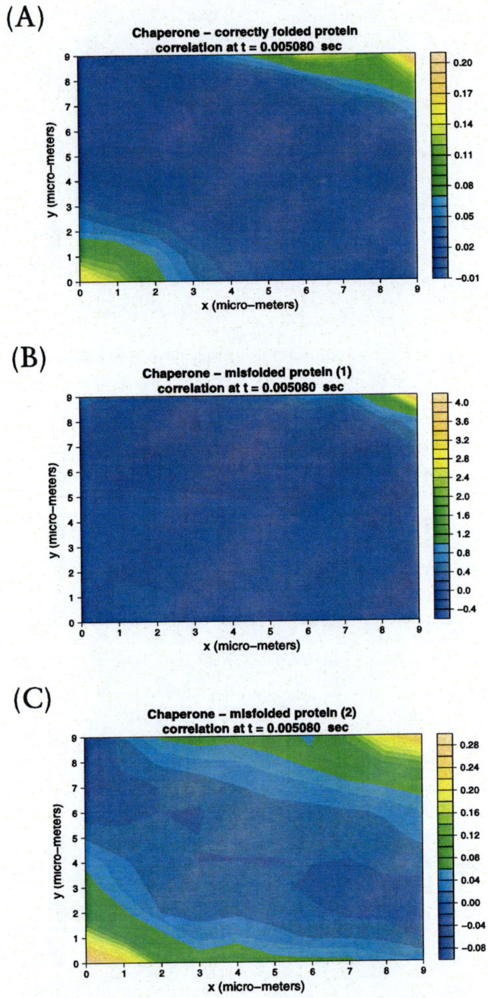

Figure 6.21 Matrices of correlation (Eq. (6.43)) between chaperones and correctly folded proteins concentrations (A), chaperones and misfolded proteins concentrations deriving from the Reaction 2 (B), and chaperones and misfolded proteins concentration deriving from Reaction 4 (C). The figures are snapshots of the system at time $t = 0.005080$ sec.

Published by Woodhead Publishing Limited, 2013

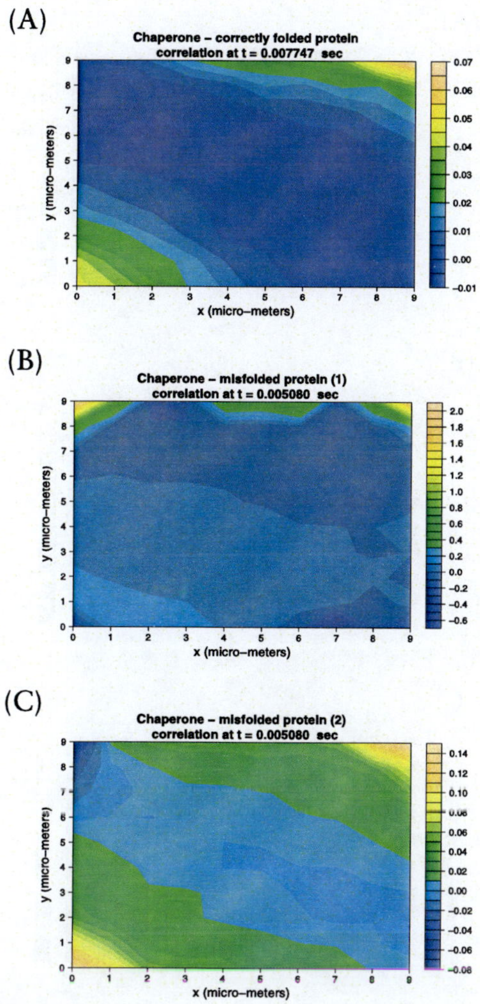

Figure 6.22 Matrices of correlation (Eq. (6.43)) between chaperones and correctly folded proteins concentrations (A), chaperones and misfolded proteins concentrations deriving from the Reaction 2 (B), and chaperones and misfolded proteins concentration deriving from Reaction 4 (C). The figures are snapshots of the system at time $t = 0.007747$ sec.

Published by Woodhead Publishing Limited, 2013

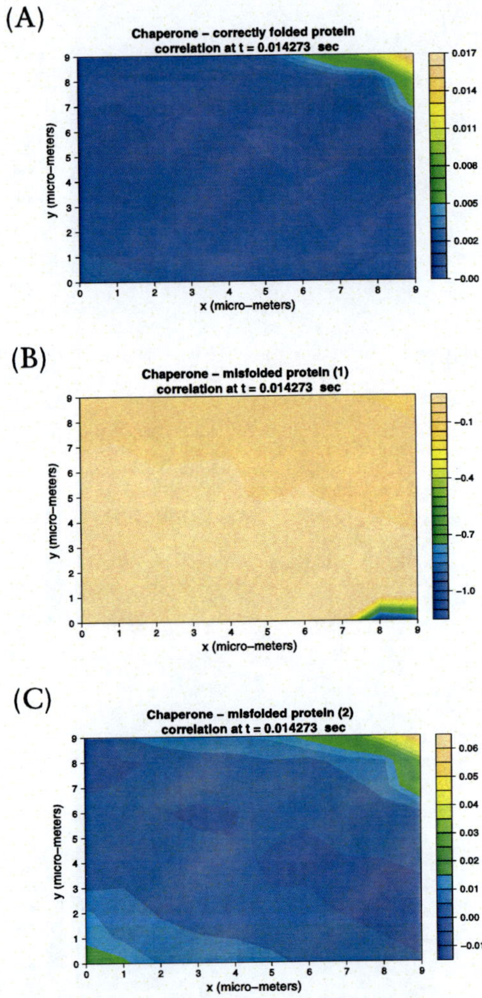

Figure 6.23 Matrices of correlation (Eq. (6.43)) between chaperones and correctly folded proteins concentrations (A), chaperones and misfolded proteins concentrations deriving from the Reaction 2 (B), and chaperones and misfolded proteins concentration deriving from Reaction 4 (C). The figures are snapshots of the system at time $t = 0.014273$ sec.

Published by Woodhead Publishing Limited, 2013

of concentrations of solute. Moreover, chaperones assist not only the efficient folding of newly-translated proteins as these proteins are being synthesized on the ribosome, but they can also maintain pre-existing proteins in a stable conformation. Chaperones can also promote the disaggregation of preformed protein aggregates. The general mechanism by which chaperones carry out their function usually involves multiple rounds of regulated binding and release of an unstable conformer of target polypeptides. These reaction are not included in this simple model.

Apart from the above limitations, the model captures the essential features of the kinetics of the chaperone-assisted protein folding (see [17, 22, 23, 24, 25, 26, 27, 28]). Both the correct and misfolded proteins appear near to the chaperones, as the proteins are released from the chaperones. The correlation between chaperones and correctly folded proteins, as well as the correlation between chaperones and misfolded proteins deriving from Reaction 4, vanish at $t \approx 0.01$. This suggest that, after that time, the proteins released from the chaperones quickly diffuse away from them and aggregates at the site where the cheperones are less abundant. The diffusion of the chaperones toward those sites causes the decrement and the subsequent stabilization of the amount of misfolded proteins.

6.6 Case study 2: modeling the formation of Bicoid gradient

One of the most fundamental problems in developmental biology is the question of morphogenesis; that is, by what mechanisms do cells self-organize into highly complex spatial distributions giving rise to tissues and organs during embryonic development? While tremendous advances have been made in the last fifty years, there are still unresolved questions on a morphogen's mechanism of transport and relationships. The first and perhaps still most elegant example of a morphogen is bicoid. The gradient formation of which acts as a suitable case study of a reaction-diffusion system;

Published by Woodhead Publishing Limited, 2013

where the medium, that is the syncytial embryo, is inhomogenous and highly dynamic, and the most pronounced changes associated with the number and the spatial distribution of the nuclei [34]. However, there seems to be brewing a controversy over some recent theories and quantitative analyses addressing the fundamental question of how the bicoid morphogen gradient is established and decoded in early Drosophila embryos. The numerical solution for the dynamics of morphogen diffusion from a localized source of production and linear morphogen degradation shows that the morphogen gradient achieves a steady state over a broad spatial region on time periods one order of magnitude larger than the morphogen's half life [35]. To demonstrate the actual participation of diffusion as a transport mechanism for the bicoid protein, several recent studies have evaluated the bicoid's diffusion rate (see the works of T. Gregor et al. in [36, 37] for a comprehensive review of the literature on this debate), all of which have yielded different values. As a step towards resolving this issue of disparities in data, a novel reaction-diffusion model has been developed by us, where the medium in which molecules react and diffuse is highly structured, and the concentration of particles not spatially homogenous. In this model we assume that the kinetics of diffusive processes are driven by multidimensional state- and time- dependent diffusion coefficients. We were able to simulate the dynamics of the gradient profile of the bicoid protein in the Drosophila Melanogaster embryo. Our procedure, implemented in our Redi software, reproduced with an accuracy of 1% the experimental spatio-temporal dynamics of bicoid, as recorded in a time-lapse experiment obtained by direct measurements of transgenic bicoid-enhanced green fluorescent protein [37]. Using our reaction-diffusion model we were able to characterize the dynamic properties of the bicoid gradient; predicting the kinetic rate of production and degradation and the range of variability of the average diffusion coefficient. We believe that these results could give a contribution to the present studies aimed at defining the

Published by Woodhead Publishing Limited, 2013

plausible biophysical mechanism responsible for the formation of the bicoid protein gradient, and towards the understanding of morphogenesis in developmental biology.

6.6.1 The Redi model of the Bicoid gradient

In Drosophila Melanogaster, bicoid is a maternally transcribed gene that organizes anterior development. Its mRNA is localized at the anterior pole of the oocyte and is translated soon after fertilization. As a consequence of the anterior localization of mRNA a gradient of the bicoid protein is formed along the antero-posterior axis, simultaneously with nuclear cleavage cycles. Because the bicoid protein is a transcription factor that functions when the early Drosophila melanogaster embryo is a syncytium, the bicoid protein confers fate directly on individual nuclei in a common cytoplasm, thus removing the need for transmembrane receptors, cytoplasmic signal transduction machinery and other possible events. With respect to the mode of dispersion, the prevailing model is that there is a point source of bicoid mRNA at the anterior pole of the embryo from which bicoid protein is synthesized and then diffuses posteriorly forming a gradient. Recent research suggests that a diffusion-based mechanism is sufficient to explain the exponentially decreasing gradient of bicoid protein observed (which gradually decreases as the distance from the center of production increases). However, the role of this diffusion-based mechanism on the movement of the bicoid is the subject of intense discussions. Lipshitz et al. [38] and Spirov et al. [39] suggest that the bicoid protein gradient is produced by a bicoid mRNA gradient, that is formed by active transport of a Stau-bicoid mRNA complex (i.e. bicoid mRNA in a complex with the mRNA-binding protein Staufen) through a microtubular network.

Numerical solution of a the Fick's law-based dynamical model of morphogen diffusion from the localized source of production and linear morphogen degradation shows that the morphogen gradient achieves a steady state over a broad

Published by Woodhead Publishing Limited, 2013

spatial region on time periods one order of magnitude larger than the morphogen half life [35].

To demonstrate the actual participation of diffusion as a transport mechanism for the bicoid, several recent studies have been used to evaluate the bicoid's diffusion rate [37]. The data, however, have yielded strikingly different values [36, 37]. On the initial study, injecting inert fluorescent dextran molecules into the anterior pole of the Drosophila syncytium and measuring the fluorescence intensity over 1 hrs at different spatial positions a few hundred microns from the injection Gregor et al. [36] generated time-evolution data curves which were fitted by computationally derived time courses from which the diffusion rate was calculated. Since dextran molecules of several molecular masses were used in the range of the bicoid protein molecular mass (55-57 kDa, [40]), the diffusion coefficient was uncovered using the Stokes-Einstein relationship, which establishes that the diffusion coefficients decrease inversely with increasing molecular radius. Thus, if the overall gradient is at a steady state and is formed by diffusion and linear degradation, based on the characteristic length of the gradient, assumed to be 100 μm, the diffusion rate was inferred to be in the order of 10 - 13 $\mu m^2/s$. However, this value does not agree with the diffusion rate inferred from the direct measurement of transgenic bicoid-enhanced green fluorescent protein levels after photobleaching at the cortical cytoplasm. The diffusion coefficient of 0.3 $\mu m^2/s$, three orders of magnitude smaller than the diffusion rate of dextran molecules and one to two orders of magnitude lower than what would be required to establish the observed bicoid gradient [37] was what was revealed. These observations are therefore inconsistent with a pure and non-generalized Fickian diffusion model, raising the open issue of how to conciliate these two disparate results.

In trying to resolve this issue of disparities in data, we propose to model the bicoid protein gradient formation as a stochastic reaction-diffusion system where diffusion is a state-dependent process. By implementing this model in Redi, we attempt to point out the range of plausible bicoid protein diffusion rates. To do this, we start by assuming

Published by Woodhead Publishing Limited, 2013

that the reaction-diffusion system involves the following events:

- bicoid protein production; localized at the anterior pole of the embryo;
- bicoid protein anisotropic diffusion;
- uniform bicoid degradation.

The experimental data used in this study have been recorded by T. Gregor et al. [37] in a time-lapse movie (available as supplementary material of [37]) of a Drosophila Embryo expressing bicoid-GFP two-photon microscopy. The video in AVI format contains 154 frames, one frame per minute from 40 min to 194 min. The best agreement between the experiments and the simulations have been obtained for the following values of the parameters: (1) molecular mass of the bicoid protein in the range of 50-60 kDa; (2) bicoid protein production rate of order of magnitude of 10^{-5} min^{-1}; and (3) degradation rate of bicoid protein of the order of magnitude of 10^{-3} min^{-1}. The initial conditions of the spatio-temporal simulation are defined by the fluorescence values reported in the first frame of the experimental video (Figure 6.24 A). The reaction-diffusion system of bicoid has been simulated in a two-dimensional reaction space of 450×200 μm^2, and the cell has been fixed to $5 \mu m$, that is the radius of a nucleus [41] (1 pixel on the image measures $1 \mu m$ in the physical space). Finally, the output is saved as a text file, visualized and then saved as RAW images (Figure 6.24 B). By using the writer `HeatMapView` through the command `/writer:Cosbi.DiffuseSim.HeatMapViewWriter,HeatMapView` a visualization of the state of the systems as a heat-colour image at each step of the simulation has been obtained (Figure 6.25). Due to space limitations, Figure 6.25 reports only 8 frames of both the experimental and the Redi simulated video that actually contain 155 frames. We can see that the Redi simulated spatio-temporal dynamics mimics quit well the observed one.

Published by Woodhead Publishing Limited, 2013

A

```
// File: bicoid.txt
var bicoid : weight 50;

bicoid -> nil [1.0E-3];
nil -> bicoid [1.0E-5];

run

bicoid [253, 33, 0, 10];
bicoid [254, 33, 0, 24];
bicoid [255, 33, 0, 24];
bicoid [256, 33, 0, 1];
bicoid [261, 33, 0, 14];
... ... ... ... ... ... ...
... ... ... ... ... ... ...
```

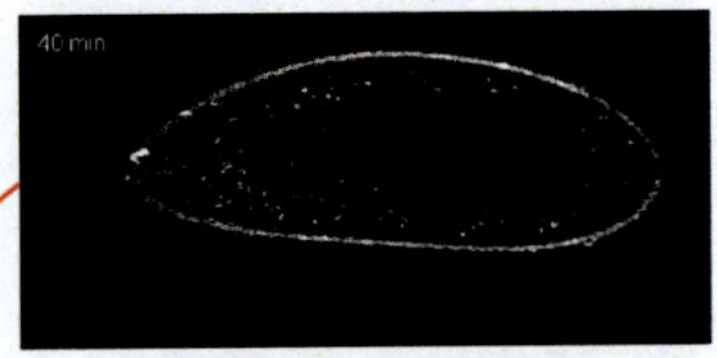

Initial conditions: first frame of the experimental video

B

```
Redi.exe /fileName:.\input_files\bicoid.txt
        /steps:1530000
        /deltaSteps:10000
       /x:450 /y:200 /z:1 /cellLength:5
      /writer:GlobalDataWriter /writer:StandardDataWriter
     /writer:Cosbi.DiffuseSim.HeatMapViewWriter,HeatMapView
    /writerSpecies:bicoid
```

Figure 6.24 **A.** The initial part of the input file to Redi. The figure shows the declaration of the species (only Bicoid Protein) and the reaction (production and degradation of the protein); the parameters of the simulation are the molecular mass of Bicoid protein, the rate constants of the production and degradation reactions. The initial conditions are specified in a sort of "matrix-like" syntax. **B.** The command line launched to run the Redi simulation.

The Figure 6.26 shows a comparison between the experimental and the simulated bicoid protein gradient. Simulations are in good agreement with the experimental observations. On the *y*-axis the average intensity/concentration of the bicoid, scaled in the range between 0 and 1, is reported, and on the *x*-axis the distance fromt he anterior

Published by Woodhead Publishing Limited, 2013

pole of the embryo is reported. In order to calculate the average intensity/concentration of the bicoid corresponding to a given distance x from the tip of the bicoid on the antero-posterior axis, the image of the embryo has been divided into vertical slices 1 pixel wide, the on each slice the average pixel intensity has been calculated and normalized in the range between 0 and 1. On the graphics of Figure 6.26 the diffusion of bicoid away from its source can be recognized as progressive widening of the gradient profile toward greater values on the x-axis. If the concentration of bicoid is mostly disposed on the anterior layer of the embryo the gradient profile is significantly different from zero for small distances x from the tip of the embryo. On the contrary, if a consistent number of bicoid molecules moved away from the anterior pole, the gradient profile will be significantly different from zero in the posterior regions of embryo (i.e. for higher values of the x coordinate).

Slight discrepancies are revealed in the spatial region within 20 μm from the anterior pole of the egg, where the experimental data are most noisy and, consequently, the approximation of first derivative given in Eq. (6.11) is less accurate. The discrepancy between the experimental video and the simulated one has been calculated frame by frame as L^1 distance. As we can see in Figure 6.27, it amounts to the 20-25% of the measured fluorescence. Nevertheless, the Euclidean metric to define the simulation accuracy provides an overestimate of the value of the difference between simulated and observed image. Slight effects of translation and dilatation in the simulated images are consequences of the numerical approximations and/or the propagation of errors through the steps of the algorithmic simulation procedure. To get around this, we used a distance measure based on Mahalanobis distance to take into account the translations and dilation in comparing the simulations with the experiments. The Mahalanobis distance between experiments and simulations is one order of magnitude less than the Euclidean distance: 2-3% of the measured fluorescence. Figure 6.28 reports the average behaviour of the Mahalanobis distance

Published by Woodhead Publishing Limited, 2013

for a simulation of bicoid diffusion where the molecular mass is 55 kDa.

Furthermore we considered, as in [36] five nominal molecular masses in the range form 10 to 150 kDa and then launched 100 simulations for each value of the molecular mass. The best agreement, estimated by a Mahalanobis distance, between observations and simulations were obtained for simulations performed with a molecular mass between 50 and 60 kDa. The measured molecular mass of the protein is seen to drop significantly within this range (see Table 6.2) and indicates that the range of variability of the diffusion coefficient is between 14 and 18 $\mu m^2\ s^{-1}$, values which were able to reproduce the observed rate of the Bicoid gradient formation.

Finally, we simulated the bicoid dynamics with fixed values of bicoid diffusivity (standard Fick's law) in and out the estimated range of variability $[14, 18]$ $\mu m^2\ s^{-1}$. We ran 100 simulations for each of the following nominal fixed values of bicoid diffusivity: 10^{-1}, 1, 10, 16, 30 $\mu m^2\ s^{-1}$. Table 6.3 shows the average Mahalanobis distance between observed and simulated dynamics. The best agreement has been achieved for a diffusivity around 10 $\mu m^2\ s^{-1}$; nevertheless, we found that using the Fick's law with constant diffusion coefficients does not allow to reproduce the observed movement of the bicoid protein. For instance, Figure 6.29 shows the simulation of the bicoid dynamics obtained using the Fick's law with a value of the diffusion coefficient equal to 0.3 $\mu m^2\ s^{-1}$. We notice that the movement of the bicoid is significantly slow with respect to the observed one, and at the time-scale of the experiments it is not characterized by the shuttling of the bicoid in and out the nuclei of the embryo. The observed movement can be obtained with a high accuracy with our generalization of the Fick's law. Moreover, with respect to the other diffusion-based models of the bicoid concentration profile, that are also able to reproduce the observations, our model does not need extra terms in the rate equation describing the spatio-temporal dynamics of the protein, and therefore can be considered a pure diffusive model.

Published by Woodhead Publishing Limited, 2013

Table 6.2 Order of magnitude of the discrepancy between experiments of simulations performed with five nominal molecular mass of Bicoid protein. The best agreement has been achieved for a molecular mass in the range from 50 to 60 kDa.

Bicoid protein molecular mass (kDa)	Average Mahalanobis distance (order of magnitude)
≤ 10	1
40	10^{-1}
50 to 60	10^{-2}
70	10^{-1}
150	1

Namely, at the best of our knowledge, the main studies aiming at modelling the bicoid movement as a diffusive process account both of the free and the nuclear bicoid dynamics. I. Hecht et. al. and M. Coppey et al. [34, 41] reproduced the observed dynamics of the bicoid by describing with two different rate equations the spatio-temporal dynamics of free bicoid and the one of nuclear bicoid concentration. T. Gregor [36] introduced an extra negative term proportional to the local bicoid concentration in the second Fick's law. These models reproduced the observed exponential decay of the bicoid concentration along the antero-posterior axis of the embryo. Our model and the Redi software were also able to reproduce both the temporal and spatial diffusion dynamics of the protein simulating the entire experimental movie.

6.7 Conclusions and future directions

The authors presented a model for the diffusion of non-charged molecules, in which the diffusion coefficients are not constant with respect to the time and space. Constant diffusion coefficients are rather more the exception than

Published by Woodhead Publishing Limited, 2013

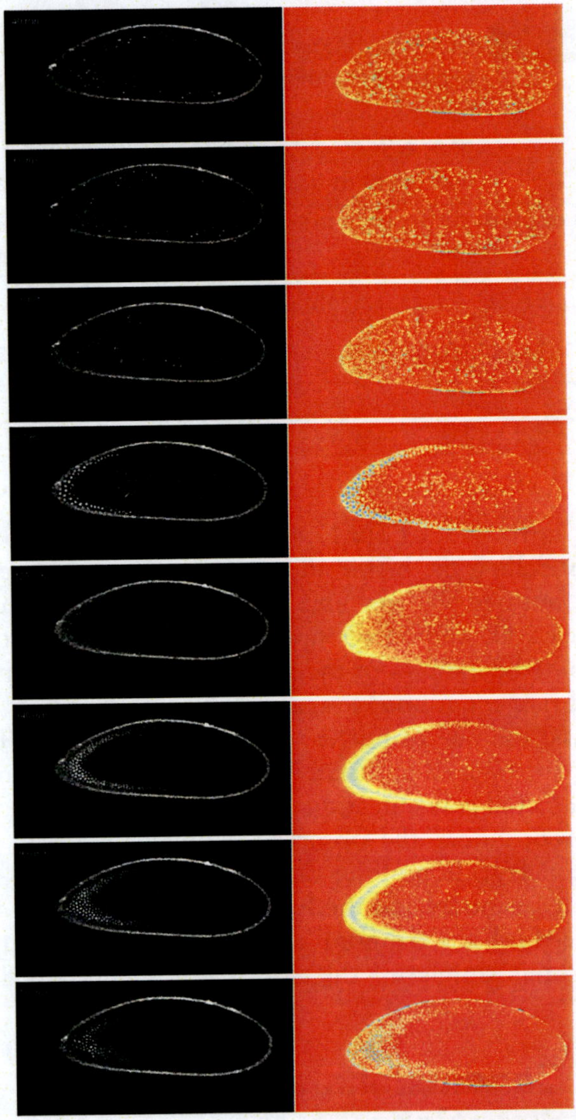

Figure 6.25 Experimental video frames (black figures on the left) and Redi simulated video frames (red heat-color maps on the right). The simulated images are an average of 100 stochastic simulations.

Published by Woodhead Publishing Limited, 2013

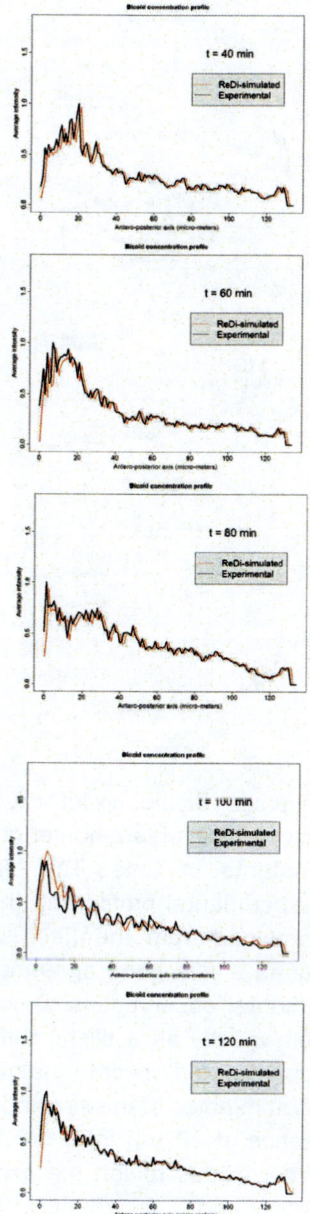

Figure 6.26 Continued

Published by Woodhead Publishing Limited, 2013

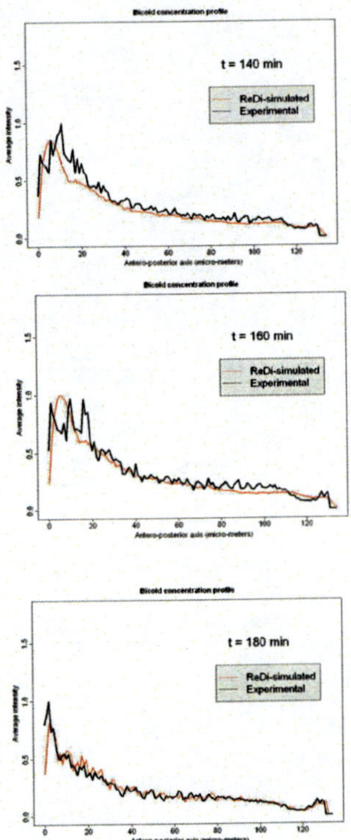

Figure 6.26 Average Bicoid protein fluorescence profiles along the antero-posterior axis at different instants of time. The black curve is the experimental profile and the red is the curve obtained from the Redi simulation. Simulations are in good agreement with the experimental observations. Discrepancies are visible mostly as a slight shift of the simulated curve with respect to the measured one in the first minutes of the simulation and within a distance of 20 μm from the anterior pole of the egg. In this region the experimental video is noisy and, consequently, the first approximation of the derivative of the concentration in Eq. (6.11) is not accurate enough.

Published by Woodhead Publishing Limited, 2013

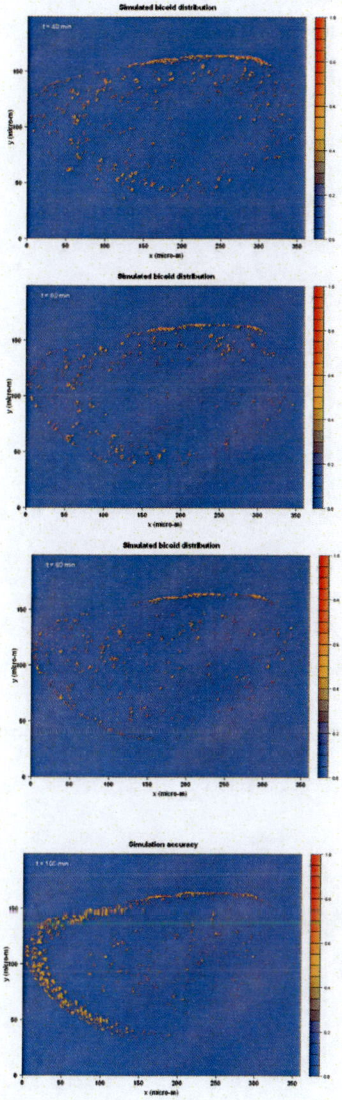

Figure 6.27 Continued

Published by Woodhead Publishing Limited, 2013

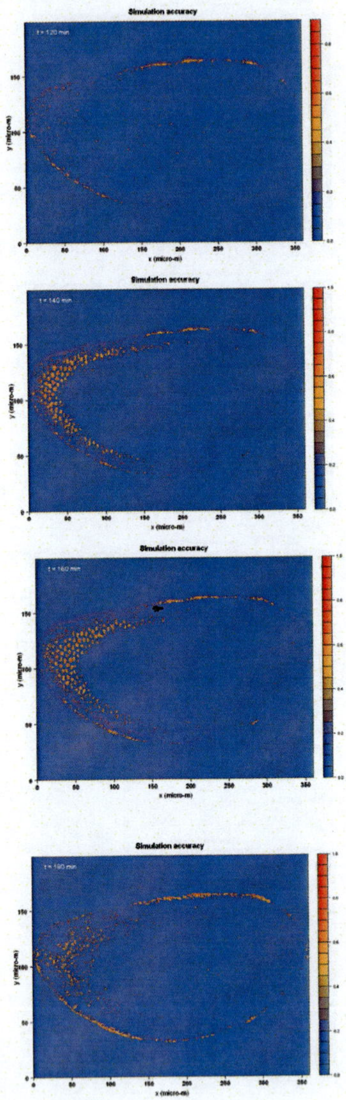

Figure 6.27 Euclidean distance between experimental frames and Redi simulated frame.

Published by Woodhead Publishing Limited, 2013

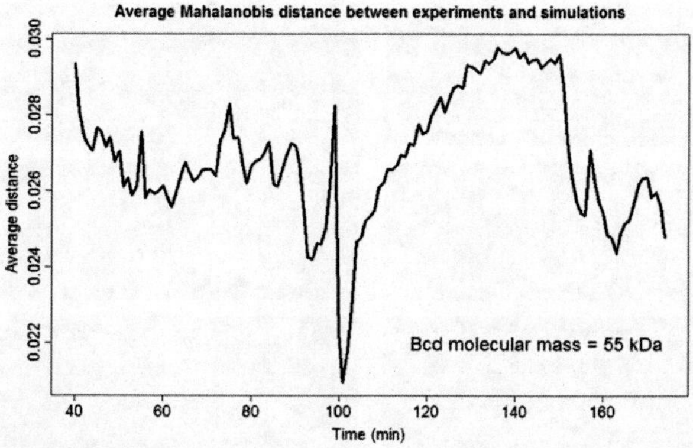

Figure 6.28 Time behavior of average Mahalanobis distance between experimental and simulated spatio-temporal dynamics of Bicoid protein gradient.

Table 6.3 Order of magnitude of the discrepancy between experiments of simulations performed with five nominal constant values of bicoid diffusivity. The best agreement has been achieved for a diffusivity around 10 $\mu m^2\ s^{-1}$.

Bicoid diffusion constant ($\mu m^2\ s^{-1}$)	Average Mahalanobis distance (order of magnitude)
10^{-1}	1
1	10^{-1}
10	10^{-2}
16	10^{-2}
30	10^{2}

the rule in living cells and, more generally in biological tissues. The authors implemented the procedure in the framework of stochastic simulation of reaction-diffusion systems and presented the results of the method on the case study of chaperone-assisted protein folding. The software tool is

Published by Woodhead Publishing Limited, 2013

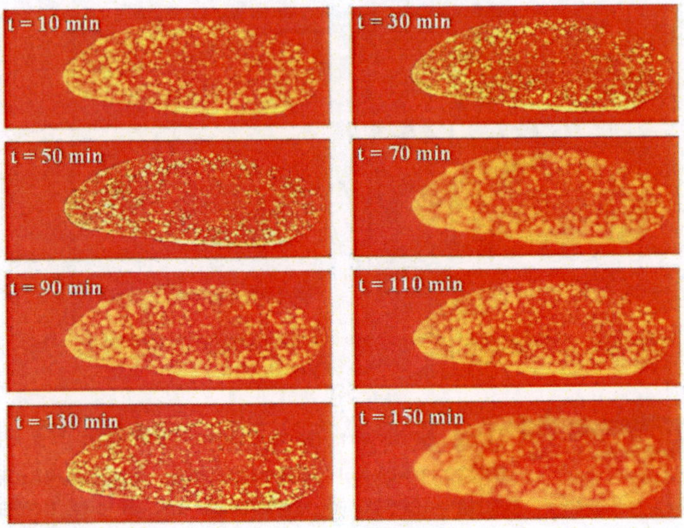

Figure 6.29 Bicoid spatio-temporal dynamics simulated with Fick's law with constant diffusion coefficient equal to 0.3 μ m ^{2} s^{-1}. The kinetics results to be much slower than the observed one and it no shuttling movement of bicoid in and out the nuclei are reproduced in the time interval from 10 to 150 min.

equipped with a 3D visualizer that shows the spatial distribution of the diffusion molecules at every step of simulations (see Figs. 6.30 and 6.31, showing the distributions from two points of view).

Unlike the previous works as [9, 20, 29], this model provides a theoretical derivation of the molecular origins of the parameters, determining the time-behavior of the diffusive phenomena. Moreover, it provides results in agreement with experimental qualitative and quantitative data. Future work will consist in a further refinement of the procedure to make it closer to the chemistry and physics of biological transport phenomena. Some future directions will consist of a more accurate calculation of the second virial coefficient for

Published by Woodhead Publishing Limited, 2013

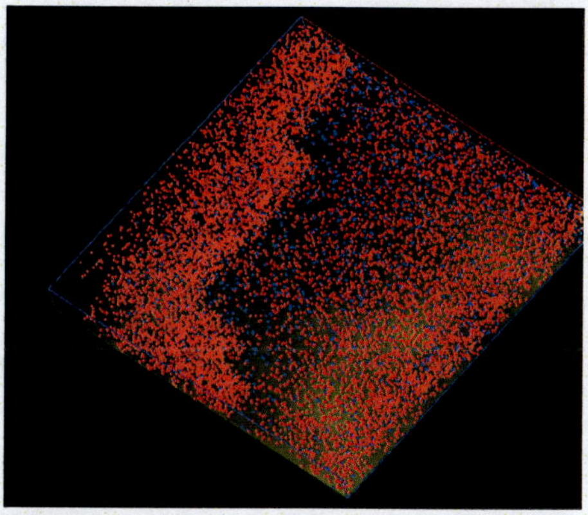

Figure 6.30 A sample view of the distribution of chaperones (bluepoints) and nascent proteins (red points), right-folded proteins (yellow points), misfolded proteins of type 1 (green points) and misfolde proiteins of type 2 (magenta points).

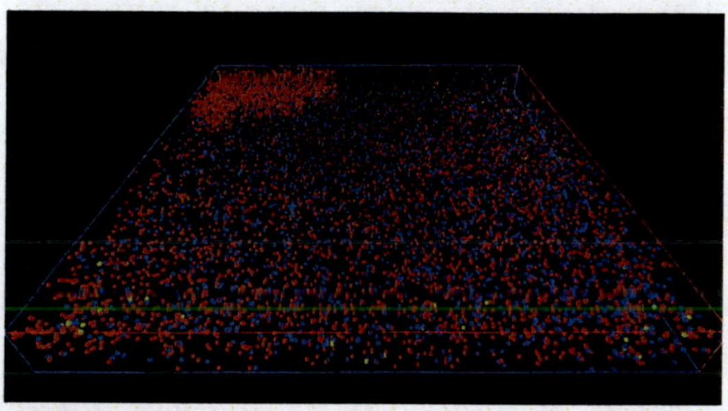

Figure 6.31 Another sample view of the distribution of chaperones (bluepoints)and nascent proteins (red points), right-folded proteins (yellow points), misfolded proteins of type 1 (green points) and misfolde proiteins of type 2 (magenta points).

Published by Woodhead Publishing Limited, 2013

biomolecules, especially for proteins. The use of the Lennard-Jones potential is a good approximation of the molecular interaction, but it is a drawback in describing protein-protein interaction is that water molecules must be included explicitly [30], complicating the computational task. The condition of solvated molecules is reflected also to the expression of the concentration-dependence of frictional coefficient, that will need to be accordingly modified.

Furthermore, more generally, as already mentioned, the cellular environment is a crowded solution. Namely, the cellular environments are packed with other biomolecules and this crowdedness may affect the stability and aggregation rates of proteins inside cells [18, 19, 31, 32, 33]. Unlike in typical biochemical experiments in which the proteins of interest are purified and diluted, the living cell is crowded with a wide variety of other proteins and macromolecules which generally occupy 20-30% of the total cell volume. This percentage is called *excluded volume*. The effects imposed by the excluded volume, that is caused by the volume excluded by the "inert" macromolecules, are called *macromolecular crowding* effects and those macromolecules are called *crowding agents*. The authors are currently extending the present study to develop a model whose simulations are of support to the investigation of excluded volume effects on the protein diffusion and folding.

References

1. J. Elf, A. Doncic, and M. Ehrenberg, "Mesoscopic reaction-diffusion in intracellular signaling," *Fluctuation and noise in biological, biophysical and biomedical systems. Procs. of SPIE*, vol. 5110, 2003.
2. P. S. Agutter and D. Wheatley, "Random walks and cell size," *BioEssays*, vol. 22, pp. 1018–1023, 2000.
3. P. Agutter, P. Malone, and D. Wheatley, "Intracellular transport mechanisms: a critique of diffusion theory," *J. Theor. Biol.*, vol. 176, pp. 261–272, 1995.
4. D. Fusco, N. Accornero, B. Lavoie, S. Shenoy, J. Blanchard, R. Singer, and E. Bertrand, "Single mrna

Published by Woodhead Publishing Limited, 2013

molecules demonstrate probabilistic movement in living mammallian cells.," *Curr. Biol.*, vol. 13, pp. 161–167, 2003.

5. B. Alberts, A. Johnson, J. Lewis, M. Raff, K. Roberts, and P. Walter, *Molecular biology of the cell.* Garland Science, 4th ed. ed., 2003.
6. P. Lecca, A. Ihekwaba, L. Dematté, and C. Priami, "Stochastic simulation of the spatio-temporal dynamics of reaction-diffusion systems: the case for the bicoid gradient," *Journal of Integrative Bioinformatics*, vol. 7, p. 150, June 2010.
7. E. R. Kandel, "The molecular biology of memory storage: a dialogue between genes and synapses," *Science*, vol. 294, pp. 1030–1038, 2001.
8. D. Gillespie, "Exact stochastic simulation of coupled chemical reactions," *Journal of Physical Chemistry*, vol. 81, pp. 2340–2361, December 1977.
9. C. J. Roussel and M. R. Roussel, "Reaction-diffusion models of development with state-dependent chemical diffusion coefficients.," *Progress in Biophysics & Molecular Biology*, 2004.
10. K. J. Laidler, J. H. meiser, and B. C. Sanctuary, *Physical chemistry*. Houghton Mifflin Company Boston New York, 2003.
11. M. P. Tombs and A. R. Peacocke, *The Osmotic Pressure of Biological Macromolecules*. Monograph on Physical Biochemistry, Oxford University Press, 1975.
12. A. Solovyova, P. Schuck, L. Costenaro, and C. Ebel, "Non ideality of sedimantation velocity of halophilic malate dehydrogenase in complex solvent," *Biophysical Journal*, vol. 81, pp. 1868–1880, 2001.
13. K. Laidler, J. Meiser, and B. Sanctuary, *Physical Chemistry*. Houghton Mifflin Company, 2003.
14. S. Harding and P. Johnson, "The concentration dependence of macromolecular parameters," *Biochemical Journal*, vol. 231, pp. 543–547, 1985.

Published by Woodhead Publishing Limited, 2013

15. R. J. E. B. van den Berg and C. M. Dobson, "Effects of macromolecular crowding on protein folding and aggregation," *The EMBO Journal*, vol. 18, p. 6927:6933, 1999.
16. J. J. Z. Hu and R. Rajagopalan, "Effects of macromolecular crowding on biochemical reaction equilibria: A molecular thermodynamic perspective," *Biophysical Journal*, vol. 93, pp. 1464–1473, 2007.
17. A. R. Kinjo and S. Takada, "Competition between protein folding and aggregation wth molecular chaperones in crowed solutions: insight from mesoscopic simulations," *Biophysical Journal*, vol. 85, pp. 3521 – 3531, 2003.
18. A. R. Kinjo and S. Takada, "Effects of macromolecular crowding on protein folding and aggregation studied bu density functional theory: statics," *Physical Review E*, vol. 66, pp. 031911: 1–9, 2002.
19. A. R. Kinjo and S. Takada, "Effects of macromolecular crowding on protein folding and aggregation studied by density functional theory: Dynamics," *Physical review. E*, vol. 66, no. 5, pp. 051902.1–051902.10, 2002.
20. D. Bernstein, "Simulating mesoscopic reaction-diffusion systems using the gillespie algorithm," *PHYSICAL REVIEW E*, vol. 71, April 2005.
21. J. Elf and M. Ehrenberg, "Spontaneous separation of bistable biochemical systems into spatial domains of opposite phases," *Syst. Biol.*, vol. 1, December 2004.
22. H. S. Chan and K. A. Dill, "A simple model of chaperonin-mediated protein folding," *PROTEINS: Structure, Function, and Genetics*, vol. 24, pp. 345–351, 1996.
23. W. A. Houry, "Chaperone-assisted protein folding," *Curr. protein Pept. Sci.*, vol. 2, no. 3, pp. 227–244, 2001.
24. J. Frydman and F. U. Hartl, "Principles of chaperone-assisted folding: differences between in vitro and in vivo mechanisms," *Science*, vol. 272, no. 5667, pp. 1497 – 1502, 1996.

Published by Woodhead Publishing Limited, 2013

25. T. Langer, J. Martin, E. Nimmesgern, and F. U. Hartl, "The pathway of chaperone-assisted protein folding," *Fresenius' Journal of Analytical Chemistry*, vol. 343, 1992.
26. J. Martin and F. U. Hartl, "The effect of macromolecular crowding on chaperonin-mediated protein folding," *Proc. Natl. Acad. Sci. USA*, vol. 94, pp. 1107–1112, 1997.
27. D. Thirulamai and G. H. Lorimer, "Chaperonin-mediated protein folding," *Ann. Rev. Biophys. Biomol. Struct.*, vol. 30, pp. 245–268, 2001.
28. D. Thirumalai and G. H. Lorimer, "Chaperonin-mediated protein folding," *Annu. Rev. Biophys. Biomol. Struct.*, vol. 30, p. 245:269, 2001.
29. S. A. Isaacson and C. S. Peskin, "Incorporating diffusion in complex geometries into stochastic chemical kinetics simulations," *SIAM Journal of Scientific computing*, pp. 47–74, 2006.
30. B. L. Neal, D. Asthagiri, and A. M. Lenhoff, "Molecular origins of osmotic second virial coefficients of proteins," *Biophysical Journal*, vol. 75, 1998.
31. G. Y. G. Ping and J. M. Yuan, "Depletion force from macromolecular crowding enhances mechanicsl stability of protein molecules," *Polymer*, vol. 27, p. 2564:2570, 2006.
32. A. P. Minton, "Molecular crowding: analysis of effects of high concetrations of inert cosolutes on biochemical equilibria and rates in terms of volume exclusion," *Methods Enzymol.*, vol. 295, p. 127:149, 1998.
33. A. P. Minton, "The influence of macromolecular crowding and macromolecular confinement on biochemical reactions in physiological media," *J. Biol. Chem.*, vol. 276, p. 10577:10580, 2001.
34. M. Coppey, A. M. Berezhkovskii, Y. Kim, A. N. Boettiger, and S. Y. Shvartsman. Modeling the bicoid gradient: diffusion and reversible nuclear trapping of a stable protein. *Developmental Biology*, 312:623–630, 2007.

Published by Woodhead Publishing Limited, 2013

35. S. Bergmann. Pre-steady state deconding of the bicoid morphogen gradient. *PloS Biol.*, 5:232, 2007.
36. T. Gregor, W. Bialek, R. R. de Ruyter van Stevenick, D. W. Tank, and E. F. Wieshaus. Diffusion and scaling during early embryonic pattern formation. *PNAS*, 102(51):18403–18407, 2005.
37. T. Gregor, E. F. Wieshaus, A. P. McGregor, W. Bialek, and D. W. Tank. Stability and nuclear dynamics of the bicoid morphogen gradient. *Cell*, 130:141–152, 2007.
38. H. D. Lipshitz. Follow the mrna: a new model for bicoid gradient formation. *Nature reviews*, 10:509–512, 2009.
39. A. Spirov, K. Fahmy, M. Schneider, E. Frei, M. Noll, and S. Baumgartner. Formation of the bicoid morphogen gradient: an mrna dictates the protein gradient. *Development*, 136:605–614, 2009.
40. W. Driever and C. Nuessein-Volhard. A gradient of bicoid protein in drosophila embryos. *Cell*, 54(1):83–93, 1988.
41. I. Hecht, W. J. Rappel, and H. Levine. Determining the scale of the bicoid morphogen gradient. *PNAS*, 106(6):1710–1715, 2009.
42. R. Hawkins, S. A. Rice. Study of Concentration Fluctuations in Model Systems. *J. Theor. Biol.*, 30: 579, 1971

Published by Woodhead Publishing Limited, 2013

7

KInfer: a tool for model calibration

Abstract Systems biology models have parameters, such as kinetic constants, decay rates and noise terms, which are unknown, difficult to measure directly or weakly constrained by existing experimental knowledge. Since systems biology models are intended to provide a mechanistic description of the system, often using ordinary differential equations, the standard approaches based on maximum likelihood or least squares methods and various optimization heuristics encounter mathematical and numerical difficulties. KInfer is a software prototype implementing a novel maximum-likelihood based method for estiamting rate constants of systems of chemical reactions from experimental time series of reagents concentrations. The only inputs required by KInfer are the list of chemical equations, or alternatively a generalized mass action, and the experimentally measured time-series of the reagents that are known to be involved in the system. Principal features of the tool are: (i) automatic generation of generalized mass action model from the chemical reactions involved in the system, (ii) automatic estimation of the initial guesses and bounds for the parameter values (iii) estimate of the propagation of the experimental errors from the input data to the parameter estimates, and (iv) estimation of the level of noise in the input data.

Published by Woodhead Publishing Limited, 2013

Keywords: parameter inference, maximum-likelihood, generalized mass-action kinetics.

7.1 Introduction

Mathematical modeling and dynamic simulation of biochemical networks are central in systems biology, as they provide new ways toward the analysis of omics data and lead to a greater understanding of the language of cells and organisms [4]. Models and simulations are systematic strategies for key issues in medicine and the pharmaceutical and biotechnological industries. For example, the model-based approaches and the *in silico* experiments on a computer provide a rational framework to guide drug development, taking into account the effects of possible new drugs on biochemical pathways and physiology. Once a mathematical model of even a small part of a biochemical network is established, then the potential benefits are noticeable: generating new hypotheses, suggesting experiments to test them, and, more generally, supporting experimental design. The construction of a mathematical model of a network consists in two tasks: 1. deciding on the model structure and 2. estimating the involved parameters values. This work is focused on the key step of parameter estimation, assuming the structure of the model as given. Parameter estimation (also known as model calibration) from experimental data is a bottleneck for a major breakthrough in computational systems biology in the present post-genomic era [5]. Parameter inference aims to find the parameters of the model which give the best fit to a set of experimental data.

Estimating the parameters of a dynamic model of a biochemical network is difficult, because often the model is non-linear and thus no general analytic result exists. Biological models are often dynamic and highly nonlinear, thus, in order to find the estimates, we must resort to nonlinear optimization techniques where a measure of the distance between model predictions and experimental data is used as the optimality criterion to be minimized. The criterion selection will

Published by Woodhead Publishing Limited, 2013

depend on the assumptions about the data disturbance and on the amount of information provided by the user. The maximum likelihood estimator maximizes the probability of the occurrence of the observed measurements. If we make the assumption that the residuals are normally distributed and independent with the same variance σ^2, then the maximum likelihood criterion is equivalent to the least squares and we aim to find which minimizes the sum of squared residuals of all the responses. This is subject to the dynamics of the system, plus possibly other algebraic constraints, and model parameters are also subject to upper and lower bounds. When estimating parameters of dynamical systems a number of difficulties may arise, like e.g. convergence to local solutions if standard local methods are used, very flat objective function in the neighborhood of the solution, overdetermined models, badly scaled model functions or non-differentiable terms in the systems dynamics. Due to the nonlinear and constrained nature of the systems dynamics, these problems are very often multimodal. Thus, traditional gradient based methods, like Levenberg-Marquardt or Gauss-Newton, may fail to identify the global solution and may converge to a local minimum when a better solution exists just a small distance away. Moreover, in the presence of a bad fit, there is no way of knowing if it is due to a wrong model formulation, or if it is simply a consequence of local convergence.

The recent literature reports many examples of new effective methods for parameter estimation both in deterministic and stochastic models. Here we briefly mention the most recent ones. Polisetty et al. [33] suggested global optimization techniques as alternative to traditional local methods. Rodrigez-Fernandez et al. [36] developed a hybrid stochastic-deterministic global optimization method. Moles et al. [30] explored several state-of-the-art deterministic and stochastic global optimization techniques and compared their accuracy

Published by Woodhead Publishing Limited, 2013

and effectiveness on nonlinear biochemical dynamic models. Tian et al. [42] presented a simulated maximum likelihood method to evaluate parameters in stochastic models described by stochastic differential equations. They proposed different types of transitional probability and a genetic optimization algorithm to search for optimal reaction rates. Chou et al. [6] developed an alternate regression method which dissects the parameter inference problem into iterative steps of linear regression. Sugimoto et al. [41] developed a computational technique based on genetic programming that simultaneously generates biochemical equations and their parameters from time series data. Reinker et al. [35] proposed the approximate maximum likelihood method and the singular value decomposition likelihood method that estimate stochastic reaction constants from molecule count data measured with errors at discrete time points. Tools for parameter fitting through regression or maximum likelihood methods can be found as integral part of simulation tools (e. g. Copasi [16]), but there exist also stand-alone, like PET [52]. Finally, Boys [2], Golitki [13] and Wilkinson [50] developed Bayesian model-based inference techniques for discrete models. Bayesian scheme offer some advantages over the maximum likelihood methods, when the volume of data is limited or the analytic form of the kinetic model makes the maximization of the likelihood difficult. Nevertheless, Bayesian approaches require to specify a prior distribution for all unknown parameters, but, in problems of calibration of biochemical models, prior knowledge is either vague, or non-existent, and that makes it very difficult to specify a unique prior distribution. Finally we note that most of the current tools for parameter estimation lacks robustness to the noise and the absence of any estimate of experimental error in their outcome. Experimental uncertainties on parameters propagate from the measurements of the concentrations of the species. Inferring the parameters with an estimate of their uncertainty is essential if we want to use these tools in the context of optimal experimental design. Furthermore, most of the current tools, based on optimization techniques do not univocally find the global optimal solution, and ask

Published by Woodhead Publishing Limited, 2013

the user to provide *a priori* the optimization algorithm with the region of parameter space in which to perform the search for the global max/minimum.

In chapter we present a recent approach – developed by P.Lecca et al. [23] – to parameter estimation, whose accuracy is robust to the experimental noise. The method is based on a probabilistic, generative model of the variations in reactant concentrations. We observe time series of concentrations for all the reactant species, gathered in N state vectors $\mathbf{X}_1, \ldots, \mathbf{X}_N$. Our method approximates the law of mass action and provides a tool to predict the values of the variables $\mathbf{X}_i$ at time t , conditioned on their values at the previous time point. The variations of the concentration of the species at different time points are conditionally independent by the Markov nature of the discrete model of the law of mass action. Assuming the observation noise to be Gaussian with variance σ^2, the probability of observing a variation D_i for the concentration $[X]_i$ of species i between time t_{k-1} and t_k is a Gaussian with variance depending on σ and with mean the expectation value of the law of mass action under the noise distribution. The likelihood for the observed increments/decrements D_i can be obptimized with respect to the kinetic rate constants of the biochemical network under consideration and with regard to the level of noise σ affecting the time course of the reactants concentration.

The discretization of the law of mass action provides a model for the variations of the species concentration, rather than a model for the time-trajectory of the species concentrations. This makes the evaluation of the expectation value of law mass action function (the integral of the transitional probability) simpler and analytically tractable. The rate coefficients and the level of noise are then obtained by maximizing the likelihood function defined by the observed variations. Our method infers the rate coefficients, the level of noise σ and an error range on the estimates of rate constants. Its probabilistic formulation handles the noise inherent in biological data, and it enables further extensions, such as a fully Bayesian treatment of the parameter inference and

Published by Woodhead Publishing Limited, 2013

an automated model selection strategies based on the comparison between marginal likelihoods of different models. Finally, the implementation of this method may be used as an interface tool, connecting the outcomes of the wet-lab activity for the concentration measurements and any software for the simulation of chemical kinetics.

The paper outlines as follows: the next section presents the mathematical model of parameter inference. Section 3 present a new procedure to estimate the variances of the parameters. KInfer is the software tool implementing the inference model. Kinfer [26, 24, 27] is free for non-commercial purposes and can be downloaded at the url

http://www.cosbi.eu/Rpty_Soft_KInfer.php.

We illustrate the results of the application of the KInfer parameter estimation method to a synthetic case study, the Maclennan-Higgins SERCA pump model, and to a real case study, the glucose metabolism of L. Lactis.

7.2 The model for inference

Consider N reactant species, $S_1, S_2, \ldots, S_N$, with concentrations $X_1, X_2, \ldots, X_N$, that evolve according to a system of rate equations

$$\frac{dX_i}{dt} = f_i(\mathbf{X}^{(i)}(t); \theta_i) \tag{7.1}$$

where θ_i , $i = 1, 2, \ldots, N$, is the vector of the rate coefficients, which are present in the expression of the function f_i . We wish to estimate the set of parameters $\Theta = \cup\theta_i$ $(i = 1, 2, \ldots, N)$, whose element θ_i is the set of rate coefficients appearing in the rate equations of i-th species, therefore

$$\theta_1 = \{\theta_{11}, \theta_{12}, \cdots, \theta_{1N_1}\}, \ldots, \theta_N = \{\theta_{N1}, \theta_{N2}, \cdots, \theta_{NN_N}\}$$

$\mathbf{X}^{(i)}$ is the vector of concentrations of chemicals that are present in the expression of the function f_i for the species i.

Published by Woodhead Publishing Limited, 2013

According to the law of mass action, the functions f_i have the general form

$$f_i(\mathbf{X}^{(i)}(t); \theta_i) = \theta_{i1} \prod_{w \in S_1 \subseteq [1,N]} X_w^{\alpha_w} + \cdots + \theta_{iN_i} \prod_{w \in S_{N_i} \subseteq [1,N]} X_w^{\alpha_w}$$

$$= \sum_{h=1}^{N_i} \left(\theta_{ih} \prod_{w \in S_h} X_w^{\alpha_w} \right) \quad (7.2)$$

where $\alpha_w \in \mathbf{R}$, and N_i is the number of parameter in the f_i rate equation The rate equations in (7.2) form the so-called Generalized Mass Action law. We assume we have noisy observations $\hat{X}_i = X_i + \epsilon$ at times $t_0, \ldots, t_M$, where $\epsilon \sim \mathcal{N}(0, \sigma^2)$ is a Gaussian noise term with mean zero and variance σ. With this choice we are assuming that the concentration measurements are not significantly affected by systematic errors, but by uncontrolled random errors and that an error is equally likely to occur in either positive or negative direction with respect to the symmetry axis of the distribution.

We also assume a number M of concentration measurements for each considered species. Approximating the rate equation (8.1) as a finite difference equation between the observation times, gives

$$X_i(t_k) = X_i(t_{k-1}) + (t_k - t_{k-1}) f_i(\mathbf{X}^{(i)}(t_{k-1}); \theta_i) \quad (7.3)$$

where $k = 1, \ldots, M$. In Eq. (7.3) the rate equation is viewed as a model of increments/decrements of reactant concentrations; i.e., given a value of the variables at time t_{k-1}, the model can be used to predict the value at the next time point t_k. Increments/decrements between different time points are conditionally independent by the Markov nature of the model (7.3). Therefore, given the Gaussian model for the noise, it is possible to estimate the probability to observe the value $\hat{X}_i(t_k)$ given the model at time t_{k-1}, $X_i(t_{k-1})$, and the set of parameters θ_i, as

Published by Woodhead Publishing Limited, 2013

$$p\left(\hat{X}_i(t_{k-1})|X_i(t_{k-1})\right) = \mathcal{N}\left(X_i(t_{k-1}) + (t_k - t_{k-1})f_i(X_i(t_{k-1}, \theta_i)), \sigma^2\right) \quad (7.4)$$

We then also have that the true value of $X_i(t_k)$ is normally distributed around the observed value $\hat{X}_i(t_k)$, so that

$$p\left(X_i(t_{k-1})|\hat{X}_i(t_{k-1}))\right) = \mathcal{N}\left(\hat{X}_i(t_{k-1}), \sigma^2\right) = \frac{1}{\sqrt{2\pi}\sigma}\exp\left[-\frac{(X_i(t_{k-1}) - \hat{X}_i(t_{k-1}))^2}{2\sigma^2}\right] \quad (7.5)$$

Therefore, the probability to observe a variation $D_i(t_k) = X_i(t_k) - X_i(t_{k-1})$ for the concentration of the i-th species between the time t_{k-1} and t_k, given the parameter vector θ_i is

$$p(D_i(t_k)|\theta_i, \sigma) = \mathcal{N}\left(E\left[f_i(\mathbf{X}^{(i)}(t_{k-1}), \theta_i)\right], 2\sigma^2\right) \quad (7.6)$$

and

$$E\left[f_i(\mathbf{X}^{(i)}(t_{k-1}, \theta_i))\right] = \int_{\Omega_{\mathbf{X}^{(i)}}} f_i(\mathbf{X}^{(i)}(t_{k-1}), \theta_i)\prod_{i=1}^{K_i}\left[p_i\left(X_i(t_{k-1})|\hat{X}_i(t_{k-1})\right)\right]d\mathbf{X}^{(i)} \quad (7.7)$$

where $\Omega_{\mathbf{X}^{(i)}}$ is the sample space of $\mathbf{X}^{(i)}$, and K_i is the number of chemical species in the expression for f_i. While the increments/decrements are conditionally independent given the starting point $X_i(t_k)$, the random variables $D_i(t_k)$ are not independent of each other. Intuitively, if $X_i(t_k)$ happens to be below its expected value because of random fluctuations, then the following increment $D_i(t_{k+1})$ can be expected to be bigger as a result, while the previous one $D_i(t_k)$ will be smaller. A simple calculation allows us to obtain the covariance matrix of the vector of increments for the i-th species. This is a banded matrix $\mathbf{C}_i \equiv \mathbf{C} = \mathrm{Cov}(\mathbf{D}_i)$ with diagonal elements given by

$$E\left[D_i^2(t_k) - E[D_i^2(t_k)]\right] = 2\sigma^2$$

Published by Woodhead Publishing Limited, 2013

and a non-zero band above and below the diagonal given by

$$E\Big[\big(D_i(t_k) - E[D_i(t_k)]\big)\big(D_i(t_{k-1}) - E[D_i(t_{k-1})]\big)\Big] = -\sigma^2$$

with all other entries zero. The likelihood for the observed increments/decrements therefore will be

$$\begin{aligned} p(\mathbf{D}|\Theta) &= \prod_{i=1}^{N} \mathcal{N}(\mathbf{D}_i|\mathbf{m}_i(\Theta), \mathbf{C}) \\ &= \left(\frac{1}{\sqrt{2\pi\,\det(\mathbf{C})}}\right)^{N} e^{\sum_{i=1}^{N} -\frac{1}{2}(\mathbf{D}_i-\mathbf{m}_i)^T\mathbf{C}^{-1}(\mathbf{D}_i-\mathbf{m}_i)} \end{aligned} \tag{7.8}$$

where $\mathbf{D} = \{\mathbf{D}_1, \ldots, \mathbf{D}_N\}$, $\mathbf{D}_i = D_i(t_1), D_i(t_2), \ldots D_i(t_M)$ $(i = 1, 2, \ldots, N)$, and $\mathbf{m}_i(t_{k-1}) \equiv E\Big[f_i(\mathbf{X}(t_{k-1}), \theta_i)\Big]$.

The Eq. (7.8) can be optimized w. r. t. the parameters $\Theta = (\theta_1, \theta_2, \ldots, \theta_N)$ of the model to yield estimates of the parameters themselves and of the noise level. The chief numerical problem of this approach is the computation of the expectations of the rate functions given by equation (7.7). Non-integer values of the coefficients α can make estimating the integral analytically difficult. We propose an approximate method in which the Gaussian noise is replaced by an approximate uniform (white) noise, with the amplitude of the uniform noise being obtained as a sample from the Gaussian cumulative distribution function. At the first order, for small σ, we can approximate the Gaussian with zero mean and variance σ with an uniform distribution defined on the interval $[-\frac{\sqrt{2\pi}\sigma}{4}, \frac{\sqrt{2\pi}\sigma}{4}]$, so that

$$\prod_{i=1}^{K_i} p_i = \prod_{i=1}^{K_i} \chi_i \tag{7.9}$$

where

$$\chi_i(X_i) = \begin{cases} \frac{2}{\sqrt{2\pi}\sigma} & \text{if } -\frac{\sqrt{2\pi}\sigma}{4} \leq X_i \leq \frac{\sqrt{2\pi}\sigma}{4} \\ 0 & \text{otherwise.} \end{cases}$$

Published by Woodhead Publishing Limited, 2013

This approximation makes the calculation of the expectation value of the rate equation (Eq. (7.7)) simpler and reduces the computational time of the procedure. Moreover, experiments not illustrated in this paper demonstrate that it does not influence the accuracy of the parameter estimates until σ is less that 30% of the concetration measurement.

Substituting Eq. (7.9) in Eq, (7.7) gives

$$E[f_i(\mathbf{X}^{(i)}(t_{k-1}),\theta)] = \left(\frac{2}{\sqrt{2\pi}\sigma}\right)^{K_i} \int_{\hat{X}-\frac{\sqrt{2\pi}\sigma}{4}}^{\hat{X}+\frac{\sqrt{2\pi}\sigma}{4}} f_i(\mathbf{X}^{(i)}(t_{k-1}),\theta_i) d\mathbf{X}^{(i)} \tag{7.10}$$

Now, substituting Eq. (7.2) in Eq. (7.10) leads to

$$\begin{aligned} &E[f_i(\mathbf{X}^{(i)}(t_{k-1}),\theta_i)] = \\ &= \left(\frac{2}{\sqrt{2\pi}\sigma}\right)^{K_i} \left\{ \sum_{h=1}^{N_i} \theta_{ih} \left[\left(\frac{\sqrt{2\pi}\sigma}{2}\right)^{\#(S-S_h)} \times \right.\right. \\ &\left.\left. \times \prod_{w\in S_h} \frac{1}{\alpha_w+1} \left(\left(\hat{X}_w + \frac{\sqrt{2\pi}\sigma}{4}\right)^{\alpha_w+1} - \left(\hat{X}_w - \frac{\sqrt{2\pi}\sigma}{4}\right)^{\alpha_w+1} \right) \right] \right\} \end{aligned} \tag{7.11}$$

where S is the set containing the indexes referring to all the K_i species appearing in f_i, and $\alpha_w \neq -1$. In case some orders are equal to -1 Eq. (7.11) takes the following form

$$\begin{aligned} E[f_i(\mathbf{X}^{(i)}(t_{k-1}),\theta_i)] &= \left(\frac{2}{\sqrt{2\pi}\sigma}\right)^{K_i} \sum_{h=1}^{N_i} \theta_{ih} \left\{ \left(\frac{\sqrt{2\pi}\sigma}{2}\right)^{\#(S-S_h)} \times \right. \\ &\times \left[\prod_{w\in S'_h} \frac{1}{\alpha_w+1} \left(\left(\hat{X}_w + \frac{\sqrt{2\pi}\sigma}{4}\right)^{\alpha_w+1} - \left(\hat{X}_w - \frac{\sqrt{2\pi}\sigma}{4}\right)^{\alpha_w+1} \right) \right] \times \\ &\left. \times \left[\prod_{w\in S''_h} \ln \frac{\hat{X}_w + \frac{\sqrt{2\pi}\sigma}{4}}{\hat{X}_w - \frac{\sqrt{2\pi}\sigma}{4}} \right] \right\} \end{aligned} \tag{7.12}$$

where S'_h is the set of indexes $\{h'_1, h'_2, \ldots, h'_s\}$ such that $\alpha_{h'} \neq -1 \;\forall h' \in S'_h$, and S''_h is the set of indexes $\{h''_1, h''_2, \ldots, h''_s\}$ such that $\alpha_{h''} = -1 \;\forall h'' \in S''_h$.

If in the Eq. (7.8), $\mathbf{m}_i$ is substituted with the expression (7.11) or (7.12), Eq. (7.8) becomes more tractable and can be optimized w. r. t. the parameters $\Theta = (\theta_1, \theta_2, \ldots, \theta_N)$ and σ. The values of the model's parameters for which $p(\mathbf{D}|\Theta)$ has

Published by Woodhead Publishing Limited, 2013

a maximum are the most likely values giving the observed kinetics.

7.2.1 Restriction of the parameter space

The search for the optimal values of rate constants can be made more efficient if we provide the algorithm of optimization of Eq. (7.8) with the initial guesses for these constants. In this way the algorithm does not waste time in exploring large regions of the parameter space or regions in which the model in Eq. (7.3) is not valid. For this purpose, we also developed and included in KInfer a procedure for the automatic calculation of the initial guesses of the parameters. Therefore, the task to direct the inference method to efficiently exploring the parameter space is not left to the user, who often does not have a precise idea about a reasonable value of the parameters.

The derivatives dX/dt at all measured time points t_k can be interpreted as slopes. Given the species i (with $i = 1, \ldots, N$), we can estimate these slopes from the data as $s_i(t_k)$, and approximate the differential equations as

$$s_i(t_k) \approx \left.\frac{dX_i}{dt}\right|_{t=t_k} \tag{7.13}$$

If the data consist of N species and the concentration of each species i is measured at M time points $(X_i(t_1), X_i(t_2), \ldots, X_i(t_M))$, we estimate $M \times N$ slopes $s_i(t_k)$ $(k = 1, \ldots, M)$. In fact, for each species we have M differential equation of the form

$$s_i(t_k) \approx f_i(X_1(t_k), X_2(t_k), \ldots, X_N(t_k); \theta_{i1}, \theta_{i2}, \ldots, \theta_{iN_i}) \tag{7.14}$$

that form a system of M algebraic equations with $M \times N_i$ unknown variables θs, as the slopes s are measurable from the data. In general $M \neq N_i$ $M \gg N_i$: more often $M \gg N_i$ so that the system of $M \times N_i$ equation results overdetermined, but $M < N_i$. The prediction intervals for the parameter can thus be obtained by computing the solution of the system

Published by Woodhead Publishing Limited, 2013

(7.14) with the least squares procedure for overdetermined systems. Note that at this stage we are not interested in a very precise estimate of the rate constants, but only in an approximate guess. Note also that the least squares method should be considered only as a method of fitting a line to a set of data, not as a method of statistical inference. We think of parameters calculate with least squares method, it might be better to call them least squares *solutions* rather than least squares *estimates*, because they are the solutions of the mathematical problem of minimizing the residual sum of squares rather then estimates derived from a statistical model. However, if we make the assumption that the residuals are normally distributed and independent with the same variance σ^2, then the maximum likelihood approach to the estimation of model parameters from data yields the classical formulae for least squares.

A system of equations similar to the system (7.14) can be written also for the experimental uncertainties Δs_i affecting the slopes s_i:

$$\Delta s_i(t_k) \approx \Delta f_i(X_1(t_k), X_2(t_k), \ldots, X_N(t_k); \theta_{i1}, \theta_{i2}, \ldots, \theta_{iN_i}) \tag{7.15}$$

where

$$\begin{aligned} \Delta s_i =& \Delta\left[\theta_{i1} \prod_{j\in S_1\subseteq[1,N]} X_j^{\alpha_j}\right] + \Delta\left[\theta_{i2} \prod_{j\in S_2\subseteq[1,N]} X_j^{\alpha_j}\right] \\ &+ \cdots + \Delta\left[\theta_{iN_i} \prod_{j\in S_{N_i}\subseteq[1,N]} X_j^{\alpha_j}\right] \end{aligned} \tag{7.16}$$

By using the standard formulas of the error propagation, a single term of the sum on the right-hand side of Eq. (7.16) is

$$\begin{aligned} &\frac{\Delta\left[\theta_{i1} \prod_{j\in S_1\subseteq[1,N]} X_j^{\alpha_j}\right]}{\left|\theta_{i1} \prod_{j\in S_1\subseteq[1,N]} X_j^{\alpha_j}\right|} = \frac{\Delta\theta_{i1}}{\theta_{i1}} + \frac{\Delta\left[\prod_{j\in S_1\subseteq[1,N]} X_j^{\alpha_j}\right]}{\left|\prod_{j\in S_1\subseteq[1,N]} X_j^{\alpha_j}\right|} \\ &= \frac{\Delta\theta_{i1}}{\theta_{i1}} + \sum_{h=1}^{\#S_1} |\alpha_h| \frac{\Delta X_h}{|X_h|} \end{aligned}$$

Published by Woodhead Publishing Limited, 2013

where $\#S_1$ is the cardinality of the set S_1. Therefore, Eq. (7.16) becomes

$$\Delta s_i = \sum_{\nu=1}^{N_i} \left\{ \left(\frac{\Delta\theta_{i\nu}}{\theta_{i\mu}} + \sum_{h=1}^{\#S_\nu} |\alpha_h| \frac{\Delta X_h}{|X_h|} \right) \cdot \left| \theta_{i\mu} \prod_{j \in S_\nu \subseteq [1,N]} X_j^{\alpha_j} \right| \right\} \tag{7.17}$$

By assuming that the measurements of times are not affected by errors, the error Δs_i is calculated from Eq. (7.3) as follows

$$\Delta s_i(t_k) = \frac{1}{t_k - t_{k-1}} \Big(\Delta X_i(t_k) - \Delta X_i(t_{k-1}) \Big)$$

where $\Delta X_i(t_k)$ is the experimental error on the measurement of concentration of species i at time t_k. Therefore $\Delta s_i(t_k)$ can be obtained from the data, and the system (7.17) can be solved to find the size of the prediction intervals $\Delta\theta$ of θs with the same procedure used for the system (7.14). These intervals are also approximate measures of the errors that propagate to the rate constants from the concentration measurements.

7.2.2 Variance of the estimated parameters

To seek the parameter matrix Θ that maximizes the function in Eq. (7.8) is equivalent to seek the parameter matrix Θ that maximizes the log-likelihood function given by

$$\begin{aligned} \ln p(\mathbf{D}|\Theta) = &-\frac{N}{2}\Big(\ln(2\pi) \ln(\det (\mathbf{C})) \Big) \\ &-\frac{1}{2}\sum_{i=1}^{N} \Big((\mathbf{D}_i - \mathbf{m}_i)^T \mathbf{C}^{-1} (\mathbf{D}_i - \mathbf{m}_i) \Big) \end{aligned} \tag{7.18}$$

Maximizing the log-likelihood function amounts to minimizing the last term of (7.18) since the other terms do not depend on Θ. The estimation problem is therefore reformulated as follows:

Published by Woodhead Publishing Limited, 2013

$$\Theta^{MLE} = \arg\min_{\Theta} \sum_{i=1}^{N} \left((\mathbf{D}_i - \mathbf{m}_i)^T \mathbf{C}^{-1} (\mathbf{D}_i - \mathbf{m}_i) \right) \quad (7.19)$$

The maximum likelihood estimate Θ^{MLE} has the following appealing asymptotic properties: it is asymptotically unbiased (i.e. $E(\Theta^{MLE}) = \Theta^*$, where Θ^* denotes the vector of the true values of Θ), consistent, asymptotically efficient and asymptotically Gaussian [3]. The latter implies that the distribution of the Θ^{MLE} converges to a normal distribution with a covariance matrix given by the Cramér-Rao bound that is also the inverse of the Fisher information matrix

$$\mathbf{F}_{\Theta^{MLE}} = \sum_{i=1}^{N} \mathbf{G}_i^T \mathbf{C}^{-1} \mathbf{G}_i \quad (7.20)$$

where $\mathbf{G}_i$ is called *sensitivity matrix*.

All the matrices $\mathbf{G}_i$ can be obtained from the sensitivity matrix $\mathbf{S}(t)$ evaluated at the sampling instants. The sensitivity matrix is a $N \times P$ time-dependent matrix, where N is the number of species and p is the length of the parameter vector Θ. It is defined as follows

$$\mathbf{S}(t) = \frac{\partial \ln \mathbf{m}(t, \Theta)}{\partial \ln \Theta} = \left(\frac{1}{\mathbf{m}(t, \Theta)} \frac{\partial \mathbf{m}(t, \Theta)}{\partial \Theta} \right) \Bigg|_{\Theta = \Theta^{MLE}} \quad (7.21)$$

The $\mathbf{G}_i$ matrices are obtained from $\mathbf{S}(t)$ as

$$\mathbf{G}_i = [s_i(t_1)^T, \ldots, s_i(t_M)^T]^T$$

where $s_i(t_k)$, $(k = 1, \ldots, M)$, is the i-th row of $\mathbf{S}(t_k)$ $(i = 1, \ldots, N)$, where

$$\mathrm{S}(t_k) = \begin{pmatrix} \frac{\partial \ln m_1(t_k)}{\partial \ln \theta_{i1}} & \cdots & \frac{\partial \ln m_1(t_k)}{\partial \ln \theta_{iN_i}} & \cdots & \frac{\partial \ln m_1(t_k)}{\partial \ln \theta_{NP}} & \\ \frac{\partial \ln m_2(t_k)}{\partial \ln \theta_{i1}} & \cdots & \frac{\partial \ln m_2(t_k)}{\partial \ln \theta_{iN_i}} & \cdots & \frac{\partial \ln m_2(t_k)}{\partial \ln \theta_{NP}} & \\ \vdots & \vdots & \vdots & \vdots & \vdots & \vdots \\ \frac{\partial \ln m_N(t_k)}{\partial \ln \theta_{i1}} & \cdots & \frac{\partial \ln m_N(t_k)}{\partial \ln \theta_{iN_i}} & \cdots & \frac{\partial \ln m_N(t_k)}{\partial \ln \theta_{NP}} \theta_{NP} & \end{pmatrix}$$

and thus

Published by Woodhead Publishing Limited, 2013

$$\mathbf{G}_i^T = \begin{pmatrix} \frac{\partial \ln m_i(t_1)}{\partial \ln \theta_{i1}} & \frac{\partial \ln m_i(t_2)}{\partial \ln \theta_{i1}} & \cdots & \frac{\partial \ln m_i(t_M)}{\partial \ln \theta_{i1}} \\ \vdots & \vdots & \vdots & \vdots \\ \frac{\partial \ln m_i(t_1)}{\partial \ln \theta_{iN_i}} & \frac{\partial \ln m_i(t_2)}{\partial \ln \theta_{iN_i}} & \cdots & \frac{\partial \ln m_i(t_M)}{\partial \ln \theta_{iNi}} \\ \frac{\partial \ln m_i(t_1)}{\partial \ln \theta_{i+1\ 1}} & \frac{\partial \ln m_i(t_2)}{\partial \ln \theta_{i+1\ 1}} & \cdots & \frac{\partial \ln m_i(t_M)}{\partial \ln \theta_{i+1\ 1}} \\ \vdots & \vdots & \vdots & \vdots \\ \frac{\partial \ln m_i(t_1)}{\partial \ln \theta_{NP}} & \frac{\partial \ln m_i(t_2)}{\partial \ln \theta_{NP}} & \cdots & \frac{\partial \ln m_i(t_M)}{\partial \ln \theta_{NP}} \end{pmatrix}$$

An element $\gamma_{ab}^{(i)}(t)$ of the $\mathbf{G}_i$ ($a = 1, 2, \ldots, M$ and $b = 1, 2, \ldots, N$), is

$$\gamma_{ab}^{(i)}(t) = \left(\frac{1}{\mathbf{m}_i}\frac{\partial \mathbf{m}_i}{\partial \theta_{ib}}\right)_{\Theta=\Theta^{MLE}} = \left(\frac{1}{\mathbf{m}_i}\right)_{\Theta=\Theta^{MLE}} \left(\frac{\sqrt{2\pi}\sigma}{2}\right)^{-\#(S_b)} \times \prod_{w \in S_b} \left[\left(\hat{X}_w(t) + \frac{\sqrt{2\pi}\sigma}{4}\right)^{\alpha_w+1} - \left(\hat{X}_w(t) - \frac{\sqrt{2\pi}\sigma}{4}\right)^{\alpha_w+1}\right] \frac{1}{\alpha_w + 1} \tag{7.22}$$

if $\alpha_w \neq -1 \ \forall \ w \in S_b$.

The square root of the p-th diagonal element of $\mathbf{F}^{-1}_{\Theta^{MLE}}$ gives an estimate of the standard deviation of the p-th component of Θ [7, 23]. Different estimates of the variance of the estimated parameters can be obtained from different covariance matrix. A rough estimate of the parameter variance can be obtained by replacing the covariance matrix C by the identity matrix, which amounts to consider uncorrelated the $\mathbf{m}_i$ vectors. A less rough estimate of this variance can be obtained replacing C by C/σ^2, where σ is the MLE of σ [7].

We end this section by noting that KInfer is the software tool implementing the procedure of inference of the parameters with their variances. KInfer is fully integrated with a design and simulation environment [8] based on the BlenX language [9] developed by our resarch group. KInfer can feed the BlenX models with rate constants for running simulation due to the clear de-coupling of qualitative and quantitative descriptions of the programming language approach.

Published by Woodhead Publishing Limited, 2013

7.3 Synthetic case study: buffering SERCA pump

Calcium oscillation in non-excitable cells act as a messenger between extracelluair stimulations and cell function, such as secretion of enzymes. The oscillations are the result of an influx of calium into the cytosol from the endoplasmic reticulum (ER) through the inositol triphosphate receptors (IP_3R) and the ryanodine receptors (RyR) followed by reuptake of calcium into the ER through the sarcoplasmic/endoplasmic reticulum calcium ATPase (SERCA) pumps. Many models have been constructed to reproduce calcium oscillations, and all these models contain a model of the SERCA pump. The SERCA pump uses the chemical energy produced from the conversion of adenosine triphosphate (ATP) into adenosine diphosphate (ADP) to transport calcium ions across the membrane from the cytosol to the ER, against a concentration gradient.

When calcium ions are transported into the ER through the SERCA pump, they are bound to pump proteins on the cytosolic side of the membrane. The protein undergoes a conformational change, which is powered by the energy released from the conversion of ATP in ADP, and the calcium ions are then released on the ER side of the membrane. Although the calcium ions are bound to the pump protein, they do not contribute to the calcium concentration inside the cytosol or to the calcium concentration inside the ER. For this reason the calcium is said to be *buffered* by the SERCA pump. Because there is a large number of of pump protein present (15-75μM/L in a cardiac ventricular cell [1]), the pump is able to bind a large amount of calcium and so the buffering effect is significant [14]. Moreover, when the SERCA pump transports calcium, the amount of calcium bound on the cytosolic side of the membrane of the membrane may not always be equal to the amount released on the ER side, as some calcium remains bound to the pump proteins.

Many different models have been proposed for describing the dynamics of the SERCA pumps. Maclennan et al. [28]

Published by Woodhead Publishing Limited, 2013

proposes a reaction cycle involving four transitions, including the binding of cytosolic calcium, the change of conformation of the pump proteins, the release of calcium on the ER side of the membrane and the return to the original conformation. Others suggest a larger number of reactions in the cycle. For example, Stockes and Green [40] give a reaction diagram with eight reactions. Dode et al. [10] give a model with six transitions. Lauger suggests a twelve-state model with twenty four reactions [22]. The SERCA model we consider here is based on the four-state diagram given in Fig. 7.1 and proposed by Higgins et al. [14]. It contains the same transitions as the reaction cycle given by Maclennan et al. in [28]. Table 7.1 report the reactions illustrated in the cycle of Fig. 7.1. X_1 denotes the pump protein on the cytosolic side with no calcium bound, X_2 denotes the pump protein on the cytosolic side with two calcium ions bound, and Y_1 and Y_2 are the analogous on the ER side. C is the calcium on the cytosolic side and C_E is the calcium on the ER side. Using the state diagram we can write down the set of reactions and convert them into a system of rate equations through the law of mass action. The parameters used in the model are given in Table 7.2: the values of k_2, k_4, k_{-2}, and k_{-4} have been determined by Higging et al. in [14], whereas the values of k_1, k_3, k_{-1} and k_{-3} are assumed to be one order of magnitude greater than the others. Higgings et al. [14] assume that these rate constants are fast and that $k_1/k_{-1} = 0.7$ $(\mu\text{M/L})^2$ and $k_3/k_{-3} = 1.11 \times 10^{-5}$ $(\mu\text{M/L})^2$.

We report the results of our inference procedure applied to two synthetic datasets: in order to be able to compare our results with those published by [14], Dode et al. [10], and Yano et al. [51], the first has been generated considering $C_E(0) = 10$ μmol/L ER (Configuration 1) and the second has been generated considering $C_E(0) = 150$ μmol/L ER (Configuration 2). In both cases $C_(0) = 5$ μmol/L CYT. The time-courses obtained from these two initial configurations has been then perturbed by a noise with variance equal to the 5% of the measurements (this value of variance reflect the typical precision achieved in the experiments). For the first configuration the average noise variances (in μmol/L)

Published by Woodhead Publishing Limited, 2013

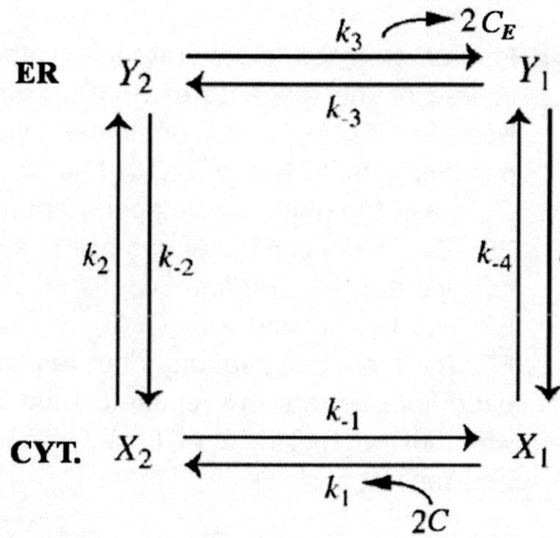

Figure 7.1 State diagram of the SERCA pumps. Adapted from [14].

Table 7.1 Set of reactions derived from the four-state SERCA pump diagram in Fig. 7.1.

Reaction
$X_1 + C + C \xrightarrow{k_1} X_2$
$X_2 \xrightarrow{k_2} Y_2$
$Y_2 \xrightarrow{k_3} Y_1 + C_E + C_E$
$Y_1 \xrightarrow{k_4} X_1$
$X_2 \xrightarrow{k_{-1}} X_1$
$Y_2 \xrightarrow{k_{-2}} X_2$
$Y_1 \xrightarrow{k_{-3}} Y_2$
$X_1 \xrightarrow{k_{-4}} Y_1$

are the following $\sigma_C = 0.0083$, $\sigma_{CE} = 0.0083$, $\sigma_{X_1} = 1.66$,

Published by Woodhead Publishing Limited, 2013

Table 7.2 Parameters used to generate the synthetic time-course of the reagents of SERCA pumps [14]. Note that X_1, X_2, and C are in the units μmoles per liter cytosol (μmol/L Cyt) and Y_1, Y_2, and C_E are in the units μmoles per liter ER (μmol/L ER). Two initial configurations of the system are considered: the first with $C_E(0) = 10$ μmol/L ER, and the second with $C_E(0) = 150$ μmol/L ER. The first configuration is not realistic, but is used here so that the results can be compared with those of Higgins et al. [14], Dode et al. [10], and Yano et al. [51]

Parameter	Value (unit)
k_1	0.1 (sec^{-1})
k_{-1}	1 (sec^{-1})
k_2	2 (sec^{-1})
k_{-2}	0.97 (sec^{-1})
k_3	0.5 (sec^{-1})
k_{-3}	0.1
k_4	0.4 (sec^{-1})
k_{-4}	1.2×10^{-3} (sec^{-1})
Reactant	**Initial concentration (unit)**
X_1	5 (μmol/L Cyt)
X_2	5 (μmol/L Cyt)
Y_1	15 (μmol/L ER)
Y_2	15 (μmol/L ER)
C	5 (μmol/L Cyt)
C_E	10 and 150 (μmol/L ER)

$\sigma_{X_2} = 0.04$, $\sigma_{Y_1} = 0.19$, and $\sigma_{Y_2} = 0.1$. For the second configuration, we obtained $\sigma_C = 8.85$, $\sigma_{CE} = 0.67$, $\sigma_{X_1} = 0.04$, $\sigma_{X_2} = 0.19$, $\sigma_{Y_1} = 0.2$, and $\sigma_{Y_2} = 0.1$. We considered these values as belonging to a possible range of values for σ, and thus we generated noisy time-course datasets perturbed by

Published by Woodhead Publishing Limited, 2013

these values of σ and applied the parameter inference procedure on all of them. The estimates of the parameters at different level of noise in the input data are reported in Tables 7.3 and 7.5, and the variances of the estimates are shown in Tables 7.4 and 7.6, for the datasets from Configuration 1 and Configigration 2, respectively. The figures 7.2-7.7 show the comparisons between the simulations obtained with the "true" parameters listed in Table 7.2, that are assumed to be the "experimental" time-courses, and those obtained with the estimated values for different levels of noise in the input data . The simulations confirm the robustness of the inference accuracy to the noise in the data and consequently the agreement with the paramters estimated by [14] and [10]. In particular, for the system in Configuration 1, the figures 7.2 and 7.3 demostrate that the endoplasmic calcium concetration and the concentration of the pump protein on Er side (Y_1) time-course are the most affected by the inaccuracies that from the input data propagate to the rate costants. The kinetics of the unperturbed and the perturbed systems are the same in the first instants of reaction and then stabilises on different equilibrium values. For the system in Configuration 2, we obtained strong agreement between estimated and "experimental" behaviors of the time-course of the endoplasmic calcium concentration (Fig. 7.5), and a bad agreement, due to the significative deviation of k_2, k_{-3} and k_{-4} from the "true" values, for the X_1 and Y_1 time courses for $\sigma = 0.67$ μmol/L (Figures 7.6 and 7.7).

Published by Woodhead Publishing Limited, 2013

Table 7.3 Estimated parameters for the simulation generated with the following initial conditions $C_E(0) = 10\ \mu$mol/L ER and $C(0) = 5\ \mu$mol/L Cyt. .

Parameter	Value ($\sigma = 0.008$)	Value ($\sigma = 0.01$)	Value ($\sigma = 0.04$)	Value ($\sigma = 0.19$)	Value ($\sigma = 1.66$)
k_1	0.0945	0.0954	0.0955	1.104	0.012
k_{-1}	0.164	0.146	0.140	0.024	0.726
k_2	3.669	3.640	3.0556	1.176	1.0526
k_{-2}	0.499	0.439	0.442	0.129	1.281
k_3	0.498	0.497	0.498	0.549	0.194
k_{-3}	0.127	0.123	0.144	0.254	0.074
k_4	0.554	0.557	0.577	0.552	0.444
k_{-4}	0.00183	0.00186	0.00256	0.0287	0.00668

Table 7.4 Variance of the estimated parameters from the time-courses simulated with the following initial conditions: $C_E(0) = 10 \mu mol/L$ ER and $C(0) = 5\ \mu mol/L$ Cyt. .

Parameter	$\sqrt{\gamma} \times 10^2$ ($\sigma = 0.008$)	$\sqrt{\gamma} \times 10^2$ ($\sigma = 0.01$)	$\sqrt{\gamma} \times 10^2$ ($\sigma = 0.04$)	$\sqrt{\gamma} \times 10^2$ ($\sigma = 0.19$)	$\sqrt{\gamma} \times 10^2$ ($\sigma = 1.66$)
k_1	0.382	0.424	2.494	23.771	99.429
k_{-1}	20.756	18.470	18.342	12.065	404.882
k_2	884	899.497	850.718	547.226	235.165
k_{-2}	465.14	641.145	448.693	70.356	453.39
k_3	333.662	324.004	328.261	541.59	114.914
k_{-3}	233.069	230.589	308.925	646.497	100.528
k_4	480.749	483.655	502.737	568.489	193.075
k_{-4}	49.432	50.891	85.66	193.197	415.83

Table 7.5 Estimated parameters for the simulation generated with the following initial conditions $C_E(0) = 150\ \mu$mol/L ER and $C(0) = 5\mu$mol/L Cyt. .

Parameter	Value ($\sigma = 0.04$)	Value ($\sigma = 0.1$)	Value ($\sigma = 0.20$)	Value ($\sigma = 0.67$)	Value ($\sigma = 8.85$)
k_1	0.0953	0.0982	0.099	0.124	0.000109
k_{-1}	0.241	0.453	0.450	1.095	0.134
k_2	2.908	0.473	0.0541	1.386	0.606
k_{-2}	0.45	0.385	0.201	0.0129	0.639
k_3	0.491	0.488	0.446	0.0977	0.448
k_{-3}	0.131	0.151	0.117	0.0198	0.0325
k_4	0.541	0.478	0.426	0.142	1.094
k_{-4}	0.00248	0.00477	0.00566	0.00381	0.0106

Table 7.6 Variance of the estimated parameters from the time-courses simulated with the following initial conditions: $C_E(0) = 150\mu$mol/L ER and $C(0) = 5\ \mu$mol/L Cyt. .

Parameter	$\sqrt{\gamma} \times 10^2$	$\sqrt{\gamma} \times 10^2$	$\sqrt{\gamma} \times 10^2$	$\sqrt{\gamma} \times 10^2$	$\sqrt{\gamma}$
	$(\sigma = 0.04)$	$(\sigma = 0.1)$	$(\sigma = 0.20)$	$(\sigma = 0.67)$	$(\sigma = 8.85)$
k_1	2.096	6.499	30.962	125.249	2090.852
k_{-1}	31.345	68.353	76.958	147.044	145.544
k_2	730.32	227.941	24.572	131.256	131.256
k_{-2}	469.598	302.94	324.685	477.676	1027.918
k_3	327.871	349.081	471.648	36.996	36.996
k_{-3}	248.41	524.356	344.394	104.64	104.64
k_4	488.769	479.545	477.328	866.457	860.835
k_{-4}	78.505	144.568	110.221	621.524	621.288

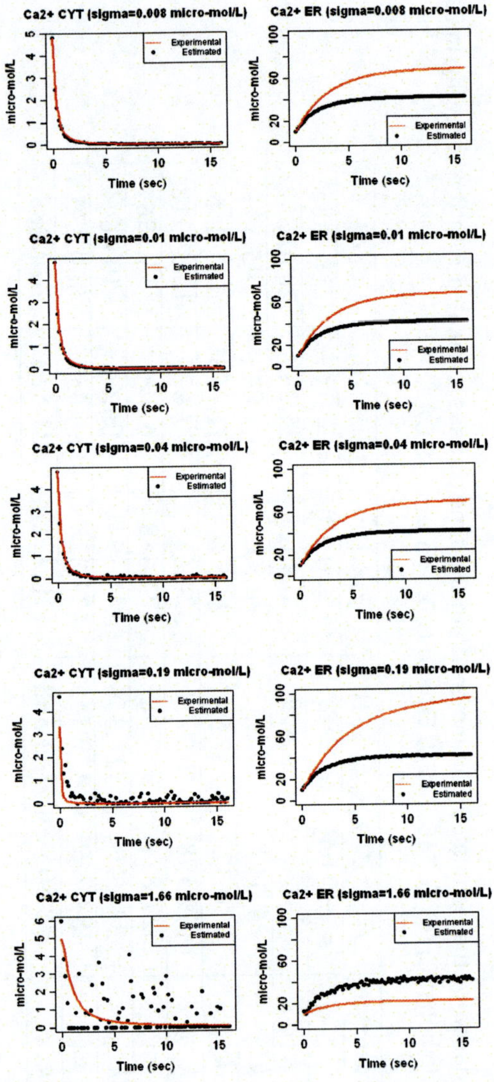

Figure 7.2 Synthetic smoothed and noisy dynamics of cytosolic and ER concentration of Ca^{2+} for $C_E(0) = 10\ \mu$mol/L ER and $C(0) = 5\ \mu$mol/L.

Published by Woodhead Publishing Limited, 2013

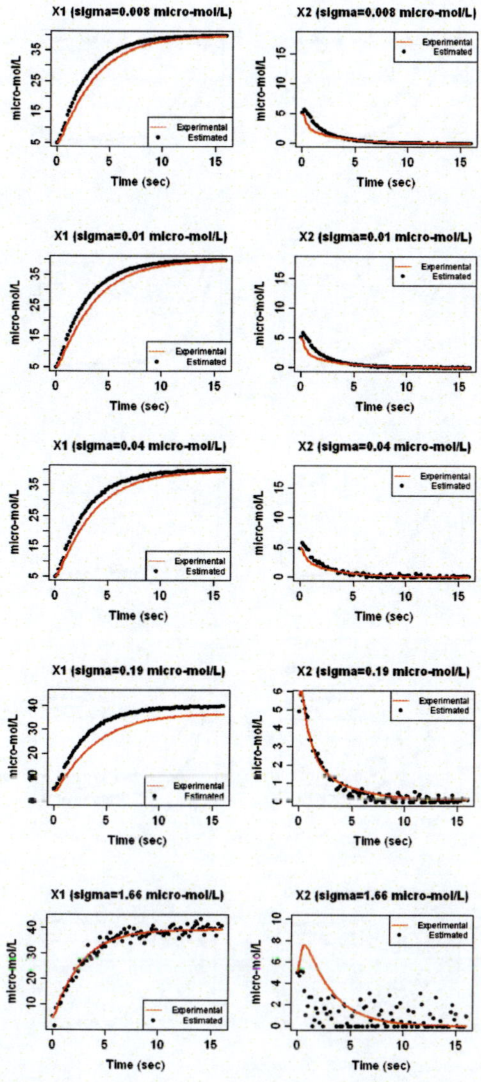

Figure 7.3 Synthetic smoothed and noisy dynamics of cytosolic and ER concentration of X_1 and X_2 for $C_E(0) = 10$ μmol/L ER and $C(0) = 5$ μmol/L.

Published by Woodhead Publishing Limited, 2013

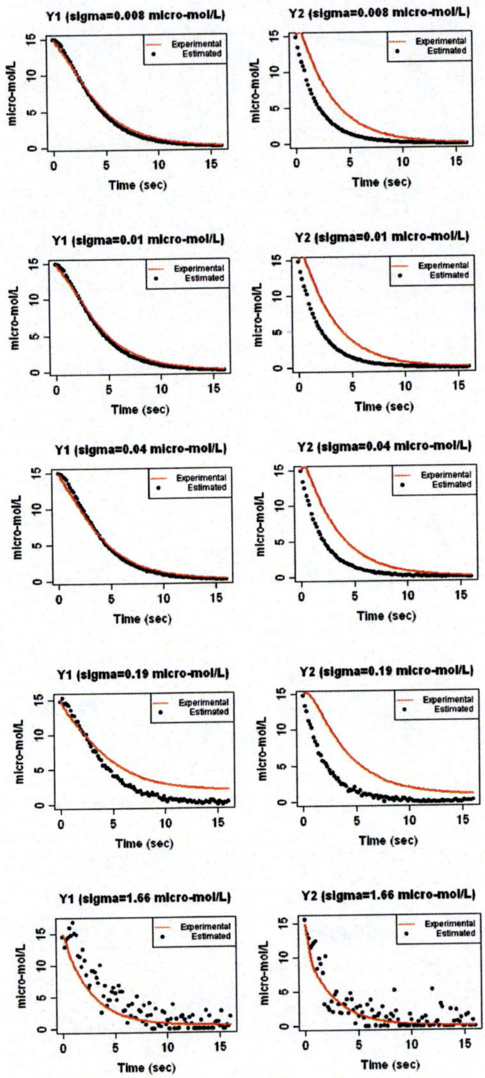

Figure 7.4 Synthetic smoothed and noisy dynamics of cytosolic and ER concentration of Y_1 and Y_2 for $C_E(0) = 10\ \mu$mol/L ER and $C(0) = 5\ \mu$mol/L.

Published by Woodhead Publishing Limited, 2013

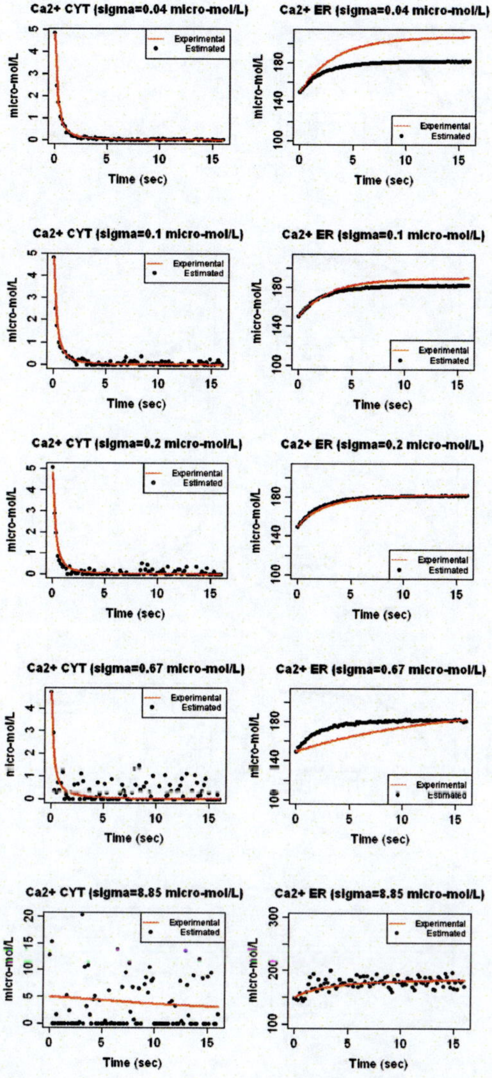

Figure 7.5 Synthetic smoothed and noisy dynamics of cytosolic and ER concentration of Ca^{2+} for $C_E(0)$ = 150 μmol/L ER and $C(0)$ = 5 μmol/L.

Published by Woodhead Publishing Limited, 2013

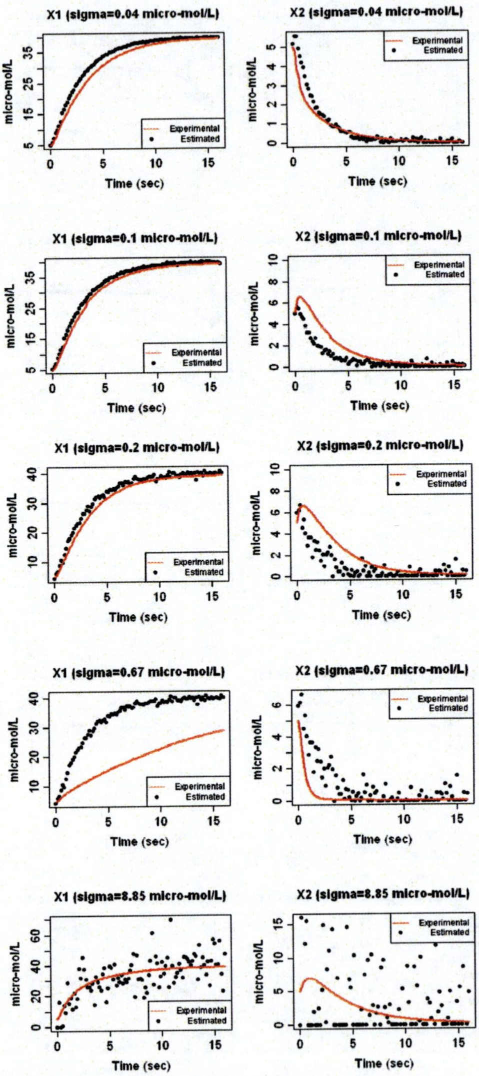

Figure 7.6 Synthetic smoothed and noisy dynamics of cytosolic and ER concentration of X_1 and X_2 for $C_E(0) = 150\ \mu$mol/L ER and $C(0) = 5\ \mu$mol/L.

Published by Woodhead Publishing Limited, 2013

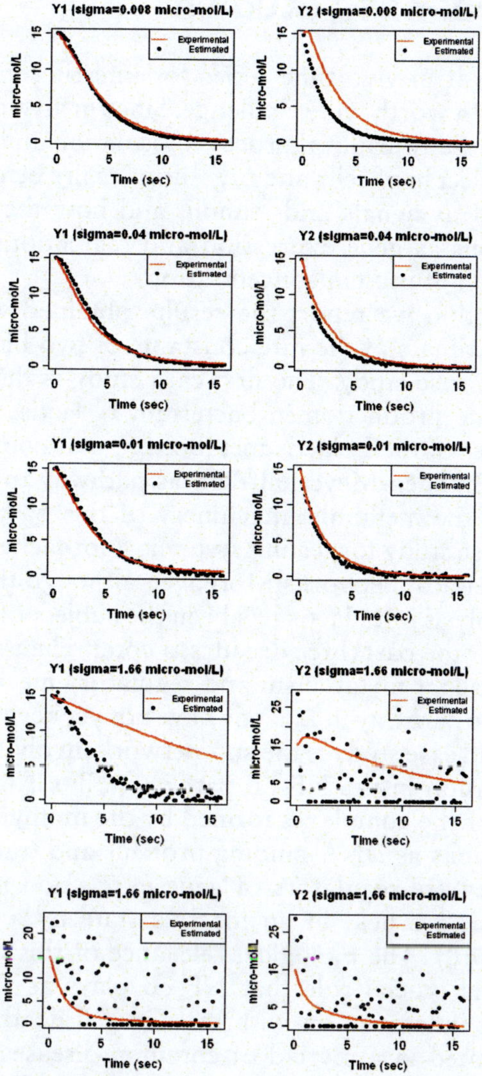

Figure 7.7 Synthetic smoothed and noisy dynamics of cytosolic and ER concentration of Y_1 and Y_2 for $C_E(0) = 150$ μmol/L ER and $C(0) = 5$ μmol/L.

Published by Woodhead Publishing Limited, 2013

7.4 Real case studies

The analysis of in vivo time series for inferring the kinetic parameters is a worthwhile challenge. Since in vivo data have not undergone any artificial process of isolation and purification, they reflect how cells and organisms really behave, how they respond to signals and stimuli, and how they orchestrate functions as gene expression and regulation, and the time evolution of protein concentration.

In this section we report the results obtained with KInfer for the estimate of the rate constants of two biologically relevant real case study. The first case study is the glycolysis and lactate production in bacterium L. lactis. The relative simplicity of the L. lactis metabolism, that converts sugars via the Embden-Meyerhalf-Parnas pathway to pyruvate [44], makes the metabolic machinery of this bacterium an attractive case study for testing systemic approaches to modelling biochemical networks.Moreover, although the regulation of glycolysis in L. lactis has been the subject of intensive research over the past three decades, a comprehensive understanding of sugar metabolism and regulatory mechanism of the glycolytic pathway in L. lactis have not yet been achieved.

The second case study is the sub-network involving the IκB phosphorylation in the NF-κB pathway. NF-κB is a collective name for the complexes formed by the multigene family which functions as DNA-binding proteins and transcription factors. They are regulators of gene expression in eukaryotic cells but are held in an inactive state by a family of inhibitors (IκB). The biological relevance of this case study is due to the crucial role that NF-κB plays as the central mediator of inflammation with roles in cell death; and has been implicated in a myriad of common diseases - such as cancer, arthritis, asthma, diabetes, atherosclerosis and septic shock, to name but a few - and in the regulation of immune responses to infection.

Published by Woodhead Publishing Limited, 2013

7.5 Glucose metabolisms of Lactococcus lactis

We applied our method to infer the rate constants of the biochemical pathway that converts glucose into lactate in the bacterium *L. lactis* [11, 44] (see Fig. 7.8). The experimental data provided by Prof. Eberhard O. Voit consist of the time-series of glucose (X_1), glucose-6-phosphate G6P (X_2), total fructose 1,6-biphosphate FBP (X_3), 3-phosphoglycerate 3-PGA (X_4), phosphoenolpyruvate PEP (X_5), pyruvate (X_6), lactate (X_7), acetate (X_8), ATP and inorganic phosphate P_i. The mathematical model of this pathway has been formulated by Voit et al. [11, 44] as in the equation system (7.23).

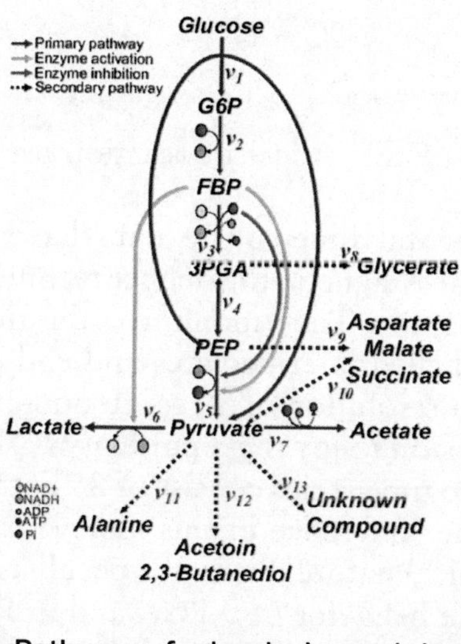

Figure 7.8 Pathway of glycolysis and lactate production in *L. lactis*. Black arrows: flow of material; grey arrows: enzyme activation and inhibition; dashed arrows indicate leakage of material into secondary pathways, that are not considered in the model presented in this paper. This figure has been adapted from [11].

Published by Woodhead Publishing Limited, 2013

$$\frac{dX_1}{dt} = -\nu_1, \quad \frac{dX_2}{dt} = \nu_1 - \nu_2$$
$$\frac{dX_3}{dt} = \nu_2 - \nu_3, \quad \frac{dX_4}{dt} = 2\nu_3 + \nu_4 \tag{7.23}$$
$$\frac{dX_5}{dt} = -\nu_4 - \nu_1 - \nu_5, \quad \frac{dX_6}{dt} = \nu_1 + \nu_5 - \nu_6 - \nu_7$$
$$\frac{dX_7}{dt} = \nu_6, \quad \frac{dX_8}{dt} = \nu_7$$

where

$$\begin{aligned}
\nu_1 &= k_1 X_1^{a_1} X_5^{a_2} \\
\nu_2 &= k_2 X_2^{a_3} \mathrm{ATP}^{a_4} \\
\nu_3 &= k_3 X_3^{a_5} \mathrm{Pi}^{a_6} \\
\nu_4 &= k_4 X_5^{a_7} - k_5 X_4^{a_8} \\
\nu_5 &= k_6 X_5^{a_9} X_3^{a_{10}} \mathrm{Pi}^{a_{11}} + k_7 X_5^{a_{12}} \\
\nu_6 &= k_8 X_6^{a_{13}} X_3^{a_{14}} \\
\nu_7 &= k_9 X_6^{a_{15}} \mathrm{Pi}^{a_{16}}
\end{aligned} \tag{7.24}$$

and the values of the orders of reaction are given in Table 7.7.

The experimental time-course data have been collected with the Nuclear Magnetic Resonance technique by Voit and co-workers [11, 44]. They monitored the time behavior of the pools of labelled intermediates and end products of the pathway with a resolution of 30 sec in non-growing L. lactis bacteria suspension following a pulse of ^{13}C-labelled glucose [44, 32]. All the time series consist of 85 data points.

The parameter inference in this model suffers from difficulties of technical nature due to the peculiarities of the data. In fact, the time behavior of 3-PGA and PEP dip down very quickly, then recover, overshoot and slowly degrades (see Fig. 7.10). Even if it is possible to model such dynamics with generalized mass action law, the search algorithm might find a set of parameters causing the time course to cross over into the negative domain. In this case, non-integer reaction orders force the integration of the equation system to produce results with imaginary values. Moreover, some variables

Published by Woodhead Publishing Limited, 2013

Table 7.7 Values of the partial orders of reaction in model (7.23)-(7.24). The values has been proposed by Goel et al. in [11].

Parameter	Value	Parameter	Value
a_1	0.4	a_9	0.53
a_2	0.81	a_{10}	1.33
a_3	0.74	a_{11}	-0.0001
a_4	0.4	a_{12}	2.30
a_5	0.88	a_{13}	0.46
a_6	0.01	a_{14}	1.04
a_7	0.43	a_{15}	1.0
a_8	0.32	a_{16}	0.46

approach to zero toward the end of the experiment. Due to the numerical inaccuracies of any integration software, these variables may become negative. Also in our specific case, the search algorithm selects a parameter combination such that the simulation of the generalized mass action model with those parameters does not identify a global fit of the experimental data over their entire time domain. namely, the integration with the XPPAUT software stops at $t \approx 6.4$ min. Therefore, to avoid termination of integration, we artificially stopped the simulation of the dynamics of the pathway at $t \approx 6.4$ min, and extrapolated with the Stineman interpolation algorithm the behavior from $t \approx 6.4$ min till $t \approx 40$ min. We maintained the values of the orders of reaction as in Table 7.7 and we estimated the kinetic rate constants of the model. Table 7.8 shows the KInfer estimates of these parameters. The rate constants estimates reproduce the experimental time behavior of the involved species, except for 3-PGA and PEP.

Published by Woodhead Publishing Limited, 2013

Table 7.8 Estimates of the kinetic rate constants of the pathway of regulation of glycolysis in L. lactis. In this experiment, the variances of the estimated parameters are two order of magnitudes bigger than the estimates and indicate the large scale of the parameters of being spread out.

Parameter	Value	Δk	σ_k
k_1	0.388	0.001	16.9
k_2	10.35	0.011	153.25
k_3	1.300496	0.000012	192.78
k_4	89.16	0.12	1602.93
k_5	87.41	0.10	1699.55
k_6	0.0538	0.0011	11.78
k_7	0.0050	0.0008	39.18
k_8	0.00824	0.00007	7.91
k_9	0.012458	0.000006	1.88

7.6 Discussion

The results of the application of our inference procedure to the calibration of models of synthetic and real biochemical networks show that the method converges to the expected solutions within the bounds of the experimental errors that propagates from concentration measurements to the kinetic rate constants. The good estimates obtained confirm the validity of the procedure applied to any kind of reaction and the validity of the discretized model of mass action law for the rate equation. Moreover, some important features missing from the existing methods for parameter inference are present in our method. The first is the automatic computation of the initial guesses of the parameters. In this way, the user is not forced to insert any *a priori* knowledge about the system, that often is quite hard to find. At the same time, the method is equipped with a rigorous procedure referring only to the experimental concentration measurements to identify

Published by Woodhead Publishing Limited, 2013

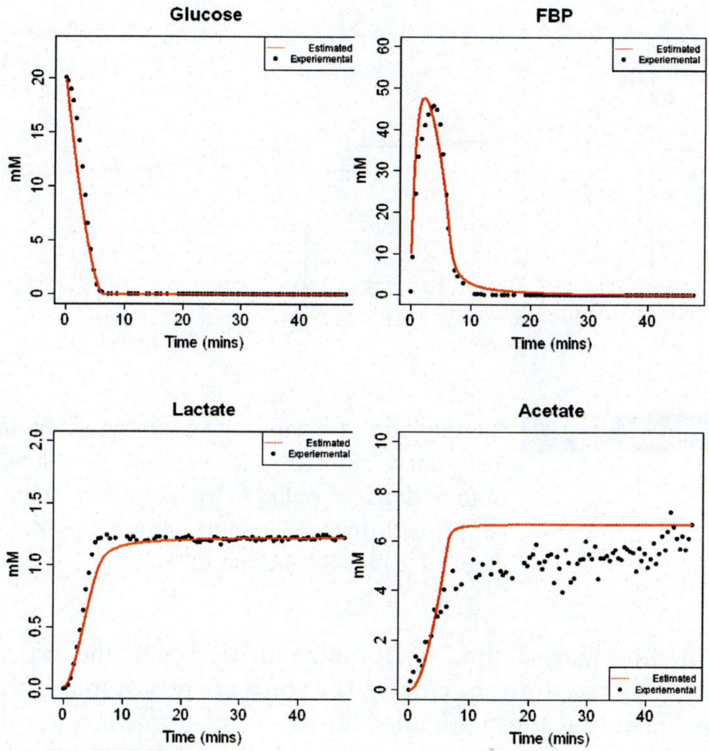

Figure 7.9 Comparison between experimental behavior (black circles) and estimated behavior obtained as a solution of equation system (7.23) with the parameters inferred by KInfer (Table 7.8).

a region of the parameter space where the optimization of the probability density function takes place. The second feature is the implementation of the experimental error propagation. The evaluation of the experimental uncertainty on the rate constants estimates is particularly useful if the parameter inference is incorporated in projects of experimental design. The size of the errors on the kinetic constants is indicative of the optimality of the experimental setup. Thus, any procedure devoted to the reduction of this error is definitely part of

Published by Woodhead Publishing Limited, 2013

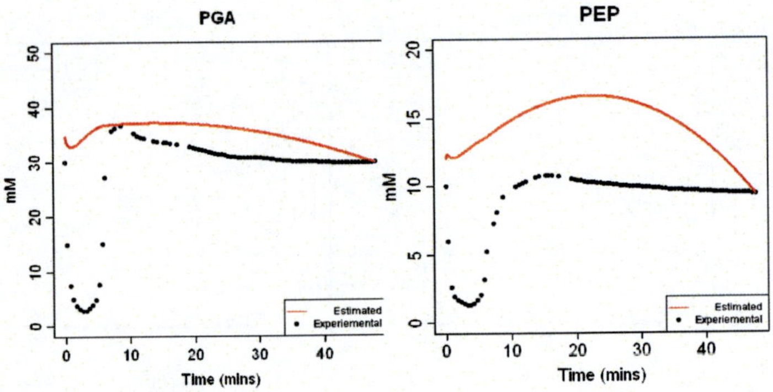

Figure 7.10 Comparison between experimental behavior (black circles) and estimated behavior obtained as a solution of equation system (7.23) with the parameters inferred by KInfer (Table 7.8) for 3-PGA and PEP.

a methodology aiming to optimize the design of the experimental configuration. Finally, the software prototype KInfer, implementing the mathematical model of inference, can be used for interfacing the outcomes of the wet-lab activity for the concentration measurements with the softwares for modelling and simulation of biochemical networks.

References

1. D. M. Bers. *Excitation-contraction coupling and cardiac contractile force.* Kluwer Academic, The Netherlands, 2nd edition, 2001.
2. R. J. Boys, D. J. Wilkinson, and T. B. Kirkwood. *Bayesian inference for a discretely observed stochastic kinetic model.* Statistics and Computing. Springer Netherlands, 2008.
3. G. Casella and R. L. Berger. *Statistical inference.* Duxbury, 2002.

Published by Woodhead Publishing Limited, 2013

4. K.H. Cho, W. Koch, and O. Wolkenhauer. Experimental design in systems biology, based on paramter sensitivity analysis using a monte carlo method: a case study for the tnfα-mediated nf-κb signal transduction pathway. *Simulation*, 79(12):726–739, 2003.
5. R. J. Cho, M. J. Campbell, E. A. Winzeler, L. Steinmetz, A. Conway, L. Wodicka, T. G.Wolfsberg, A. E. Garielian, D. Landsman, D. J. Lockhart, and R. W: Davis. A genome-wide transcriptional analysis of the mitotic cell cycle. *Molecular Cell*, 2:65–73, 1998.
6. I. C. Chou, H. Martens, and E. O. Voit. Parameter estimation in biochemical systems models with alternating regression. *Theoretical Biology and Medical Modelling*, 3(25), 2006.
7. B. David and G. Bastin. A maximum likelihood parameter estimation method for nonlinear dynamicsl systems. In *38th Conference on Decision & Control*, Phoenix - Arizona USA, December 1999.
8. L. Dematté, C. Priami, and A. Romanel. The beta workbench: a computational tool to study the dynamics of biological systems. *Briefings in bioinformatics*, 9(5):437–449, 2008.
9. L. Dematté, C. Priami, and A. Romanel. The blenx language: A tutorial. *LNCS*, 5016:313–365, 2008.
10. L. B. Dode, K. Vilsen, F. van Baelen, F. Wuytack, J. D. Clausen, and J. P. Andersen. Dissection of the functional differences between sarco(endo)plasmic reticulum ca^{2+}-atpase (serca) 1 and 3 isoforms by steady state and transient kinetic analyses. *J. Biol. Chem.*, 277:45579–45591, 2002.
11. G. Goel, I-C. Chou, and E. O. Voit. System estimation from metabolic time-series data. *Bioinformatics*, 24(21):2505–2511, 2008.
12. D. E. Goldberg. *Genetic algorithms in search, optimization and machine learning*. Addison-Wesley, Massachusetts, 1989.
13. A. Golightly and D. J. Wilkinson. Bayesian inference for nonlinear multivariate diffusion models observed

Published by Woodhead Publishing Limited, 2013

with error. computational statistics and data analysis. *Computational statistics and data analysis*, 52(3):1674–1693, 2008.

14. E. R. Higgins, M. B. Cannel, and J. Sneyd. A buffering serca pump in models of calcium dynamics. *Biophysical Journal*, 91:151–163, 2006.
15. W. S. Hlavacek and M. A. Savageau. Rules for coupled expression of regulator and effector genes in inducible circuits. *J. Mol. Biol.*, 255:121–139, 1996.
16. S. Hoops, S. Sahle, R. Gauges, C. Lee, J. Pahle, and N. Simus. Copasi - a complex pathway simulator. bioinformatics. *Bioinformatics*, 22:3067–3074, 2006.
17. I-C.Chou and E. O. Voit. Recent developments in parameter estimation and structure identification of biochemical and genomic systems. *Mathematical Biosciences*, 219(2):57–83, 2009.
18. A. E. C. Ihekwaba, D. S. Broomhead, R. Grimley, N. Benson, and D. B. Kell. Iκbα. *Systems Biology*, 1:99–103, 2004.
19. A. E. C. Ihekwaba, D. S. Broomhead, R. Grimley, N N. Benson, M. R. H. White, and D. B. Kell. Iκbα. *IEE Proceedings Systems Biology*, 152:153–160, 2005.
20. A. E. C. Ihekwaba, S. J. Wilkinson, D. S. Broomhead, D. Waithe, R. Grimley, N. Benson, and D. B. Kell. Bridging the gap between in silico and cell based analysis of the nf-κb signalling pathway by in vitro studies of ikk2. *FEBS Journal*, 27:1678–1690, 2007.
21. S. Kikuchi, D. Tominaga, M. Arita, K. Takahashi, and M. Tomita. Dynamic modeling of genetic networks using genetic algorithm and s-system. *Bioinformatics*, 10(5):643–650, 2003.
22. P. Lauger. *Electrogenic ions pumps*. Addison-Wesley, L. Nadels and D. Stein eds, 1991.
23. P. Lecca, A. Palmisano, A. Ihekwaba, and C. Priami. Calibration of dynamic models of biological systems with kinfer. *European biophysics journal*, 29(6):1019–1039, 2010.

Published by Woodhead Publishing Limited, 2013

24. P. Lecca, A. Palmisano, and C. Priami. Inferring rate coefficients of biochemical reactions from noisy data with kinfer. Technical Report 17, The Microsoft Research - University of Trento Centre for Computational and Systems Biology, http://www.cosbi.eu/Rpty_Tech.php, 2008.
25. P. Lecca, A. Palmisano, and C. Priami. Deducing chemical reaction rate constants and their regions of confidence from noisy measurements of time series of concentration. In *11th Int. Conference on Computer Modelling and Simulation (UKSim 2009), Cambridge - England*, 2009. In press.
26. P. Lecca, A. Palmisano, C. Priami, and G. Sanguinetti. A new method for inferring rate coefficients from experimental time-consecutive measurements of reactant concentrations. In *Int. Conf. on Systems Biology, Long Beach, California*, www.icsb07.org, 2007.
27. P. Lecca, A. Palmisano, C. Priami, and G. Sanguinetti. A new probabilistic generative model of parameter inference in biochemical networks. In *Proceedings of the 2009 ACM Symposium on Applied Computing '09*, Hawaii USA, 2009.
28. D. H. MacLennan, W. J. Rice, and N. M. Green. The mechanism of the ca^{2+} transport by sarco(endo)plasmic reticulum ca^{2+}-atpase. *J. Biol. Chem.*, 272:28815–28818, 1997.
29. A. Marin-Sanguino, E. O. Voit, C. Gonzalez-Alcon, and N. V. Torres. Optimization of biotechnological systems through geometric programming. *Theoretical Biology and Medical Modelling*, 4(38), 2007.
30. G. C. Moles, P. Mendes, and J. R. Banga. Parameter estimation in biochemical pathways: a comparison of global optimization methods. *Genome Res.*, 13:2467–2474, 2003.
31. D. E. Nelson, A. E. C. Ihekwaba, M. Elliott, J. R. Johnson, C. A. Gibney, B. E. Foreman, G. Nelson, V. See, C. A. Horton, D. G. Spiller, S. W. Edwards, H. P. McDowell, J. F. Unitt, E. Sullivan, R. Grimley, N. Benson, D. Broomhead, D. B. Kell, and M. R. H. White.

Published by Woodhead Publishing Limited, 2013

Oscillations in nf-κb signaling control the dynamics of gene expression. *Science*, 306:704–708, 2004.

32. A. R. Neves, A. Ramos, H. Costa andI. I. I. van Swam, J. Hugenholtz, V. W. de Vos M. Kleerebezem, and H. Santos. Effect of different nadh oxidase levels on glucose metabolism by lactococcus lactis: kinetics of intracellular metabolite pols determined by in vivo nuclear magnetic resonance. *Appl. Environ. Microbiol.*, 68:6332–6342, 2002.
33. P. K. Polisetty and E. O. Voit. Identification of metabolic system parameters using global optimization methods. *Theoretical Biology and Medical Modelling*, 3(4), 2006.
34. S. Ramsey, D. Orrell, and H. Bolouri. Dizzy: stochastic simulation of large-scale genetic regulatory networks. *J. Bioinform. Comput. Biol.*, 3(2):415–436, 2005.
35. S. Reinker, R. M. Altman, and J. Timmer. Parameter estimation in stochastic biochemical reactions. In *IEEE Proc. Syst. Biol*, volume 153, 2006.
36. M. Rodrigez-Fernandez, P. Mendes, and J. Banga. A hybrid approach for efficient and robust parameter estimation in biochemical pathways. *BioSystems*, 83:248–265, 2006.
37. M. Savageau. *Coupled circuits of gene regulation.* Sequence specificity in transcription and translation. R. Calendar and L. Cold eds, Alan R. Liss, New York, 1985.
38. M. A. Savageau and P. J. Sands. *Completely uncoupled and perfectly coupled circuits for inducible gene regulation.* Canonical non-linear modeling: S-system approach to understanding complexity. E. O. Voit eds, Van Nostrand Reinhold, New York, 1985.
39. M. A. Savageau and E. O. Voit. Power-law approach to modeling biological-systems theory. *J. Ferment. Technol.*, 60:221–228, 1982.
40. D. L. Stokes and N. M. Green. Structure and function of the calcium pumps. *Annu. rev. Biophys. Biomol. Struct.*, 32:445–668, 2003.

Published by Woodhead Publishing Limited, 2013

41. M. Sugimoto, S. Kikuchi, and M. Tomita. Reverse engineering of biochemical equations from time-course data by means of genetic programming. *BioSystems*, 80:155–164, 2005.
42. T. Tian, S. Xu, and K. Burrage. Simulated maximum likelihood method for estimating kinetic rates in gene expression. *Bioinformatics*, 23(1):84–91, 2007.
43. E. O. Voit and J. Almeida. Decoupling dynamical systems for pathway identification from metabolic profiles. *Bioinformatics*, 20:1670–1681, 2004.
44. E. O. Voit, J. Almeida, S. Marino, R. Lall, G. Goel, A. R. Neves, and H. Santos. Regulation of glycolysis in lactococcus lactis: an unfinisched systems biological case study. *IEE Proc.-Syst. Biol.*, 153(4):286–298, 2006.
45. V. Vyshemirsky and M. A. Girolami. Bayesian ranking of biochemical system models. *Bioinformatics*, 24(6):833–839, 2008.
46. V. Vyshemirsky and M. A. Girolami. Biobayes: a software package for bayesian inference in systems biology. *Bioinformatics*, 24(17):1933–1934, 2008.
47. V. Vyshemirsky and M. A. Girolami. Biobayes: Bayesian inference for systems biology, 2008.
48. XPPAUT web page. X-windows phaseplane plus auto, 2008.
49. D. Wilkinson. *Stochastic Modelling for Systems Biology*. Chapman and Hall/CRC, 2006.
50. D. J. Wilkinson. Bayesian methods in bioinformatics and computational systems biology. *Briefings in bioinformatics*, 1(8), 2007.
51. K. Yano, O. H. Petersen, and A. V. Tepikin. Dual sensitivity of sarcoplasmic/endoplasmic ca^{2+}-atpase to cytosolic and endoplasmic reticulum ca^{2+} as a mechanism of modulating cytosolic ca^{2+} oscillations. *Biochem. J.*, 383:353–360, 2004.
52. J. W. Zwolak, J. J. Tyson, and L. T. Watson. Estimating rate constants in cell cycle models. In A. Tentner (ed.), editor, *Proc. High Performance constants in cell cycle models, San Diego*, pages 53–57, 2001.

Published by Woodhead Publishing Limited, 2013

8

Modelling living systems with BlenX

Abstract In recent years the experimental and computational research approaches in life sciences have been abandoning the reductionist vision to adopt a system-level point of view. Unlike the reductionist approach, the framework of systems theory proposes an integrative planning out to model complex biological phenomena. The integrative modelling is the main aspect of systems biology. This emerging discipline describes the activity of biological entities, such as biochemical networks, cells, tissues, organs, and organisms, as the result of the properties and mutual interactions of the single components of these systems. In particular, systems biology integrates knowledge about structure and functions of the components of a system obtained by the past reductionist investigation methodologies with the current knowledge about the dynamical processes concerning those components. The development of a language suitable to describe in system-level description of biological processes and enable an incremental modelling approach able to express the modularity of biological systems is a challenge. In this chapter we present the programming language BlenX developed by the team of The Microsoft Research - University of Trento Centre for Computational Systems Biology, Italy, to face this challenge.

Published by Woodhead Publishing Limited, 2013

Keywords: languages for systems biology, BlenX.

8.1 Deterministic vs stochastic approach in systems biology

Systems of living entities are composed of several interacting elements. This implies that mathematical models can be designed at various observation and representation scales. The microscopic scale corresponds to model, by integro-differential equations the time evolution of the state of each single variable of the system. If the system is composed by a large number of elements it is possible to obtain suitable local in space averages of their state in an elementary space volume ideally tending to zero [2]. In this case the modelling can be developed at the macroscopic scale, which describes the time behavior of locally averaged quantities called *macroscopic variables*. Moreover, generally the modelling is deterministic, i.e. it follows deterministic causality principles: unless some external noise is added, once a cause is given, the effect is determined. The macroscopic modelling scale can still be applied when the number of system components is sufficiently large and a sufficiently small volume still contains enough elements to allow the averaging process mentioned above.

It is generally believed that understanding the properties and the time evolution of a system follows from a detailed knowledge of the state of each of its elements. Consider as an illustrative example a system composed by a certain number of particles (proteins, molecules, ions, functional complex, etc). At microscopic molecular level the states of the particles evolve according to the laws of classical mechanics that describes with a system of first order differential equations the time behavior of the position and velocity of each particle of the system. If the initial values position and velocity are known, the system of differential equations can be solved and the macroscopic properties of the physical system can be obtained as averages involving the microscopic information contained in such a solution. However, it is very hard

Published by Woodhead Publishing Limited, 2013

to implement such a program. Even in principle, it is impossible to predict the exact molecular population levels at some future time unless we know the exact positions and velocities of all the particles of the systems. D. Gillespie in [6] points out that a reacting system of classical molecules is a deterministic process in the position-momentum phase space, but it is not a deterministic process in the multidimensional subspace of the species population numbers.

An alternative to the deterministic approach is the stochastic representation, where the state of the whole system is described by a suitable probability distribution function over the macroscopic state of the interacting system. In this chapter, we will focus on discrete-space continuous-time stochastic modelling, because living systems either at the molecular scale or at the ecological scale are composed by a discrete number of particles and individuals, respectively [8]. At the molecular scale the adoption of stochastic representation is recommended when the number of molecules is small, whereas at the large scale typical of ecological systems it is recommended when the network of interactions among species is inherently affected by factors of random noise.

The deterministic and the stochastic essence of a natural process depends on the properties of the components of the system and on the physics of the characteristic interactions among them. For example, chemical reactions are due to random collisions between interacting particles. Another example: random motion of genetic particles imbues the cellular environment with intrinsic noise that frequently causes cell to cell variability and even significant phenotypic differences within a clonal cell population. Extending our glance to ecosystems, if population sizes are small, then models should be stochastic: the effects of fluctuations due of population size must be explicitly analyzed. Nowadays, stochastic models in ecology have begun to be systematically studied because of their relevance to biological conservation.

The difference between the deterministic and the stochastic nature of a biological or physical process also requires different modelling languages. In life science differential equations

Published by Woodhead Publishing Limited, 2013

are appropriate for continuous time, continuous space modelling of systems composed by a large number of elements. The stochasticity manifests itself when the number of the system components is small and it is exalted when the system includes parallel and/or concurrent interactions. Biological processes are often stochastic, parallel and concurrent. Therefore, living systems require a descriptive approach substantially different from differential equations. It has to be able to represent parallelism and concurrency of the interactions, that at the microscopic scale derive from the multiple functionalities of the proteins and molecular functional complexes, whereas at the ecological scale are the engine of Darwinian selection.

In this chapter we first describe the BlenX language and then we present two time-continuous discrete stochastic models specified in BlenX language [4] and simulated with Beta WB simulator [3]: (i) a model of ubiquitin-proteasome system, and (ii) a simple predatory-prey model. BlenX implements a stochastic process calculus explicitly developed to represent biochemical entities and their interactions at the micro- and meso-scale. BlenX is part of the software platform *CoSBiLab*, on which our group is currently working and that implements a new conceptual modeling, analysis and simulation approach - primarily inspired by *algorithmic systems biology* [12] - to biological processes. Algorithmic systems biology grounds on the belief that algorithms and computer-science formalisms - like processes calculi - can help not only in modelling well established knowledge but also in coherently extracting the key biological principles that underlie the experimental observations [12].

The BlenX language offers to the modeler the possibility to address parallelism and concurrency of interactions, to express causality of the interactions, to represent multifunctionality of living entities. Moreover, BlenX formalisms are quantitative, interaction-driven, composable, scalable and modular, and thus able to represent not only the main static features of modularity and compositionality of a living system, but also the principal characteristic of its quantitative time evolution.

Published by Woodhead Publishing Limited, 2013

8.2 The BlenX language

BlenX is a programming language implementing the Beta-binders calculus. Here we provide a descriptive user-point-of-view introduction to the fundamental units, operators, and "actions" of this language. We refer the reader to [3, 4, 1] for a detailed technical description of the language.

In computer science, the process calculi (or process algebras) are a diverse family of related approaches to formally modelling concurrent systems. Beta-binders in particular is an extension of the stochastic π-calculus [11]. Beta-binders calculus, as the other members of the family of process algebras, is based on the notion of *communication* described through a set of temporally ordered *actions*. The fundamental units of the calculus are the interlocutors of this communication, represented by *computational processes*. Just as in a conversation, the main actions that a computational process can take are sending and receiving messages. To denote a chain of events, the action prefix operator is used, which is written as an infix dot. For instance, `a!.b?.P` denotes a process that may offer action on `a`, then offers an action on `b`, and finally behaves as process `P`. `a` and `b` are the *channels* through which the communication take place. The behaviour of the process `a!.b?.P` consists of sending a signal over the channel `a` (`a!`) and waiting for a message over a channel `b` (`b?`). The processes can be composed in parallel. Parallel composition (denoted by the infix operator |, for instance `P|Q`) allows the description of processes which may run independently in parallel and also synchronize on *complementary actions* (by complementary action we mean a *send* and a *receive* over the same channel). Communication between processes is always binary and synchronous. The `rep` operator replicates copies of the process passed as argument. Only guarded replication is used, i.e. the process argument of this operator must be prefixed by an action that forbids any other action of the process untile the first action has been executed. In addiction to the parallel composition processes can be also composed through a *non deterministic choice*, indicated with

Published by Woodhead Publishing Limited, 2013

the summation operator "+". The sum of processes P and Q, P + Q behaves either as P or as Q and selection of an alternative discards the other forever. To represent a deadlock situation, where the process is unable to perform any sort of action or co-action, the `nil` operator is used.

Beta-binders calculus adds to these simple syntactical elements *boxes*, also called *bio-processes*, that can be intuitively pictured as shapes encapsulating processes. Formally, the boxes are defined by unique identifiers that express the interaction capabilities of the processes encapsulated. These identifiers, called *binders* can be pictured as interaction sites put in charge of allowing the inter-boxes communication. Consider Fig. 8.1. A binder is a pair (x, A), written as x, A, where x is the name used by the internal process P to perform send/receive actions, while the binder identifier A, called *type*, expresses the interaction capabilities at the site x. The usefulness of the type of binder can be understood if we consider the interaction between boxes. The type A is a syntactical structure through which it is possible to quantitatively express the affinity of the interaction between boxes. Two boxes are likely to interact if their interface contains binders whose types are affine, i.e. binders whose affinity is non null.

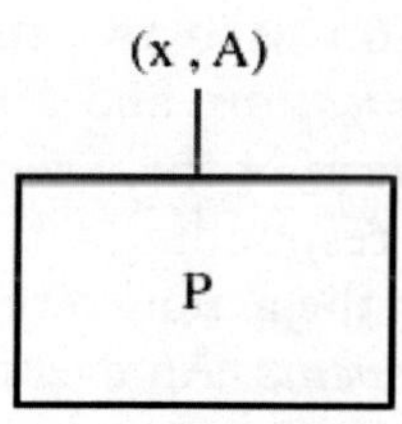

Figure 8.1 A pictorial view of a box. The sites of interaction are represented as binders on the box surface. In this figure, the box has only one binder identified by its name x and its type A, and an internal process P.

Published by Woodhead Publishing Limited, 2013

The BlenX language adds to the actions defined in the original Beta-binders calculus new actions that extend the possibilities to define and control the way in which a box evolves. The evolution of the interfaces of a box is driven by suitable actions that are defined by the processes inside the box. Such actions are named `hide`, `unhide`, `ch` (i.e. "change") and `expose`. These actions corresponds to the following transformation of the status of a binder: `hide` disables any communication through the binder, by hiding it from the tentatives of communication with binders of other boxes; `unhide` takes the opposit action: it enable communication through the binder, by undisclosing it to the view of other boxes; `change` changes the type of the binder; and `expose` adds a new binder to the box interface. The evolution of the processes inside a box that do not directly affects the interface can be defined by the following actions: send/receive action between processes (called *intra-communication*; and `delay` action that imposes a delay of a certain amount of time before the execution of subsequent actions.

The execution of the actions can be controlled with `if-then` statements, used to express conditions that need to be satisfied before executing the actions defined in the statement. Finally, a box can be eliminated from the system by executing the action `die`.

Boxes can interact in different ways: they can join (`join` is the action verb), they can form complexes, they can send/receive information each to other through dedicated binders (*inter-communication*, and a box can split into two boxes (`split` is the verb of this event). New boxes can be created (`new` is the corresponding action verb), boxes can be eliminated (`delete` is the action verb). `join`, `split`, `new`, `delete` are verbs of *events*. An event is the composition of a condition and an action verb, namely, events are used to express actions that are enabled by global conditions.

Boxes can be interpreted as biological entities, i.e. components that interact in a model to accomplish some biological function: proteins, enzymes, organic or inorganic compounds as well as cells or tissues. Binders of boxes are models of molecules interaction sites, protein sensing and effecting

Published by Woodhead Publishing Limited, 2013

domains. The biochemical interactions between the biological entities are abstracted as communications between boxes and join events, whereas conformational changes, allosteric reactions, and zero-th order degradation or production are established respectively by the processes inside the box, by `split` actions, by `delete` and `new` actions. With regard to conformational change and allosteric reaction, for instance, the internal structure of a box can codify for the mechanism that transforms an input signal into a protein conformational change, which can result in the activation or deactivation of another domain.

In order to obtain quantitative simulations of a BlenX model, a specific speed (or rate constant) is associated to each action. This attribute is a generalization of the rate constant of a biochemical interaction. The affinity between two binder types is a number that can quantify chemical affinity in a reaction, but also the degree of structural complementarity in key-lock reaction mechanisms. The dynamics of a BlenX model is governed by the values of these rate constants and is stochastically defined an efficient adaptation of the Gillespie algorithm [6]. The physical basis of the algorithm is the collision of molecules within a reaction vessel. It is assumed that collisions are frequent, but collisions with the proper orientation and energy are infrequent. Therefore, all reactions within the Gillespie framework must involve at most two molecules. Reactions involving three molecules are assumed to be extremely rare and are modeled as a sequence of binary reactions. It is also assumed that the reaction environment is well mixed. The algorithm executes four main steps: 1. initialization of the number of molecules in the system, reactions constants, and random number generators; 2. (Monte Carlo step) generation of random numbers to sample from a uniform and an exponential probability density respectively the next reaction to occur as well as the time interval. The probability of a given reaction to be chosen is proportional to the number of substrate molecules. 3. Update: the time step is increased by the randomly generated time in step 2., and the molecule count is updated on the basis of the reaction that occurred. 4. Iteration: the algorithm execute all the

Published by Woodhead Publishing Limited, 2013

steps from Step 1 unless the number of reactants is zero or the simulation time has been exceeded.

In order to enable the reader to catch the potentialities and teh essence of BlenX, we show in Fig. 8.2 the result of the execution of an inter-communication, that will be largely used in our models. The process inside the first box can receive a message on channel `x`, that is bound to an active binder of the box (`(x:1, A)`). The process inside the second box sends a message through the active binder (`(x:1, A)`) through the action `y!()`. The empty brackets in these actions mean that, in this model, there is no need to specify the object of communication. The exchange of information between the two box is permitted only if the binders involved in the communication are compatible, i.e. if their affinity is non null. Once the communication has occurred, their change of state if reflected by the modification of the internal processes: the execution of the inter-communication results in the "disappearance" of the channels `x` and `y` and with the exposure of the subsequent action, that in this example is simply the deadlock process.

In Fig. 8.3 a small model of the interaction between a nascent protein and a chaperone, and between misfolded protein and proteasome.

A nascent protein (NP) is a box. Its interface is defined by two binders `(y:1, P)` and `(prot:1, PTSP)`. The number `1` after the name of the binder indicate the value of th especific speed of the activity involving the binder. The internal structure is specified by the process in (8.1). This process expresses a non deterministic choice ("+") between the process `y?().ch(100,y, DR).hide(1,ubi).nil` and process `y?().ch(100,y, DW).prot?().die(1).nil`, Let call NP_1 the first process and NP_2 the second process. NP_1 can receive a message on channel `y` (`y?()`), then it can change - at a specific speed set to 100 - the type of binder `(y:1, P)` into `(y:1, DR)` (`ch(100,y, DR)`), and finally it ends with a deadlock (`nil`), i.e. a process that can do nothing. NP_2 can receive a message on channel `y` (`y?()`), then it can change - at a specific speed set to 100 - the type of binder `(y:1, P)` into `(y:1, DW)` (`ch(100,y,`

Published by Woodhead Publishing Limited, 2013

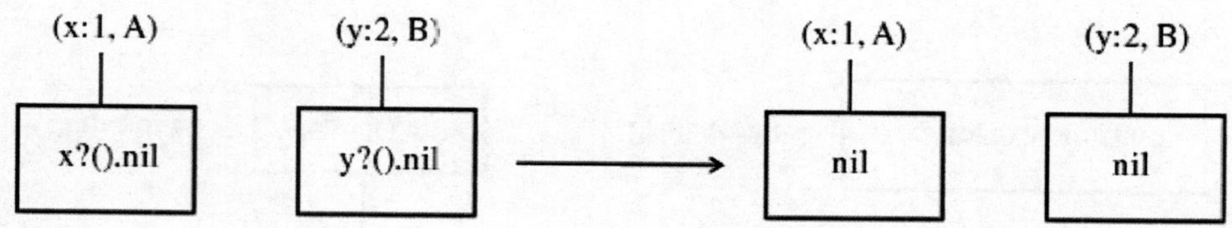

Figure 8.2 Graphical rpresentation of an inter-communcation.

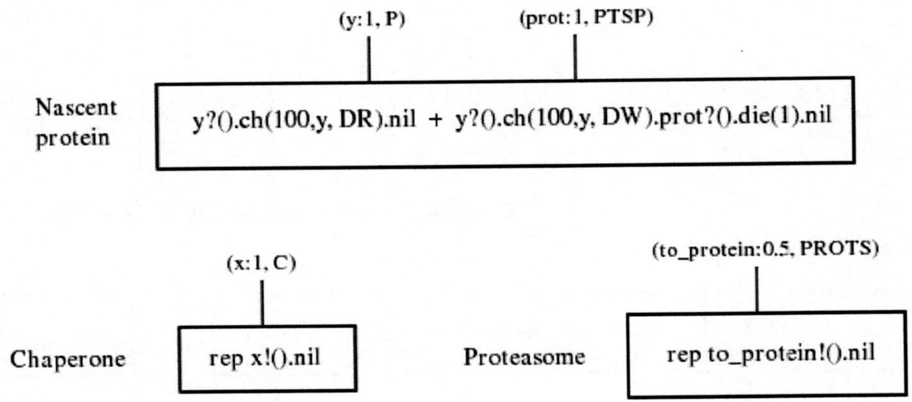

Figure 8.3 Pictorial scheme of a model of chaperone-protein interaction. The system includes a nascent protein, a molecular chaperone, and a proteasome. After the interaction with chaperone through the binders `x` and `y`, the protein can result correctly folded or misfolded. The type `P` of the protein binder changes to `DR` if the protein assumes the healthy 3D shape, whereas it changes to `DW` if it assumes the faulty shape. In the second case it is ready to undergo an interaction with the proteasome through the binders `prot` and `to_protein`. More details in the text.

DR)), then it can receive on channel prot (prot?()) and finally it can perform a die action that eliminate the box with a specific speed equal to 1 (die(1)).

```
y?().ch(100,y, DR).nil
+ y?().ch(100,y, DW).prot?().die(1).nil
```
(8.1)

The chaperone is represented by a box with one binder, (x:1, C) and an internal process that replicates the sending action of channel x. Similarly, the proteasome is represented by a box with one binder, to_protein:0.5, PROT, and an internal process replicating the sending action on channel to_protein.

Since a nascent protein has to be sequestered by the chaperone to be folded, an interaction between box Nascent protein and Chaperone has to be enabled. The way to enable a possible interaction is to define a non null affinity between the binders (y:1, P) and (x:1, C). Given a non null affinities between the binder, an inter-communication thorough these binders on the complementary channels y and x is enabled.

In stochastic regime, according to the Gillespie algorithm [6], the eventuality that the inter-communication involves process NP_1 or process NP_2 is quantitatively determined by the number of available chaperones, nascent proteins and the rates of interaction associated to channels x and y.

The non deterministic choice $NP_1 + NP_2$ reflect the possibility that the interaction between chaperone and misfolded protein through the binders (y:1, P) and (x:1, C) can result in a healthy protein or a faulty protein. In fact, after the inter-intercommunication is fired, a change action is enabled to change the type of (y:1, P) into (y:1, DR) to denote a right-folded protein *or* into (y:1, DW) to denoted a "wrong" protein. If the process NP_1 is stochastically selected for the inter-communication the interaction between chaperone and nascent protein ends in a healthy protein that does not undergo any further interaction with the other components of the systems (the process terminates in a deadlock nil). If the process NP_2 is stochastically selected for

Published by Woodhead Publishing Limited, 2013

the inter-communication the interaction between chaperone and nascent protein produces a faulty protein that is going to interact with the proteasome though the binders (prot:1, PTSP) and (to_protein:0.5, PROT) on the channels prot and to_protein, provided that in the model the affinity between (prot:1, PTSP) and (to_protein:0.5, PROT) has been defined non null. After this interaction has occurred the misfolded proteins is eliminated from the system with the execution of a die action.

The internal behaviors of the chaperone and proteasome is defined by a replicated process, to express the fact that after the interaction with nascent protein and misfolded protein, respectively, they remain unaltered.

The BlenX code for this example is reported in Tables 8.1 and 8.2, that shows the program file and the files for the definition of binder types and affinities, respectively. The code in Table 8.2 says that couple of binders having non null affinities are (P,C), (DW,C), (PTSP, PROTS), and the value of their affinities are 1, 1, and 10, respectively.

A possible trajectory of the time evolution of the protein-chaperone-proteaseome system is sketched in Figs. 8.4 and 8.5. After the interaction with the chaperone, the nascent protein is still not correctly folded (the box represent the protein has changed (y, P) into (y, DW)) and undergoes an interaction with proteasome at the end of which the protein is degraded and is eliminated from the system (in the the BlenX code, its corresponding box dies and becomes the deadlock box).

8.3 The ubiquitin-proteasome system

The ubiquitin-proteasome system is the major pathway that mediates the degradation of unwanted intracellular soluble proteins (i.e., mutant, misfolded, denatured, misplaced, or damaged) in the cytoplasm, nucleus, and endoplasmic reticulum of eukaryocytic cells. The process whereby the ubiquitin-proteasome system clears these unwanted proteins

Published by Woodhead Publishing Limited, 2013

Table 8.1 The BlenX model coding for the interaction between nascent protein and chaperon and between misfolded protein and proteasome. The system is defined as the parallel composition (||) of three boxes: `protein` (line 4), `chaperone` (line 8), and `proteasome` (line 11). The absolute simulation time is set to 100 (line 3), and the initial amounts of the model components is set to 1000 (lines 14-15-16).

```
1    // File: example.prog
2
3    [time=100]
4    let protein : bproc =
6     #(y:1, P), #(prot:1, PTSP)
6     [y?().ch(100,y, DR).nil
7     + y?().ch(100,y, DW).prot?().die(1).nil];
8
8    let chaperone : bproc =
9      #(x:1, C) [rep x!().nil];
10
11   let proteasome : bproc =
11    #(to_protein:0.5, PROTS)
12    [rep to_protein!().nil];
13
14   run 1000 protein
15       || 1000 chaperone
16       || 1000 proteasome
```

mainly involves two steps: (i) labeling of unwanted/damaged proteins with chains of activated ubiquitin molecules transported by parkin proteins; (ii) transport of ubiquitinated proteins to the proteasome by chaperone molecules (e.g., heat shock proteins).

Multiple molecules of ubiquitin, a small highly-conserved polypeptide, attached to the target protein, constitute the signal for proteasome attack. Mutant variants of α-synuclein protein can interfere with normal ubiquitin-proteasome system function, by inhibiting the signal transmitted by the

Published by Woodhead Publishing Limited, 2013

Table 8.2 The binder definition file stores all the binders identifiers and the affinities between binders associated with a particular identifier.

```
// example.types

{P, PTSP, C, PROTS, DW, DR}
%%
{
 (P,C,1),
 (DW,C,1),
 (PTSP, PROTS,10)
}
```

ubiquitinated misfoeld protein to protesome [7, 10, 13]. The switch off of the signal sent by aberrant ubiquitinate proteins to the regulatory complex of proteasome causes proteolytic stress, protein accumulation and aggregation, and finally cell death.

Our BlenX model of the ubiquitin-protesome system consists of the parallel composition of seven boxes representing the nascent protein, the chaperone, that parkin and the ubiquitine (Fig. 8.6); the stress factor, the α-synuclein protein and the proteasome (Fig. 8.7).

The box of the nascent protein has three binders `(y:1, P)`, `(ubi:10, U)`, and `(prot:1, PTSP)` through which the communication with chaperone, ubiquitin and proteasome are respectively enabled. The process inside the nascent protein box is given by

```
  y?().ch(100,y, DR).hide(1,ubi).nil
+ y?().ch(100,y, DW).ubi?().prot?().die(1).nil
```
(8.2)

where a non deterministic choice models the eventuality of a correctly folded protein or a faulty protein as a result of the interaction between chaperone and nascent protein. In the case in which the nascent protein is correctly shaped (i.e. the first term of the summation is selected the action `y!()`

Published by Woodhead Publishing Limited, 2013

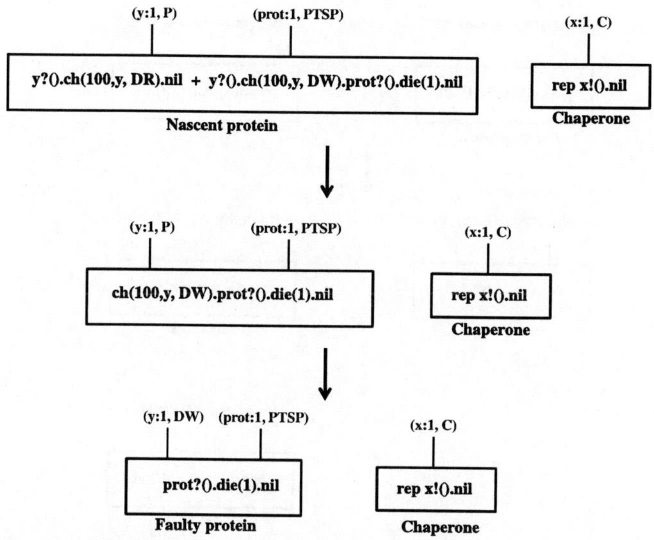

Figure 8.4 Inter-communication between nascent protein and chaperone represents the biochemical interaction between these two entities. In this figure a porssible trajectory of the system is shown: the interaction results in a faulty protein.

and `ch(100,y, DR)` are executed), the internal process first disables with a `hide` action any further communication through the binder `(ubi:10, U)`, that is dedicated to the ubiquitylation's activities, and then terminates in a deadlock, meaning that the correctly faulty protein does not perform any other reaction with the other components of the system. On the contrary, in the case in which the nascent protein did not assume the healthy shape, it is going to be ubiquitinated (i.e it is willing to receive through channel `ubi` bound to the bidner `(ubi:10, U)`) and than degraded by the proteasome (i.e. if first communicates with the proteasome box thorugh `prot?()` and then "dies" (due to the execution of `die(1).nil`).

Published by Woodhead Publishing Limited, 2013

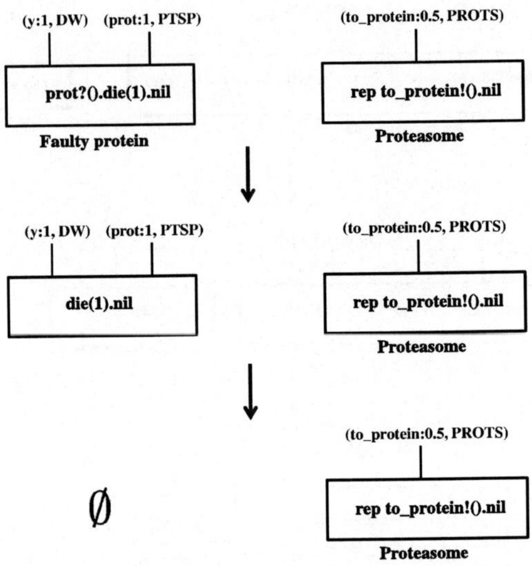

Figure 8.5 The faulty protein undergoes an inter-communcation with the proteasome and then it becomes the deadlock process, i.e. it degrades.

The parkin is represented as a box having a binder devoted to communicate with ubiquitin: `(to_ubiquitin:0.5, T_UB)`. The internal process is a send action on the channel `to_ubiquitin`. The ubiquitin molecule is a box having three binders: `(from_parkin:0.5, F_PARK)`, `h(u:1, UB)`, and `(actp:1, UB2)` allowing communications with parkin, misfolded protein and proteasome, respectively. At the beginning the binder `(u:1, UB)` is hidden (an "h" prefixes the definition of the binder), and thus it is invisible to any another system's boxes.

The box representing the stress factor contains a replicated send action through the binder `(st:1, STR)`. The α-synuclein box has four binders: `h(s:0.8, SY)`, `(t:0.8, ST)`, `(with_prot:0.1, VS_UB)`, `(degrad:0.1, KILL)` for the communications with the proteasome, the stress factor, the ubiquitin, respectively.

Published by Woodhead Publishing Limited, 2013

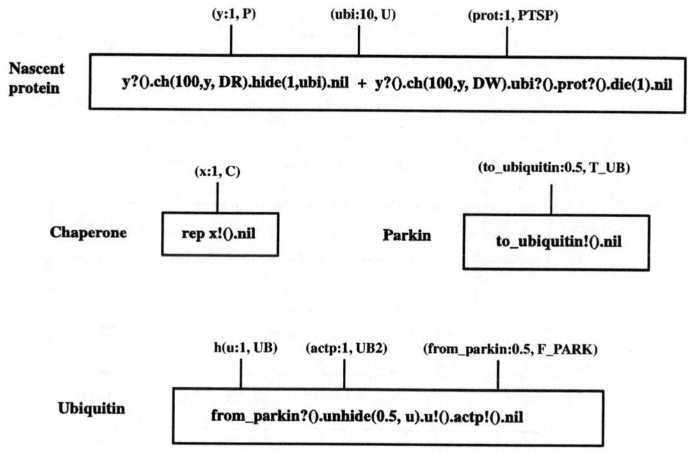

Figure 8.6 Sketch of the boxes representing nascent protein, molecular chaperone, parkin and ubiquitin molecules (see in the text the detailed description).

As shown in Fig. 8.7, the protein α-synuclein first interacts with the stress factor through the binders `(t:0.8, ST)` and `(st:1, STR)`, respectively. Then, since we assume also the possibility that this interaction does not change the protein, the internal process of the α-synuclein box can evolve also into a deadloack process. If the stress factor changes the structure of the α-synclein, the Fig. 8.7 discloses the binder `h(s:0.8, SY)` by performing an `unhide` action. Now, a second non deterministic choice is enabled. The first term of the sum describes the effective interaction of the mutant α-synclein and the proteasome through the binders `(s:0.8, SY)` and `(syn:0.8, SNC)`, that results in the inactivation of the proteasome. Formally, this inactivation is described by the masking of binder `(actv:0.8, SNC)` performed by the execution of the action `hide(1, actv)`. On the contrary, the second term of the sum describes first

Published by Woodhead Publishing Limited, 2013

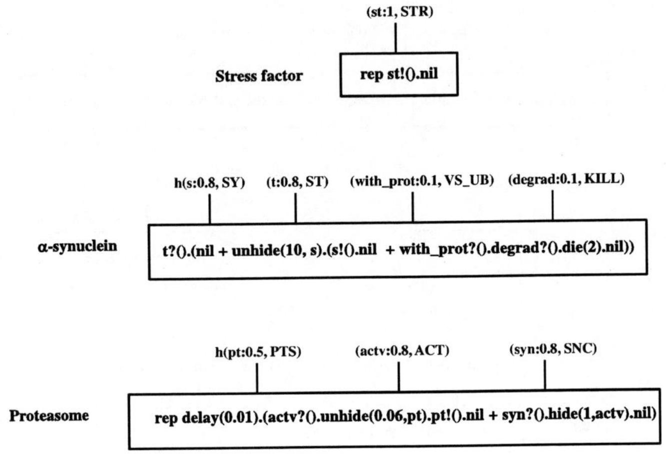

Figure 8.7 Sketch of the boxes representing enzymes stress factor, synuclein, and proteasome (see in the text the detailed description).

the activation of the proteasome by the mutant ubiquitinated α-synuclein (i.e. the communication through the binders (`with_prot:0.1, VS_UB`) and (`actv:0.8, ACT`), and its subsequent degradation realized by the executions of the inter-communication between binders `degrad:0.1, KILL`) and (`pt:0.5, PTS`) of α-synuclein and proteasome, respectively. At the end of this inter-communication, the action `die` in the internal process of α-synuclein is finally enabled and determine the deletion of the box of mutant α-synuclein.

The Fig. 8.8 shows the time behavior of the number of misfolded proteins, healthy proteins, mutant variant of α-synuclein, and proteasomes. The values of the rate constants, and of the affinities of interaction producing these behavior are reported in the "pictorial" codes in Figs. 8.6-8.7. All

Published by Woodhead Publishing Limited, 2013

the rates are expressed in units of μsec^{-}1, that is the typical scale of α-synuclein folding [5]. Their values are fictitious, but their orders of magnitude respect the typical temporal scale of the processes involved in ubiquitin-proteasome system functioning. The initial values for the numbers of molecules of the species have been chosen as follows: 5×10^3 nascent proteins, 10^4 molecules of parkin and ubiquitin, 10^4 proteasomes. They are fictitious values chosen used with the only purposes of experimenting *in silico* possible kinetics and dynamics of the systems. The number of bio-processes representing the stress factor has been varied over the range from 10 to 10^5 without obtaining significant changes in the dynamics. This number is an indicator of the intensity of the perturbation with which the stress factor causes the formation of aberrant molecules of α-synuclein. In Fig. 8.8 we can see that the number of mutant α-synuclein proteins has a fast linear growth within the first 5 μsec, then it decreases and reach a stable value at 55 μsec. Simultaneously, the number of available proteasomes decreases proportionally to the number of mutant α-synucleins and it vanishes as soon as the number of mutant α-synuclein becomes constant. The decrement of the number of α-synuclein proteins is due to the action of the protesomes that attack and degrade them. In particular, within the first 9 μsec the curve of the number of proteasome and the one of mutant α-synuclein are superimposed, i.e. the decrement of these two species occur at the same rate. We see also that the number of faulty proteins experiences a rapid linear growth within the first 5 μsec, then it slowly decreases and stabilizes as soon as the number of available proteasome is approximately zero (around 45 μsec). This behavior correctly reflect the obvious impossibility of degrading faulty proteins if the system does not have a sufficient number of available proteasomes. Finally, the simulations of our model show that the number of healthy proteins also linearly grows during the first 5 μsec. Then, since these proteins are not involved in any other processes, their number remains constant for the rest of the time.

Published by Woodhead Publishing Limited, 2013

This model also includes a mechanism of production of free molecules of ubiquitin and parkin to guarantee the reactions of ubiquitination of the new formed nascent proteins, if the number of free ubiquitin and parkin molecules drops below a critical threshold, that does not allow the sustenance of the ubiquitation process. Free molecules of ubiquitin are returned to the systems as a consequence of possible unbinding reactions that break the complexes ubiquitin-parkin. The production of parkin and ubiquitin molecules is modeled in BlenX with *events*:

```
when (ubiquitin: |ubiquitin| = 0: inf) new(10000);
   when (parkin: |parkin| = 0 :inf) new(10000);
```

and the eventuality that the complex ubiquitin-parkin breaks to free new molecules of ubiquitin and parkin is modeled by assigning a non-null value to the rate constant of unbinding reaction in the affinity definition for the complex ubiquitin-parkin (0.34 is the value of the rate constant of the unbinding reaction), as follows

```
(F_PARK, T_UB, 0.7, 0.34, 1.2).
```

8.4 A predator-prey model

Here, the main parts of a simple predator/prey developed in collaboration with our student is presented [9]. The code is listed in Table 8.3. The reader is referred to [9] for the complete model and the simulation results.

Consider a top-predator, for instance the transient orca. The predator is defined as a box with the communication channel `eat` and the additional duplication channel `dupl`.

Orca's `eat` channel is affine to the `eat` channel of the prey species to enable an inter-communication between prey and predator. The internal process `ptransorca` is composed of the three subprocesses linked by the non-deterministic choice operator + (see lines 8-9-10 in Table 8.3).

The process `eat!()` creates an inter-communication over the `eat` channel with the corresponding *eat* channel of

Published by Woodhead Publishing Limited, 2013

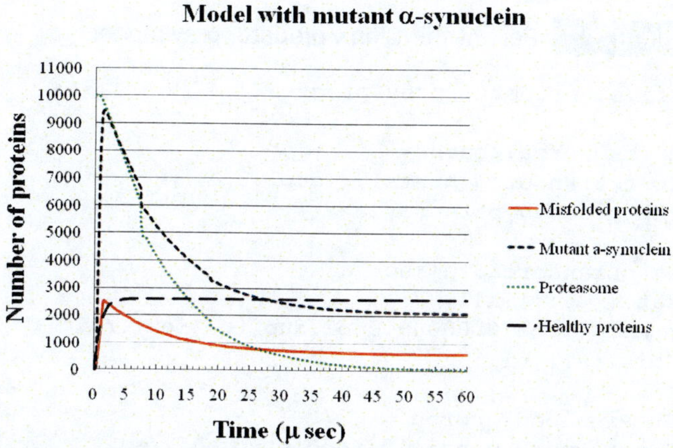

Figure 8.8 Simulated time-behavior of the number of misfolded proteins, mutant α-synuclein, proteasomes and healthy proteins. Each curve is the average of 100 simulation runs.

the prey species when the predator "hunts" and "eats" the prey. Then, one instance of the prey pinniped disappears from the system. The eat-process has two possibilities to be executed, with two different successive behaviours: (i) `eat!().ch(0.00005,dupl,duplication).nil`. In this case the sequence of the processes is "eat" and then "reproduce". The change command `ch` changes the binder type `A` into the binder type `dupl` (with a propensity rate of 0.00005) which causes a split action. (ii) `eat!().y!().nil`. This second possibility is the sequence of the processes "eat" and then "go back to life". With the command `rep` the `ptransorca` process replicates a copy of itself and starts from the beginning, without changes in the interface or in the internal behaviour.

Finally, the death of the predator is modeled as at line 14 of Table 8.3. The third process and internal evolution is the abstraction for the death of the transient orca box, implemented as *die*(*inf*).*nil*.

Published by Woodhead Publishing Limited, 2013

Table 8.3 Part of the BlenX predator-prey model.

```
// file: predator_prey.prog
let Transorca: bproc=
 #(eat,transorca_hunts),#(dupl:0,A)
[ rep proc<<y,ptransorca>> ];
 let ptransorca: pproc=
  eat!().y!().nil
  + eat!().ch(0.00005,dupl,duplication).nil
+ die(inf).nil;
let Pinniped: bproc =
 #(eat,hunts_pinni),#(food,pinni_lifes),
    #(dupl:0,A)
[ repproc<<x,ppinni >> ];
let ppinni: pproc=
 food!().x!().nil
 + food!().ch(0.0043,dupl,duplication).nil
 + eat?().die(inf)
 + die(inf).nil;
...
```

Now, consider the prey species or intermediate species pinniped. Similarly to the predator, the prey is defined as a box with the communication channels `eat`, `food` and the additional duplication channel `dupl`. The `food` channel creates an inter-communication with the corresponding `eat` channel of another prey species (e.g. the salmon), or creates an inter-communication with the corresponding *food* channel of the food species (e.g. macroalges). After this communication, one instance of the hunted species disappears from the system. The addition of the specific replication rate, which is unique for every species is implemented identically as in the predator. The food-process has two possibilities to be executed, with two different results: the duplication of the box or "going on with life" without changes neither in interface nor in the internal processes. The death without being hunted

Published by Woodhead Publishing Limited, 2013

is also implemented in the prey as in the predator. The *delay* command retards the execution of the `die` command, which causes the deletion of the box from the system.

8.5 Conclusions

As all the other sciences, computational systems biology makes progress in three different fields: (i) solution of new problems, (ii) elaboration of new methods, and (iii) development of new symbolisms. This chapter focused particularly on this last aspect and presents a language developed by the CoSBi team for modelling and simulating complex living systems. The language is born from the convergence of computer science and biology. Namely, the language spoken by biologists to describe the mechanisms of the dynamics of a biological pathway is similar to the formal languages spoken by computer scientists to describe the functioning of mobile communicating systems. Living systems and computational and communication systems of devices share many features, principally the parallelism and concurrency of the interactions driving the time evolution of these systems. We believe that taking this convergence far biology can benefit from the use of formal languages for the purposes of modeling, simulations and ultimately, understanding of living system dynamics. At the same time, computer science can be inspired by the way in which nature build complex systems and interaction networks.

References

1. http://www.cosbi.eu/index.php/research/prototypes/overview.
2. N. Bellomo, *Modelling complex living systems. A kinetic theory and stochastic game approach*. Birkhaeuser, 2008.
3. Dematté, L., C. Priami, & A. Romanel, The beta workbench: a computational tool to study the dynamics of biological systems. *Briefings in Bioinformatics 9*(5), 437–448, 2008.

Published by Woodhead Publishing Limited, 2013

4. Dematté, L., C. Priami, & A. Romanel (2008b). The blenx language: A tutorial. *SFM 2008 LNCS 5016*, 313–365.
5. F. C. M. Ferreon, Y. Gambin, E. A. Lemke, & A. A. Deniz, Interpaly of α-synuclein binding and confromational switching probed by single-molecule fluorescence. *PNAS 106*, 5645–5650, 2009.
6. D. T. Gillespie, Exact stochastic simulation of coupled chemical reactions. *The Journal of Physical Chemistry 81*(25), 2340–2361, 1977.
7. A. E. Lang & A. M. Lozano, Parkinson's disease - first of two parts. *The New England Journal of Medicine 339*(15), 1044–1053, 1998.
8. P. Lecca, Modelling, designing and simulating living systems with BlenX, 6th International Conference on Technology and Medical Sciences, 3–13, Taylor & Francis Group, London, 2011.
9. C. M. Livi, Modelling and simulating ecological networks with blenx. the food web of prince william sound: a case study. Master's thesis, Alma Mater Studiorum - University of Bologna, Italy, 2009.
10. C. W. Olanow & K. S. P. McNaught, Ubiquitin-proteasome system and parkinson's disease. *Movement Disorders 21*(11), 1806–1823, 2006.
11. C. Priami, Stochastic π-calculus. *The Computer Journal 38*(6), 578–589, 1995.
12. C. Priami, Algorithmic systems biology. *Communications of the ACM 52*, 80–88, 2009.
13. M. Vila & S. Przedborski, Genetic clues to the pathogenesis of Parkinson's disease. *Nature Medicine*, 58–62, 2004.

Published by Woodhead Publishing Limited, 2013

9

Simulation of ecodynamics: key nodes in food webs

Abstract Ecological systems are composed by a several components (e.g. populations) and multiple interactions among them. A major challenge for systems ecology is to understand the general properties of ecological interaction networks and to identify their key elements. Beyond structural analyses, a dynamical view is essential here: both deterministic and stochastic approaches have been suggested in the literature, the former ones being more developed. Here we discuss the relevance of stochastic ecosystem simulations and present a case study. We argue that the major role of stochastic system models is studying and quantifying the essential role of variability in nature.

Keywords: systems ecology, food web, network dynamics, keystone species, stochastic simulation.

9.1 Systems ecology

Modern systems biology is a relatively young science, aided by large databases and supercomputers. However, systems theory had made its impact on biology much earlier, in the fifties and the sixties [3, 65], and one of the most influenced fields was ecology. Systems thinking in ecology, in these golden years, needed no databases and computers, but resulted in conceptual diagrams and a new view focusing

Published by Woodhead Publishing Limited, 2013

on systems, their components and the mutual interdependency among them. The systems ecological view suggested that components cannot be understood only if also the system is studied, and the system cannot be understood only if also the components are studied. Major advances included higher-level descriptions of ecosystems [45], explicit studies on the role of indirect effects [50] and input-output analysis, borrowed from economy [23]. System-level indicators (e.g. cycling index [17]; stress [68]) were introduced and intensely studied. Modern syntheses (e.g. [24]) provide really interesting alternatives to conventional ecological thinking. This new paradigm of ecology studies system components, feedbacks and dynamical trajectories, instead of populations, competition and niches.

9.2 Ecological interaction networks

The most important problems of ecology are studied in several contexts, according to various scientific schools. Such a unifying topic is ecological networks. In traditional population and community ecology, food webs depict who eats whom [8, 54, 52], the summary of the antagonistic (+/-) prey-predator relationships between species. Other interaction types of various sign combinations are also studied, including competition (-/-), facilitation (+/0) and mutualism (+/+). The key questions are how populations influence each other in ecological networks, which interactions are most important and, in the case of prey-predator interactions, how do food webs look in general? All these questions serve to better understand the evolution and ecology of individual species (e.g. extinction). Similar questions in systems ecology are identifying the design principles of trophic flow networks [36], how cycles and feedbacks are built in these systems [17] and how inputs and outputs indicate system maturity [45]. Here, flow networks contain a much smaller number of trophic components, and these are typically highly aggregated groups of species (like phytoplankton, large benthic feeders, small pelagic fish or dissolved organic carbon). All

Published by Woodhead Publishing Limited, 2013

these questions serve to better understand large-scale organization of ecosystems (e.g. nutrient cycles, energetics). An interesting theoretical but also applicable problem, appearing in each of the two contexts, is what are the key elements of ecological interaction networks? Keystone species need to be quantified and predicted [42] and this is a major challenge, especially because of the importance of indirect interactions in ecosystems [41, 71, 72]. Different network indices (and their combinations [55]) have already been used to predict important species based on their key positions in networks [28]. However, the real task is to understand the relationship between structure and dynamics [53, 27, 29].

9.3 Pattern and process

Ecological patterns constrain processes and processes produce patterns. How regular they are depends on a number of factors, and leads to very different views.

9.3.1 *Laws and universal rules of nature*

According to early, mostly descriptive studies on the available food webs, many of their typical structural properties had been identified [8, 9]. These include the number of nodes (n; representing species or larger trophic groups), the number of links (L; representing trophic interactions), link density L/n and connectance ($C = 2L/n(n-1)$; the ratio of realized and possible links in a directed food web graph). Some other, still simple network statistics had been given historically but not really used in recent analyses: these include the number of prey species (PY, with out-degree $D_{out} > 0$), the number of predators (PR, with in-degree $D_{in} > 0$), the number of basal species (i.e. producers, B, with $D_{in} = 0$), the number of top species (i.e. top predators, T, with $D_{out} = 0$) and the number of intermediate species (I, if neither D_{in} nor D_{out} equals 0). Further, the ratios of certain types of species ($PY : PR$, $B : I : T$) and links ($BI : II : IT : BT$) have also been registered and some rules of thumb have been set. For example,

Published by Woodhead Publishing Limited, 2013

ranges of connectance values were suggested to be associated with stability [59]. Recently, there is much less hope in finding such elegant laws describing nature. Apart from finding simple and nice structural patterns in food webs, the dynamics of these networks can also be regarded as the outcome of some relatively simple rules. Ordinary differential (sometimes difference) equations are the most acknowledged way to describe population dynamics and the set of coupled equations provide the definition of community dynamics (of the interaction network). For example, the behavior of species *i* in a community of *n* interacting species can be characterized by the Lotka-Volterra equation

$$\dot{x}_I = a_i + \sum_{j=1}^{n} b_{ij}x_j, \quad i = 1, 2, \ldots, n$$

where x_i is the population density of species *i*; a_i is the per capita rate of reproduction (for growth, $a_i > 0$, for decay, $a_i < 0$); and b_{ij} indicates the per capita effect of species *j* on the per capita reproduction rate of species *i*. Some level of randomness can be added to this model in order to make its behavior more interesting. For example, a_i and/or b_{ij} can follow some pre-defined probabilistic distribution. The dynamics of such a system can be numerically simulated and a species can be considered extinct, for example, if its density falls below a critical threshold during a given time [27]. More complicated equations may reflect some additional aspects of reality. For example, the Holling III-type functional response [25] takes into account the possibility of prey switching (when the predator changes its feeding preferences according to the density of the prey species), as well as the searching and handling times of predators. Beyond analyzing the mathematical properties of theoretical models, there is an increasing interest in understanding large, real databases. It is only recently that the dynamic simulation of large, real food webs has become feasible, due to growth in computational capacity and available methodology. A deterministic approach of high popularity and applicability is implemented in the EwE (Ecopath module with

Published by Woodhead Publishing Limited, 2013

Ecosim) software package [7], mostly adapted for studying marine ecosystems. Here, different, standard, measurable parameters compose the master equation of the static Ecopath module, expressing the mass balance of the ecosystem (the net production of trophic component i equals the various sources of its mortality). The Ecosim module uses differential equations to dynamize this static master equation. These parameters include the biomass of species i, the production/biomass ratio, the ecotrophic efficiency (the fraction of the total production of a group utilized in the system), the fisheries catch per unit area and time, the food consumption per unit biomass of species; the contribution of species i to the diet of species j, the biomass accumulation of species and the net migration of species i (emigration minus immigration). Ecosim provides a deterministic description of ecodynamics with several measurable parameters that make it applicable in everyday fisheries ecology [69, 48, 46, 34, 10]. All the above approaches to describing population dynamics suggest that simple and elegant equations are able to describe community dynamics, considering only a limited amount of noise in the system (in the form of randomized parameters). Deterministic approaches still dominate ecological thinking (see [4], the best exception being the field of meta-population biology). This view suggests that abundance patterns are either biotic and deterministic (density-dependent interactions) or a biotic and stochastic (e.g. weather, disasters). In this context, another interpretation of stochasticity is that it may be responsible for unexplained phenomena and cause the "noise" on the ecological "signal". However, another research line suggests that noise and variability should be studied explicitly instead of being neglected or forgotten.

9.3.2 *Noise and disorder*

According to [62], "What physicists view as noise is music to the ecologist". The key message here is that biological systems are not "just" noisy, but inherently noisy. Variability and stochastic behavior are not imperfections but very

Published by Woodhead Publishing Limited, 2013

important components of evolutionary systems. Homogenous systems do not evolve, as heterogeneity is one of the three essential components of evolvability [70]. Perfectly controlled, fully deterministic systems produce constant behavior, thus these systems are neither variable nor adaptable. The question is whether noise and stochasticity are large enough to make deterministic models useless or are they still useful even if inherently contra-intuitive. Stochasticity and the sensitivity to the actual environment are sometimes essential for survival (cf. phenotypic plasticity, [1]): certain traits should be less genetically and more environmentally controlled (e.g. clutch size). An obvious external source of variability is a biotic, environmental gradients. Along a gradient, identical genotypes produce a series of different phenotypes (reaction norm, [2] and interspecific interactions may also produce variability according to a gradient (interaction norm, [66]). Thus, even if we know the components of the system, both their individual properties and their interaction parameters do vary, depending on the actual, local environmental conditions. A major, internal source of variability is complexity: myriads of simple, local rules make up large and complex systems, but never in exactly the same way. In a complex ecosystem, predictability has its serious limits, since nothing guarantees the invariance of the outcome resulting from multiple, intercrossed processes. There is increasing interest in stochastic modeling in ecology. In conservation biology, explicitly considering variability is needed [16] and could add a lot to our understanding. For example, considering the variability of population dynamics completes our information on rarity: a rare but dynamically less variable species can be in less danger than a more abundant but more variable species (even if the population is large, wide fluctuations can make it extinct). Considering stochastic dynamical models calls for individual-based approaches (IBM), since well-defined mass actions do not work here. IBM is already well-established in ecology [12, 30, 19, 20, 21]. However, although it is very intuitive to describe a complex system at the level of individual agents and local rules of behavior (e.g. [33, 47], it is much harder to evaluate these models than

Published by Woodhead Publishing Limited, 2013

deterministic ones. Simulations and sensitivity analyses still need to be more developed methodologically.

9.4 Food web dynamics: simulation and sensitivity analysis

9.4.1 *Algorithmic challenges*

The multiplicity of system components and interactions, as well as the large number of parameters in stochastic IBM pose huge computational problems [49, 40]. Recently, ecology (and biology in general) had a continuously increasing algorithmic toolkit, including novel topological network indices [28], cellular automata for spatial modelling [44], genetic algorithms [61], formal logic comparing different models [60], Petri nets [22], swarm intelligence [14] and process-based modelling [6].

Various process algebras have already been used in ecology, for simulating social insect colonies [67, 64], population biology and epidemiology [43, 39] and community ecology [35]. Biological organisms in BlenX [13, 58] are represented by boxes, composed by a set of interfaces (defining potential interacting partners, like a prey or a pollinator) and an internal program. The internal program transforms an input signal into demographic (e.g. reproduction) or behavioural (e.g. changed prey preference) changes. System behaviour emerges out of lower-level processes. BlenX uses the Gillespie algorithm [18] for stochastic simulations, just like some other ecological simulation models (e.g. [57, 31, 37]). In BlenX, the kinetics of the system is adopted from modelling molecular kinetics in the cell. Simple rates are assigned to single-individual (cf. mono-molecular) interactions (like death), while the kinetics of pairwise (bi-molecular) interactions follow mass-action:

$$A + B \xrightarrow{k_1} 2A$$

and

Published by Woodhead Publishing Limited, 2013

$$A + B \xrightarrow{k_2} A$$

meaning that there is a rate for "eat and reproduce" (k_1) and another for "eat" (k_2). In the first case, the predator A eats the prey B and produces another predator individual ($2A$), while the prey disappears. In the second case, the predator eats the prey and the prey disappears. The k_1/k_2 ratio determines how many prey B individuals are needed for the reproduction of A. This is a simple description of a prey-predator interaction but simple kinetic rules seem to apply quite well in several ecological situations (e.g. [63]).

9.4.2 Ecodynamics simulated in BlenX: a case study

A deterministic Ecosim model of the Prince William Sound ecosystem (Alaska) was built and analyzed by Okey (2004). The trophic network is composed of 48 living components (Figure 9.1) and sensitivity analysis was performed in order to quantify the relative importance of these groups. Different dynamical importance indices were used to quantify the key groups. For example, the KI keystone index predicts that "transient orca" is the most important group of the ecosystem, followed by "avian predators" and "porpoise" (Table 9.1), while the least important groups include "macrophytes" and various groups of meiofauna [46]. Given this ranking, it is of interest how do similarly motivated but stochastic simulations rank the same groups. Based on this food web model, we have constructed a stochastic counterpart in BlenX [35] and used a stochastic simulator for sensitivity analysis. We simulated the community importance of species, offering quantitative tools for conservation practice. The biomass and trophic flow data of the original EwE model have been translated to number of individuals (based on species-specific body size data, www.fishbase.org) and interaction rates, respectively. Death rates were chosen from a realistic range (from 0.001 to 0.1, typically 0.01) to fine-tune the model to quasi-equilibrium. With all the parameters used (number of individuals, reaction rates, birth and

Published by Woodhead Publishing Limited, 2013

death rates), conventional modelling tools would be computationally expensive. Process algebra reduces computational costs and opens future possibilities for using a number of other parameters (e.g., age structure and genetic variance; not presented here for simplicity). An important feature of our stochastic approach is that it makes extinction possible, in contrast with ODEs where the curves can only approach the X axis. In the reference simulation, based on *s* runs with the same, realistic, initial conditions and parameters, we can measure the variability of system behaviour (by the variance of population size after *t* steps). In a simple, illustrative case, we only change the initial number of individuals of each species, one by one. For each disturbed species we run the model s times and measure the response of every other species (in terms of both change in the mean and change in the variance of population size). Figure 9.3 shows an example: dark bars show the mean and variance of population sizes based on s reference simulations and light bars show the same after disturbing species 12 (dividing its initial population size by 2). Based on these response values, matrices can be generated, informing us how large is the response of species in column j after disturbing species in row *i* (we use the Hurlbert-response [26]. Row sums provide a measure of community importance: they express how big is the community-wide effect of disturbing an i focal species. Table 9.1 show these community importance values for each group, based on both the mean and the variance. Column sums provide a measure of community sensitivity: they express how sensitive is a focal species j to disturbing all others. Community importance indicators, like these quantitative, stochastic simulation-based importance measures are strongly needed in conservation biology [42]. The final outcome can be a species importance rank based on a dynamical model parameterized with realistic values [35]. As the structural (positional) importance of species can be measured simply by network analysis, this simulation framework can be used for approaching the classical "structure to dynamics" problem of ecology [51, 52, 56, 11, 15]. Results based on these stochastic simulations suggest that the most important

Published by Woodhead Publishing Limited, 2013

group of the system is "nearshore demersals" (Figure 9.4, Table 9.1), that the relative importance of groups is more balanced and that more important groups tend to be closer to the middle of the food web (at intermediate trophic levels). This somewhat contrasts the deterministic EwE model, where, according to different indices, key groups are either close to the top or close to the bottom of the food web. The key nodes based on stochastic simulations (mean) have intermediate importance based on EwE (Figure 9.4.2), the relationship being negative for variance (Figure 9.4.2). Thus, deterministic simulations predict key groups that are not necessarily important based on stochastic and vice versa. It is easier to compare the mean- and variance-based stochastic approaches. If group i has a large effect on the mean population size of other *j* groups, it is not automatic that it also generates behavioural variability (Figure 9.6). Rather, based on the negative correlation, trophic groups have larger effect either on the mean or on the variance. This result, if it will be better justified by future studies, challenges current conservation policy. If variability in population dynamics is considered essential for survival, different species could be indicated as key groups in ecosystem dynamics. Species having small effects on the mean population size of others but a large influence on their variability are not on the horizon of conservationists currently.

9.4.3 Predictions and applications

Stochastic simulation focusing on individuals and local, parallel processes is especially promising in conservation practice. Here, what we mostly want to understand is the behaviour and extinction risk of rare species. The weakest side of traditional models (e.g., deterministic simulations) is describing these species by means of average population features, while individual-level variability can be of key importance under certain conditions. As population size becomes smaller, genetics and demography, environmental variability and also unique interactions can be of major importance [32]. Also, below some critical population size, interaction

Published by Woodhead Publishing Limited, 2013

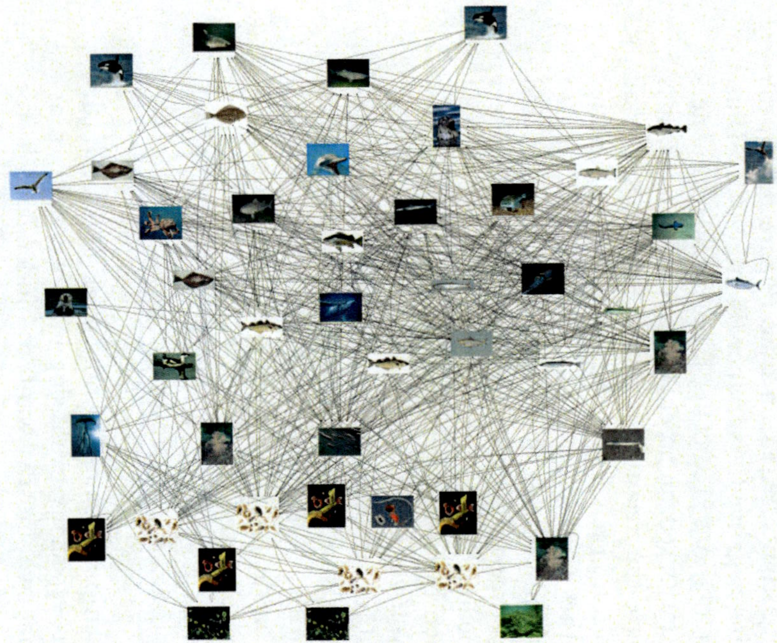

Figure 9.1 The Prince William Sound food web. Following the conventions, lower groups are consumed by the higher ones (arrows not shown for simplicity). Some pictures are identical, these are groups differing only in location, size or age (e.g. "shallow large infauna" and "deep large infauna" differ for location).

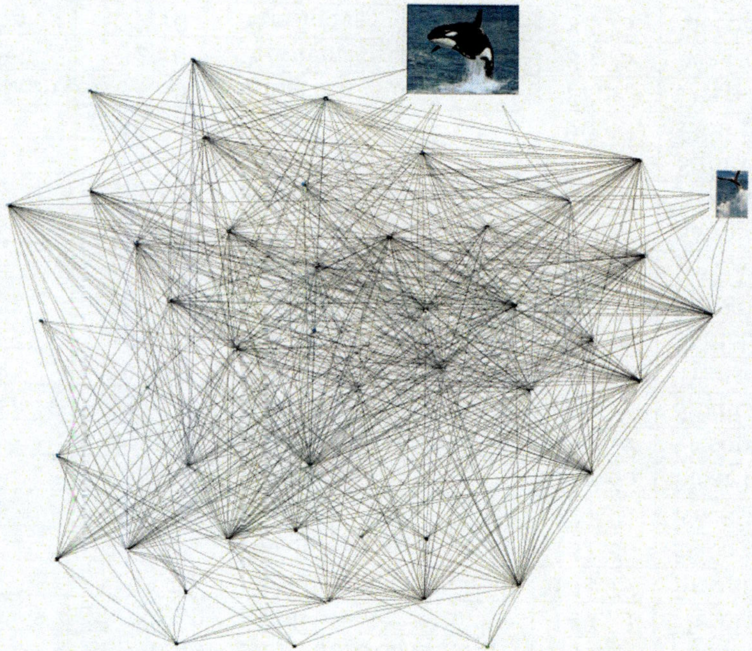

Figure 9.2 The Prince William Sound food web: the size of trophic groups is proportional to their relative importance based on deterministic dynamical simulations (KI). The (outstandingly) most important group is "Transient orca".

Table 9.1 Relative importance ranks for the trophic groups of the Prince William Sound ecosystem. based on deterministic (KI) and stochastic (IH(M): based on response in the mean, IH(V): based on response in the variance) dynamical models.

	KI		IH(M)		IH(V)
Transient orca	66330.80	Nearshore demersals	231.09	Nearshore demersals	396.69
Avian raptors	28064.80	Adult arrowtooth	147.71	Herbivorous zooplank	331.94
Porpoise	1299.70	Herbivorous zooplank	141.79	Juvenile herring	57.32
Seabirds	869.70	Juvenile herring	131.80	Deep epibenthos	56.07
Sea otter	812.10	Seabirds	121.91	Rockfish	55.59
Invert-eating birds	689.20	Sleeper shark	117.45	Near herbiv zooplank	55.26
Juvenile pollock (0)	385.00	Salmon shark	115.22	Octopods	55.15
Juvenile salmon (0-1)	367.10	Juvenile pollock (0)	103.66	Shallow sm epibenthos	54.78
Pinnipeds	293.40	Juv. Arrowtooth	103.21	Transient orca	54.75
Sleeper shark	265.30	Adult Pollock (1+)	98.94	Sablefish	54.33
Salmon shark	161.50	Near phytoplankton	98.05	Jellies	54.13
Octopods	143.90	Capelin	94.51	Juv. Arrowtooth	54.07
Juv. Arrowtooth	110.60	Shallow sm infauna	93.31	Sandlance	53.23
Baleen Whales	66.30	Deep lg infauna	92.59	Juvenile salmon (0-1)	53.12
Resident orca	54.40	Pacific cod	91.97	Squid	52.73
Spiny dogfish	38.20	Jellies	90.49	Omnivorous zooplank	52.03
Offshore phytoplankton	34.90	Lingcod	88.97	Deep lg infauna	52.00
Deep demersal fishes	32.80	Resident orca	88.75	Macrophytes	51.59
Sablefish	27.90	Pinnipeds	88.75	Sleeper shark	51.53

Table 9.1 Continued

	KI		IH(M)		IH(V)
Pacific cod	26.10	Macrophytes	87.24	Shallow lg infauna	51.06
Lingcod	25.90	Adult herring	84.56	Deep demersal fishes	51.00
Adult arrowtooth	23.90	Sea otter	84.51	Offshore phytoplankton	50.67
Adult herring	21.70	Porpoise	82.18	Adult herring	50.58
Eulachon	14.10	Deep sm infauna	80.51	Sea otter	50.49
Shallow lg epibenthos	12.30	Invert-eating birds	79.90	Seabirds	50.44
Nearshore demersals	10.90	Juvenile salmon (0-1)	79.40	Halibut	50.43
Capelin	10.30	Baleen Whales	79.10	Near phytoplankton	50.10
Squid	8.80	Adult salmon	78.41	Shallow sm infauna	50.06
Adult salmon	7.80	Sandlance	77.68	Eulachon	49.69
Halibut	7.40	Near omnivorous zoops	77.25	Adult Pollock (1+)	49.65
Sandlance	6.20	Octopods	77.00	Avian raptors	49.51
Near herbiv zooplank	6.00	Deep demersal fishes	75.17	Lingcod	49.35
Herbivorous zooplank	6.00	Shallow lg epibenthos	75.01	Capelin	49.09
Near phytoplankton	5.70	Shallow lg infauna	72.59	Deep sm infauna	48.90
Adult Pollock (1+)	5.40	Spiny dogfish	70.29	Meiofauna	48.55
Rockfish	4.90	Deep epibenthos	66.93	Baleen Whales	47.83

Table 9.1 Continued

	KI		IH(M)		IH(V)
Juvenile herring	4.50	Rockfish	66.83	Pacific cod	47.78
Deep epibenthos	4.40	Shallow sm epibenthos	66.51	Spiny dogfish	47.71
Jellies	4.20	Near herbiv zooplank	63.48	Adult salmon	47.58
Near omnivorous zoops	3.40	Halibut	61.96	Shallow lg epibenthos	47.47
Shallow sm infauna	3.20	Sablefish	61.79	Adult arrowtooth	47.08
Omnivorous zooplank	2.10	Meiofauna	60.89	Near omnivorous zoops	46.87
Shallow sm epibenthos	1.90	Omnivorous zooplank	60.59	Pinnipeds	46.82
Deep sm infauna	1.40	Eulachon	57.33	Juvenile pollock (0)	45.97
Shallow lg infauna	1.40	Avian raptors	56.25	Invert-eating birds	45.49
Meiofauna	0.60	Transient orca	53.77	Resident orca	45.39
Deep lg infauna	0.60	Squid	52.26	Salmon shark	44.27
Macrophytes	0.20	Offshore phytoplankton	48.99	Porpoise	40.80

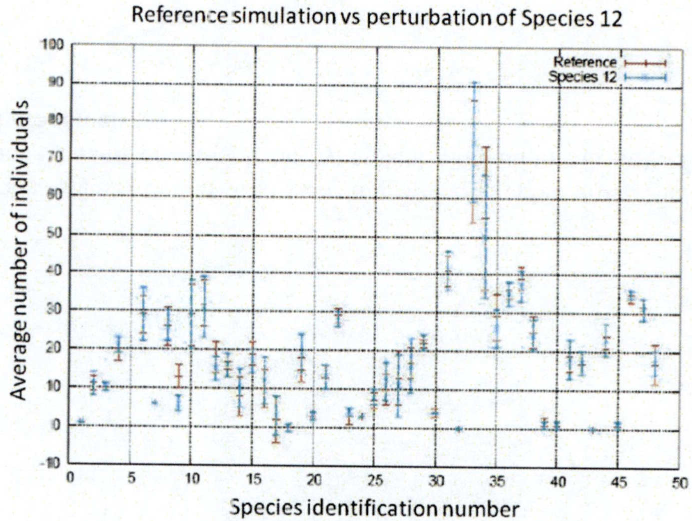

Figure 9.3 One example for performing sensitivity analysis in stochastic food web simulations. The red bars show the average number of individuals (y axis) of each trophic group of the simulated ecosystem (x axis) and its variance. This is the reference simulation. Then we perturb one of the species (here, species # 12) by dividing its initial population size by two. The simulations are run again and the average (and variation) of population size values are shown by the blue bars. The difference between the red and the blue bars (in terms of both average and variation) characterize the response of individual species to perturbing species 12, while their sum provides a community response, i.e. the community importance measure of species 12. Note that some species give larger response in the mean (species 9), while others do in the variance (species 44).

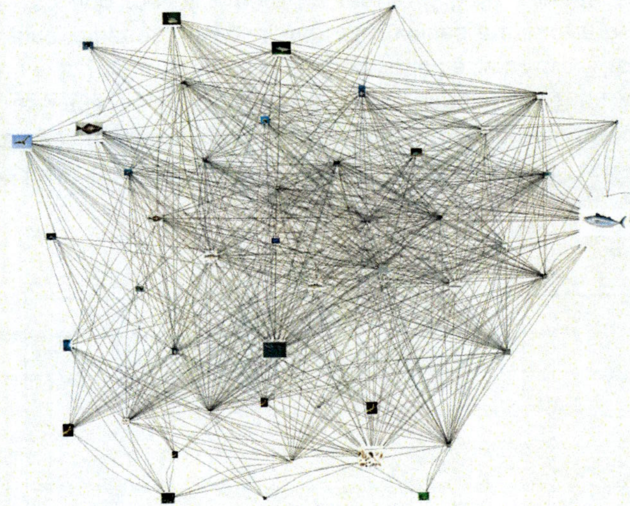

Figure 9.4 The Prince William Sound food web: the size of trophic groups is proportional to their relative importance based on stochastic dynamical simulations (IH(M)). The most important group is "Nearshoredemersals".

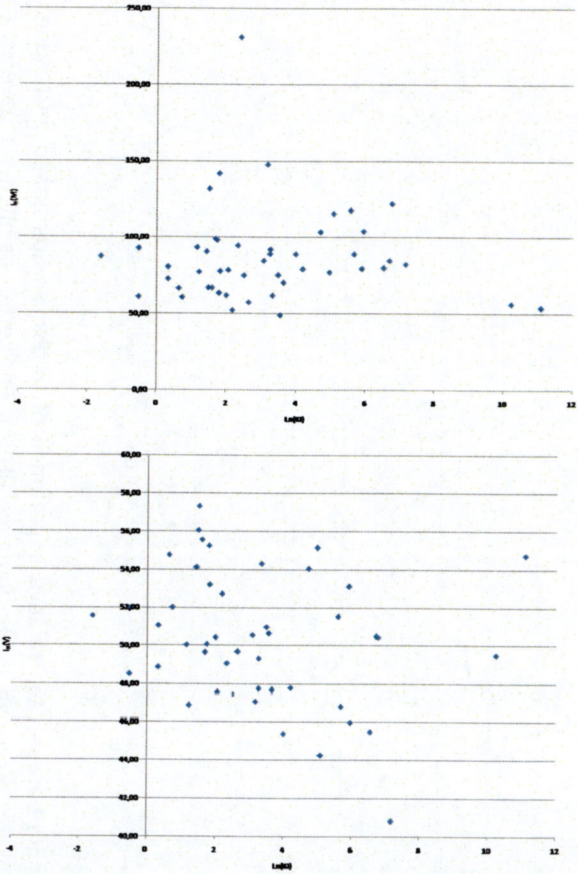

Figure 9.5 The relationship between the relative importance of trophic groups based on response in the mean (a: IH(M) in a) and in the variance (IH(V) in b) and the logarithm of the KI importance index based on deterministic dynamical simulations (Ln(KI)). For better visibility, we have excluded two outlier values.

Published by Woodhead Publishing Limited, 2013

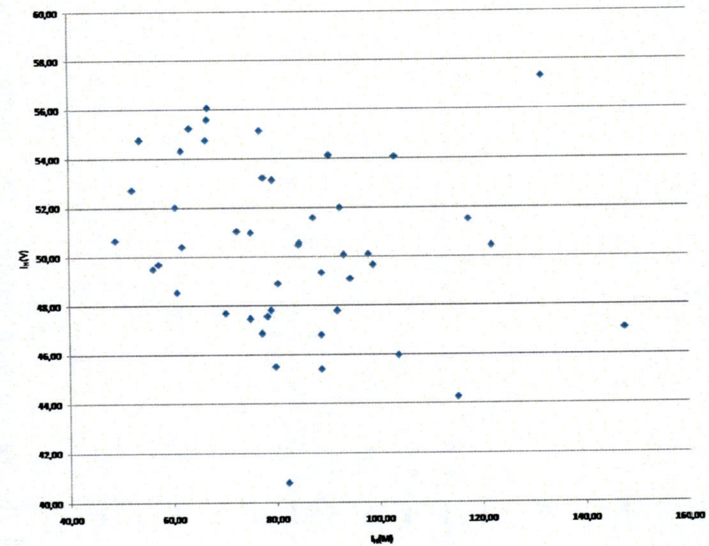

Figure 9.6 The relationship between the relative importance of trophic groups based on response in the mean (IH(M)) and response in the variance (IH(V)). For better visibility, we have excluded two outlier values.

patterns and rates may change, and the modeller has to be able to redefine or update parameters, and even model structure. It is fairly plausible if rules are local and model specification is bottom-up. Stochastic, individual-based or process-based simulations make it possible to study these aspects of systems dynamics. In practice, these tools may serve to plan systems-based conservation strategies [5] or defining optimum programs for managing multispecies fisheries [38, 72].

Notes

Carmen Maria Livi, Nerta Gjata, Marco Scotti and Thomas A. Okey are acknowledged for help.

References

1. A.A. Agrawal, Phenotypic plasticity in the interactions and evolution of species. Science 294:321-326, 2001.
2. M.J. Angilletta, R. S. Wilson, C. A. Navas, R. S. James , Tradeoffs and the evolution of thermal reaction norms, Trends. Ecol. Evol. 18, 234-240, 2003.
3. W.R. Ashby, Design for a Brain.Wilcy, Oxford, 1952.
4. M. Begon, C. R. Townsend, J. R. Harper, Ecology:from individuals to ecosystems. Fourth Edition. Blackwell, Oxford, 2006.
5. F. Berkes, Rethinking community-based conservation.Conserv. Biol. 18, 621-630, 2004.
6. S. R. Borrett, W. Bridewell, P. Langley, K. R. Arrigo, A method for representing and developing process models. Ecological Complexity 4:1-12, 2007.
7. V. Christensen and C. J. Walters, Ecopath with Ecosim: methods, capabilities and limitations. Ecological Modelling, 172:109-139, 2004.
8. J.E. Cohen,Food Webs and Niche Space. Princeton University Press, Princeton, 1978.
9. J. E. Cohen and F. Briand, Trophic links of community food webs. Proc. Natl. Acad. Sci. 81:4105–4109, 1984.

Published by Woodhead Publishing Limited, 2013

10. M. Coll, I. Palomera, S. Tudela, Decadal changes in a NW Mediterranean Sea food web in relation to fishing exploitation. Ecological Modelling 220, 2088-2102, 2009.
11. P. C. deRuiter, A. M. Neutel, J. C. Moore, Energetics, patterns of interaction strengths, and stability of real ecosystems. Science 1995;269:1257-1260, 2005.
12. D.L. DeAngelis, L. J. Gross, Individual-based Models and Approaches in Ecology. Chapman and Hall, New York, 1992.
13. , L. Dematt´, C. Priami, A. Romanel, O. Soyer, Evolving BlenX programs to simulate the evolution of biological networks.Theor. Comp. Sci. 408, 83-96, 2008.
14. M. Dorigo and T. Stutzle, Ant Colony Optimization.MIT Press, 2004.
15. J. A. Dunne, The network structure of food webs. In: Pascual, M. and Dunne, J.A. (Eds.), Ecological Networks: Linking Structure to Dynamics in Food Webs, Oxford University Press, pages 27-86, 2006.
16. A. Feest, T. D. Aldred, K. Jedamzik, Biodiversity quality: a paradigm for biodiversity. Ecological Indicators 10, 1077-1082, 2010.
17. J. T. Finn, Measures of ecosystem structure and function derived from analysis of flows. J. Theor. Biol. 1976;56:363–380, 1976.
18. D. T. Gillespie, Exact stochastic simulation of coupled chemical reactions.J. Phys. Chem. 81, 2340-2361, 1977.
19. V. Grimm, E. Revilla, U. Berger, F. Jeltsch, W. M. Mooij, S. F. Railsback, H.-H. Thulke, J. Weiner, T. Wiegand and D. L. DeAngelis, Pattern-oriented modeling of agent-based complex systems: lessons from ecology. Science 310, 987-991, 2005.
20. V. Grimm, U. Berger, F. Bastiansen, S. Eliassen, V. Ginot, J. Giske, J. Goss-Custard, T. Grand, S. K. Heinz, G. Huse, A standard protocol for describing individual-based and agent-based models. Ecol. Model. 198, 115-126, 2006.

Published by Woodhead Publishing Limited, 2013

21. V. Grimm and S. F. Railsback, Individual-based Modeling and Ecology. Princeton University Press, Princeton, New Jersey, 2005.
22. A. Gronewold and M. Sonnenschein, Event-based modelling of ecological systems with asynchronous cellular automata. Ecol. Model. 108:37-52, 1998.
23. B. Hannon, The structure of ecosystems. J. Theor. Biol. 41:535-546, 1973.
24. M. Higashi and T. P. Burns (eds.), Theoretical Studies of Ecosystems - the Network Perspective. Cambridge University Press, Cambridge, 1991.
25. C. S. Holling, Some characteristics of simple types of predation and parasitism. The Canadian Entomologist 91, 385-398, 1959.
26. S. H. Hurlbert, Functional importance vskeystoneness: reformulating some questions in theoretical biocenology. Australian Journal of Ecology 22, 369-382, 1997.
27. F. Jordán, I. Scheuring, G. Vida, Species positions and extinction dynamics in simple food webs. Journal of Theoretical Biology, 215:441-448, 2002.
28. F. Jordán, I. Scheuring, Network Ecology: topological constraints on ecosystems dynamics. Physics of Life Reviews 1:139-172, 2004.
29. F. Jordán, T. A. Okey, B. Bauer, B. and S. Libralato, Identifying important species: a comparison of structural and functional indices. Ecological Modelling, 216: 75-80, 2008.
30. O. P. Judson, The rise of the individual-based model in ecology.Trends. Ecol. Evol. 9, 9-14, 1994.
31. C. Kazanci, L. Matambaa, E.W. Tollnerb, Cycling in ecosystems: an individual based approach. Ecol. Model. 220, 2908-2914, 2009.
32. R. Lande, Genetics and demography in biological conservation. Science 241, 1455-1460, 1988.
33. S. A. Levin, Ecosystems and the biosphere as complex adaptive systems. Ecosystems 1, 431-436, 1998.
34. S. Libralato, V. Christensen, D. Pauly, A method for identifying keystone species in food web models. Ecological Modelling 195, 153-171, 2006.

Published by Woodhead Publishing Limited, 2013

35. C. M. Livi, F. Jordán, P. Lecca, T. A. Okey, Identifying key species in ecosystems with stochastic sensitivity analysis. Ecological Modelling 222, 2542-2551, 2011.
36. R. Margalef, Perspectives in ecological theory. University of Chicago Press, Chicago, 1968.
37. L. Matamba, C. Kazancia, J. R. Schramskib, M. Blessingc, P. Alexanderd, B. C. Pattene, Throughflow analysis: A stochastic approach. Ecol. Model. 220:174-3181, 2009.
38. R. M. May, J. R. Beddington and C. W. Clark, S. J. Holt, R. M. Laws, Management of multispecies fisheries. Science 1979;205:267-277, 1979.
39. C. McCaig, R. Norman, C. Shankland, From individuals to populations: a symbolic process algebra approach to epidemiology. Mathematics in Computer Science 2:535-556, 2009.
40. A. J. McKane and B. Drossel, Models of food-web evolution. In: Pascual M, Dunne JA (eds) Ecological Networks: Linking Structure to Dynamics in Food Webs, Oxford University Press, Oxford, pages 223-243, 2006.
41. B. A. Menge, Indirect effects in marine rocky intertidal interaction webs: Patterns and importance. Ecol. Monogr. 65:21-74, 1995.
42. L. S. Mills, M. L. Soulé, D. F. Doak, The keystone-species concept in ecology and conservation. BioScience, 43:219-224, 1993.
43. R. Norman, C. Shankland, Developing the use of process algebra in the derivation and analysis of mathematical models of infectious disease. Lecture Notes in Comp. Sci. 280:404-414, 2004.
44. B. Oborny, Á. Kun, T. Czárán, Szilárd Bokros, The effect of clonal integration on plant competition for mosaic habitat space. Ecology 81:3291-3304, 2000.
45. E. P. Odum, The strategy of ecosystem development. Science 164:262-270, 1969.
46. T. A. Okey, Shifted community states in four marine ecosystems: some potential mechanisms. PhD thesis, University of British Columbia, Vancouver, 2004

Published by Woodhead Publishing Limited, 2013

47. T. Okuyama, Local interactions between predators and prey call into question commonly used functional responses. Ecol. Model. 220:1182-1188, 2009.
48. M. Ortiz and M. Wolff,Dynamical simulation of mass-balance trophic models for benthic communities of north-central Chile: assessment of resilience time under alternative management scenarios. Ecological Modelling, 148:277-291, 2002.
49. M. Pascual, Computational ecology: from the complex to the simple and back. PLoSComput. Biol. 1, e18, 2005.
50. B. C. Patten, An introduction to the cybernetics of the ecosystem: the trophic-dynamic aspect. Ecology, 40:221-231, 1959.
51. B. C. Patten, Network ecology: indirect determination of the life/environment relationship in ecosystems. In: Higashi M, Burns TP (eds) Theoretical studies of ecosystems - the network perspective, Cambridge University Press, Cambridge, pages 117-154, 1991
52. S. L. Pimm, The Balance of Nature? University of Chicago Press, Chicago, 1991.
53. S. L. Pimm, Food web design and the effect of species deletion, Oikos 35:139-149, 1980.
54. S. L. Pimm, Food Webs. Chapman and Hall, London, 1982.
55. M. J. O. Pocock, O. Johnson, D. Wasiuk, Succinctly assessing the topological importance of species in flower-pollinator networks, Ecological Complexity 8:265-272, 2011.
56. G. A. Polis, K. O. Winemiller (eds), Food webs: integration of patterns and dynamics. Chapman and Hall, London, 1996.
57. C. R. Powell and R. P. Boland, The effects of stochastic population dynamics on food web structure. J. Theor. Biol. 257:170-180, 2009.
58. C. Priami, Algorithmic systems biology. Communications of ACM 52:80-89, 2009.
59. M. Rejmnek and P. Stary, Connectance in real biotic communities and critical values for stability of model ecosystems. Nature 280:311-313, 1979.

Published by Woodhead Publishing Limited, 2013

60. D. Robertson, A. Bundy, R. Muetzelfeldt, M. Haggith, M. Uschold, Eco-logic: Logic-based Approaches to Ecological Modelling, MIT Press, 1991.
61. Ruiz-Moreno, M. Pascual, R. Riolo, Exploring network space with genetic algorithms: modularity, resilience and reactivity. In: M. Pascual, J. A. Dunne (eds) Ecological Networks: Linking Structure to Dynamics in Food Webs, Oxford University Press, Oxford, pages 187-208, 2006.
62. D. Simberloff, A succession of paradigms in ecology: essentialism to materialism and probabilism. Synthese 43:3-39, 1980.
63. J. D. Stevens, R. Bonfil, N. K. Dulvy, P. A. Walker, The effects of fishing on sharks, rays, and chimaeras (chondrichthyans), and the implications for marine ecosystems. ICES J. Mar. Sci. 57:476-494, 2000.
64. D. J. T. Sumpter, D. S. Broomhead, Relating individual behaviour to population dynamics. Proc. R. Soc. Lond. B. 268:925-932, 2001.
65. R. Thom, Structural Stability and Morphogenesis, W.A.Benjamin, Inc., Massachusetts, 1975.
66. J. N. Thompson, Variation in interspecific interactions. Ann. Rev. Ecol. Syst. 19:65-87, 1988.
67. C. Tofts, Algorithms for task allocation in ants (A study on temporal polyethism: Theory). Bull. Math. Biol. 55:891-918, 1993.
68. R. E. Ulanowicz, Growth and Development: Ecosystems Phenomenology. Springer, Berlin, 1986.
69. M. Vasconcellos, S. Mackinson, K. Sloman, D. Pauly, The stability of trophic mass-balance models of marine ecosystems: a comparative analysis. Ecol. Model. 100:125-134, 1997.
70. G. P. Wagner and L. Altenberg, Complex adaptations and the evolution of evolvability. Evolution 50:967-976, 1996.
71. P. Yodzis, Diffuse effects in food webs. Ecology 81:261-266, 2000.
72. P. Yodzis, Must top predators be culled for the sake of fisheries? Trends EcolEvol 16:78-84, 2001.

Published by Woodhead Publishing Limited, 2013

Index

Published by Woodhead Publishing Limited, 2013

Published by Woodhead Publishing Limited, 2013

Published by Woodhead Publishing Limited, 2013